Schizophrenia

Evolution and Synthesis

Strüngmann Forum Reports

Julia Lupp, series editor

The Ernst Strüngmann Forum is made possible through the generous support of the Ernst Strüngmann Foundation, inaugurated by Dr. Andreas and Dr. Thomas Strüngmann.

This Forum was supported by funds from the
Deutsche Forschungsgemeinschaft (German Science Foundation)

Schizophrenia

Evolution and Synthesis

Edited by

Steven M. Silverstein, Bita Moghaddam, and Til Wykes

Program Advisory Committee:
Anil K. Malhotra, John McGrath, Andreas Meyer-Lindenberg,
Bita Moghaddam, Steven M. Silverstein, Til Wykes

The MIT Press
Cambridge, Massachusetts
London, England

© 2013 Massachusetts Institute of Technology and
the Frankfurt Institute for Advanced Studies

This volume is the result of the 13th Ernst Strüngmann Forum,
held July 22–27, 2012, in Frankfurt am Main, Germany.

Series Editor: J. Lupp
Assistant Editor: M. Turner
Photographs: U. Dettmar
Design and realization: BerlinScienceWorks

All rights reserved. No part of this book may be reproduced in any form
by electronic or mechanical means (including photocopying, recording,
or information storage and retrieval) without permission in writing from
the publisher.

MIT Press books may be purchased at special quantity discounts
for business or sales promotional use. For information, please email
special_sales@mitpress.mit.edu or write to Special Sales Department,
The MIT Press, 55 Hayward Street, Cambridge, MA 02142.

The book was set in TimesNewRoman and Arial.
Printed and bound in the United States of America.

Library of Congress Cataloging-in-Publication Data

Schizophrenia : evolution and synthesis / edited by Steven M. Silverstein,
Bita Moghaddam, and Til Wykes.
   pages   cm. — (Strüngmann Forum reports)
Includes bibliographical references and index.
ISBN 978-0-262-01962-0 (hardcover : alk. paper)
1. Schizophrenia. 2. Schizophrenia—Etiology. 3. Clinical medicine—Decision making. I. Silverstein, Steven M. II. Moghaddam, Bita. III. Wykes, Til.
RC514.S3353   2013
616.89'8—dc23
                                                          2013013617

10 9 8 7 6 5 4 3 2 1

Contents

The Ernst Strüngmann Forum   vii

List of Contributors   ix

1   **Schizophrenia: The Nature of the Problems and the Need for Evolution and Synthesis in Our Approaches**   1
Steven M. Silverstein, Bita Moghaddam, and Til Wykes

Heterogeneity

2   **What Kind of a Thing Is Schizophrenia? Specific Causation and General Failure Modes**   25
Angus W. MacDonald III

3   **How the Diagnosis of Schizophrenia Impeded the Advance of Knowledge (and What to Do About It)**   49
William T. Carpenter Jr.

4   **What Dimensions of Heterogeneity Are Relevant for Treatment Outcome?**   63
Leanne M. Williams and Chloe Gott

5   **Which Aspects of Heterogeneity Are Useful to Translational Success?**   77
Aiden Corvin, Robert W. Buchanan, William T. Carpenter Jr., James L. Kennedy, Matcheri S. Keshavan, Angus W. MacDonald III, Louis Sass, and Michèle Wessa

Risk and Resilience

6   **How Should Resilience Factors Be Incorporated in Treatment Development?**   93
Peter B. Jones

7   **Insights into New Treatments for Early Psychosis from Genetic, Neurodevelopment, and Cognitive Neuroscience Research**   101
Kristin S. Cadenhead and Camilo de la Fuente-Sandoval

8   **From Epidemiology to Mechanisms of Illness**   127
John McGrath and Andreas Meyer-Lindenberg

9   **How Can Risk and Resilience Factors Be Leveraged to Optimize Discovery Pathways?**   137
Craig Morgan, Michael O'Donovan, Robert A. Bittner, Kristin S. Cadenhead, Peter B. Jones, John McGrath, Steven M. Silverstein, Heike Tost, Peter Uhlhaas, and Aristotle Voineskos

Models

10 **Human Cell Models for Schizophrenia** 167
Ashley M. Wilson and Akira Sawa

11 **How Can Animal Models Be Better Utilized?** 183
Patricio O'Donnell

12 **How Can Computational Models Be Better Utilized for Understanding and Treating Schizophrenia?** 195
Daniel Durstewitz and Jeremy K. Seamans

13 **How Can Models Be Better Utilized to Enhance Outcome? A Framework for Advancing the Use of Models in Schizophrenia** 209
Kevin J. Mitchell, Patricio O'Donnell, Daniel Durstewitz, André A. Fenton, Jay A. Gingrich, Joshua A. Gordon, Wolfgang Kelsch, Bita Moghaddam, William A. Phillips, and Akira Sawa

Development and Treatment

14 **Why Kraepelin Was Right: Schizophrenia as a Cognitive Disorder** 227
René S. Kahn

15 **What Will the Next Generation of Psychosocial Treatments Look Like?** 235
Kim T. Mueser

16 **Creative Solutions to Overcoming Barriers in Treatment Utilization: An International Perspective** 261
Wulf Rössler

17 **What Is Necessary to Enhance Development and Utilization of Treatment?** 273
Vera A. Morgan, Richard Keefe, René S. Kahn, Anil K. Malhotra, Andreas Meyer-Lindenberg, Kim T. Mueser, Karoly Nikolich, Wulf Rössler, William Spaulding, Sharmili Sritharan, and Til Wykes

Bibliography 307

Subject Index 381

The Ernst Strüngmann Forum

Founded on the tenets of scientific independence and the inquisitive nature of the human mind, the Ernst Strüngmann Forum is dedicated to the continual expansion of knowledge. Through its innovative communication process, the Ernst Strüngmann Forum provides a creative environment within which experts scrutinize high-priority issues from multiple vantage points.

This process first begins with the identification of themes. By nature, a theme constitutes a problem area that transcends classic disciplinary boundaries. It is of high-priority interest, requiring concentrated, multidisciplinary input to address the issues involved. Proposals are received from leading scientists active in their field and are selected by an independent Scientific Advisory Board. Once approved, a steering committee is convened to refine the scientific parameters of the proposal and select the participants. Approximately one year later, the central meeting, or Forum, is held to which circa forty experts are invited.

Preliminary discussion for this theme began in 2010, when Steven Silverstein brought the initial idea to our attention. Together with Bita Moghaddam and Til Wykes, the resulting proposal was approved by the Scientific Advisory Board and from June 27–29, 2011 the steering committee was convened. The committee, comprised of Anil Malhotra, John McGrath, Andreas Meyer-Lindenberg, Bita Moghaddam, Steven Silverstein, and Til Wykes, identified the key issues for debate and selected the participants for the Forum, which took place in Frankfurt am Main, from July 22–27, 2012.

A Forum is a dynamic think tank. The activities and discourse that accompany it begin well before participants arrive in Frankfurt and conclude with the publication of this volume. Throughout each stage, focused dialog is the means by which participants examine the issues anew. Often, this requires relinquishing long-established ideas and overcoming disciplinary idiosyncrasies, which otherwise could inhibit joint examination. When this is accomplished, however, new insights begin to emerge.

This volume conveys the synergy that arose out of myriad discussions between diverse experts, each of whom assumed an active role. It contains two types of contributions. The first provides background information to key aspects of the overall theme. Originally written in advance of the Forum, these chapters have been extensively reviewed and revised to provide current understanding on these topics. The second (Chapters 5, 9, 13, and 17) summarizes the extensive group discussions that transpired. These chapters should not be viewed as consensus documents nor are they proceedings. Instead, their goal is to transfer the essence of the discussions, expose the open questions that still remain, and highlight areas in need of future enquiry.

An endeavor of this kind creates its own unique group dynamics and puts demands on everyone who participates. Each invitee contributed not only their time and congenial personality, but a willingness to probe beyond that which is evident. For this, I extend my gratitude to all.

A special word of thanks goes to the steering committee, the authors of the background papers, the reviewers of the papers, and the moderators of the individual working groups: Robert Buchanan, Michael O'Donovan, Patricio O'Donnell, and Richard Keefe. To draft a report during the week of the Forum and bring it to its final form in the months thereafter is never a simple matter. For their efforts and tenacity, I am especially grateful to Aiden Corvin, Craig Morgan, Kevin Mitchell, and Vera Morgan—the rapporteurs of the discussion groups. Most importantly, I extend my sincere appreciation to Steven Silverstein, Bita Moghaddam, and Til Wykes. As chairpersons of this 13th Strüngmann Forum, their commitment ensured a most vibrant intellectual gathering.

A communication process of this nature relies on institutional stability and an environment that encourages free thought. The generous support of the Ernst Strüngmann Foundation, established by Dr. Andreas and Dr. Thomas Strüngmann in honor of their father, enables the Ernst Strüngmann Forum to conduct its work in the service of science. The Science Advisory Board guides this work and ensures the scientific independence of the Ernst Strüngmann Forum. Supplemental financial support for this theme was received from the German Science Foundation, and the Frankfurt Institute of Advance Studies provided the backdrop for this intellectual exercise.

Long-held views are never easy to put aside. Yet, when this is achieved, when the edges of the unknown begin to appear and gaps in knowledge are able to be defined, the act of formulating strategies to fill such gaps becomes a most invigorating exercise. We hope that this volume will convey a sense of this lively endeavor. Most importantly, we hope that this joint examination of schizophrenia will lead to a novel conceptualization of the disorder and accelerate advances in treatment development and prevention efforts.

Julia Lupp, Program Director
Ernst Strüngmann Forum
Frankfurt Institute for Advanced Studies (FIAS)
Ruth-Moufang-Str. 1, 60438 Frankfurt am Main, Germany
http://esforum.de

List of Contributors

Robert A. Bittner   Department of Psychiatry, Psychosomatic Medicine and Psychotherapy, Goethe University, 60528 Frankfurt am Main, Germany
Robert W. Buchanan   Maryland Psychiatric Research Center, Baltimore, MD 21228, U.S.A.
Kristin S. Cadenhead   Department of Psychiatry, University of California, San Diego, La Jolla, CA 92014, U.S.A.
William T. Carpenter Jr.   University of Maryland School of Medicine, Maryland Psychiatric Research Center, Baltimore, MD 21228, U.S.A.
Aiden Corvin   Department of Psychiatry, Trinity Centre for Health Sciences, St. James's Hospital, Dublin 8, Ireland
Camilo de la Fuente-Sandoval   Laboratory of Experimental Psychiatry and Neuropsychiatry Department, Instituto Nacional de Neurología y Neurocirugía, Mexico City, 14269, Mexico
Daniel Durstewitz   Bernstein Center for Computational Neuroscience, Psychiatry, Central Institute of Mental Health, Medical Faculty Mannheim of Heidelberg University, Germany
André A. Fenton   Center of Neural Science, New York University, New York, NY 10003, U.S.A.
Jay A. Gingrich   New York State Psychiatric Institute, Columbia University, New York, NY 10032, U.S.A.
Joshua A. Gordon   New York State Psychiatric Institute, Columbia University, New York, NY 10032, U.S.A.
Chloe Gott   Brain Dynamics Centre, Acacia House, The University of Sydney, NSW 2145, Australia
Peter B. Jones   Department of Psychiatry, University of Cambridge, Herchel Smith Building for Brain and Mind Sciences, Cambridge CB2 0SX, U.K.
René S. Kahn   Department of Psychiatry, University of Utrecht, CX Utrecht, The Netherlands
Richard Keefe   Psychiatry and Behavioral Sciences, Duke University Medical Center, Durham, NC 27710, U.S.A.
Wolfgang Kelsch   Emmy Noether Group, University Heidelberg, 69120 Heidelberg, Germany
James L. Kennedy   Centre for Addiction and Mental Health, University of Toronto, Toronto, ON M5T 1R8, Canada
Matcheri S. Keshavan   Beth Israel Deaconess Medical Center, Harvard Medical School, Boston, MA 02215, U.S.A.
Angus W. MacDonald III   Department of Psychology, University of Minnesota, Minneapolis, MN 55455, U.S.A.

Anil K. Malhotra   The Zucker Hillside Hospital, Glen Oaks, NY 11004, U.S.A.
John McGrath   Queensland Centre for Mental Health Research, The Park Centre for Mental Health, Wacol, Q4076, Australia
Andreas Meyer-Lindenberg   Central Institute of Mental Health, J5, 68159 Mannheim, Germany
Kevin J. Mitchell   Genetics, Trinity College Dublin, Dublin 2, Ireland
Bita Moghaddam   Department of Neuroscience, University of Pittsburgh, Pittsburgh, PA 15260, U.S.A.
Vera A. Morgan   School of Psychiatry and Clinical Neurosciences, The University of Western Australia, Perth, Western Australia 6000, Australia
Craig Morgan   Institute of Psychiatry, King's College London, London, SE5 8AF, U.K.
Kim T. Mueser   Center for Psychiatric Rehabilitation, Boston University, Boston, MA 02215, U.S.A.
Karoly Nikolich   Department of Psychiatry and Stanford Institute of Neuroinnovation and Translational Neuroscience, Stanford University, Stanford, CA 94305, U.S.A.
Patricio O'Donnell   School of Medicine, University of Maryland, Baltimore, MD 21201, U.S.A.
Michael O'Donovan   Department of Psychological Medicine and Neurology, Cardiff University, Cardiff CF14 4XN, U.K.
William A. Phillips   Department of Psychology, University of Stirling, Stirling, FK9 4LA, Scotland, U.K.
Wulf Rössler   Department of General and Social Psychiatry, University Hospital Zurich, 8021 Zurich, Switzerland
Louis Sass   Graduate School of Applied and Professional Psychology, Rutgers University, Piscataway, New Jersey 08854, U.S.A.
Akira Sawa   Johns Hopkins Schizophrenia Center, Department of Psychiatry Molecular Psychiatry Program, Department of Mental Health, Johns Hopkins University School of Medicine and Bloomberg School of Public Health, Baltimore, MD 21287, U.S.A.
Jeremy K. Seamans   Department of Psychiatry and The Brain Research Centre, University of British Columbia, Vancouver, BC, Canada
Steven M. Silverstein   Division of Schizophrenia Research, University of Medicine and Dentistry of New Jersey, Piscataway, NJ 08854, U.S.A.
William Spaulding   Department of Psychology, University of Nebraska Lincoln, Lincoln, NE 68588, U.S.A.
Sharmili Sritharan   Max Planck Institute for Brain Research, 60528 Frankfurt am Main, Germany
Heike Tost   Central Institute of Mental Health, BCCN, J5, 68159 Mannheim, Germany
Peter Uhlhaas   Department of Neurophysiology, Max Planck Institute for Brain Research, 60528 Frankfurt am Main, Germany

List of Contributors

Aristotle Voineskos   Centre for Addiction and Mental Health, University of Toronto, Toronto, ON M5T 1R8, Canada
Michèle Wessa   Department of Clinical Psychology and Neuropsychology, Institute of Psychology, Johannes Gutenberg University, 55122 Mainz, Germany
Leanne M. Williams   Brain Dynamics Centre, Acacia House, The University of Sydney, NSW 2145, Australia
Ashley M. Wilson   Department of Psychiatry and Behavioral Sciences, Johns Hopkins University School of Medicine, Baltimore, MD 21287
Til Wykes   Psychology PO77, Institute of Psychiatry, London SE5 8AF, U.K.

1

Schizophrenia

The Nature of the Problems and the Need for Evolution and Synthesis in Our Approaches

Steven M. Silverstein, Bita Moghaddam, and Til Wykes

Overview

What is schizophrenia? What are its causes? Can it be cured? Can it be prevented? These fundamental issues have confronted the field of schizophrenia research and treatment for over 100 years. Our ability to improve the lives of people with the disorder, however, has not improved at nearly the same rate as the accumulation of new knowledge about it and technological advances to study it. Paradigm shifts may thus be needed to accelerate progress. This was the aim of the Ernst Strüngmann Forum, "Schizophrenia: Evolution and Synthesis," to which a group of researchers were invited to explore novel ways of conceptualizing the disorder, integrating data across levels of analysis, and accelerating advances in treatment development and prevention efforts.

In this introductory chapter, we introduce the questions and issues that motivated the Forum, in terms of fundamental problems facing the field of schizophrenia research and treatment, and discuss the specific issues identified for debate and the questions which served as starting points for deliberation. We briefly summarize the debate and conclusions of each of the four thematic groups and highlight issues that emerged during the final plenary discussion.

Rationale and Motivation for Challenging Current Paradigms in Schizophrenia Research and Treatment

Schizophrenia is a diagnostic term which describes a serious mental disorder that affects approximately 1% of the population worldwide; current global prevalence is calculated at over 20 million people (McGrath et al. 2008). Common clinical features of the condition include hallucinations, delusions, bizarre behavior, affective dysregulation and/or blunted affect, difficulties in social cognition and interpersonal functioning as well as cognitive impairment. Schizophrenia is typically diagnosed in late adolescence or early adulthood; it is often associated with lifelong disability, especially when appropriate services are not provided, and accounts for high levels of expenditures. In the United States, for example, it is estimated that as many as 10% of all mentally disabled persons are diagnosed with schizophrenia (Rupp and Keith 1993), and the diagnosis accounts for 75% of all mental health spending and approximately 40% of all publicly funded disability payments (Martin and Miller 1998). Among people with the diagnosis, 80–85% are typically unemployed at any given time; those who do obtain a job typically work for a few hours per week and quit or are fired after several weeks or months (Silverstein and Bellack 2008).

Schizophrenia imposes an immense financial burden on individuals, families, and societies. In the United States alone, the cost of treating people diagnosed with schizophrenia has been estimated to be USD 62.7 billion (~ EUR 50 billion) per year, including direct treatment costs and lost business productivity due to patient and family caretaker work absence (Wu et al. 2005). European studies also indicate high costs for treatment, although estimates are lower in southern European countries that use primarily older, less expensive medications, and where patients tend to live with families instead of in residential facilities. For example, Salize et al. (2009) calculate that the mean total cost per year, per patient, was EUR 36,978 in Zürich, EUR 16,868 in Mannheim, but only EUR 2,958 in Granada. These European cost estimates, however, represent only the direct costs of treatment; they do not include indirect costs such as lost work productivity of patients and families, or legal costs, which typically double the overall cost estimate. In the most recent comprehensive analysis of costs, Andrews et al. (2012), in a report prepared for the U.K. Schizophrenia Commission, estimated that the average annual cost per person with schizophrenia to society is GBP 60,000 and to the public sector GBP 36,000. In short, by any standard, schizophrenia is a major individual, family, and public health problem.

In recent years, numerous advances in research technology (e.g., in molecular biology and brain imaging) have resulted in an accumulation of new findings about schizophrenia. Despite this, the general sense in the field is that we are no closer to an integrated understanding of the disorder or to better methods to treat it (e.g., Insel 2009). Progress has not been made on a number of critical issues. For example, diagnosis is still made relatively late in the course of the neurodevelopmental trajectory—typically when persistent psychotic symptoms emerge, but many years after cognitive, academic, and social decline has begun. Our ability to predict who will develop the condition is poor, and etiology is essentially unknown. These issues, together with poorly developed prevention and the fact that we still do not know whether schizophrenia represents one or more disorders, means that treatment is by trial and error. Even more shocking is that although medical illness-related mortality has decreased significantly in the general population, and life span has increased significantly for people with medical diseases (e.g., diabetes, heart disease, cancer), mortality for people with schizophrenia has not decreased over the past 100 years. Moreover, the average life span for a person with the condition is 25 years less than for people without it, and this has not changed for at least 50 years. In fact, treatment outcomes in some domains are arguably equivalent to what they were 100 years ago, the effect size of the difference between active treatments and placebo has *decreased*, and few patients are able to work or live independently (see Insel 2009, 2010; Kemp et al. 2010). Despite psychopharmacological developments over the past 20 years, increased effectiveness has not been demonstrated over medications that were developed in the 1950s and 1960s (Davies et al. 2007; Lewis and Lieberman 2008), treatment noncompliance is high (Lieberman et al. 2005), and several major pharmaceutical companies are eliminating new drug development efforts that target psychotic disorders. Similarly, despite many psychosocial treatment developments over the past 20 years, meta-analyses of some widely used interventions indicate small or near-zero effect sizes (e.g., Lynch et al. 2010), with inverse relationships between study quality and effect size (e.g., Wykes et al. 2008).

Fifteen years ago, many researchers thought that genetics, in the form of a relatively small number of genetic abnormalities, would provide the answers to guide treatment. It now appears, however, that the number of genome "lesions" may be over one million, and thus it is becoming increasingly difficult to develop and maintain an understanding of the genetic basis of schizophrenia. Moreover, many genetic findings have not been replicated. The extent to which this is due to greater than expected human variation, heterogeneity, and/or false positives is unknown. Another technique that offered much promise 15 years ago, and which spawned a great deal of investment, was functional magnetic resonance imaging (fMRI). These studies have added to our appreciation of the complexity of the pathophysiology of the condition, by demonstrating that schizophrenia is not the sum of multiple localized and independent brain dysfunctions but rather the result of altered connectivity between and within brain regions, as well as altered coordination and modulation of brain activity (Phillips and Silverstein 2003). Imaging findings have also contributed to the appreciation of significant heterogeneity within the disorder as well as to the sobering realization of the considerable overlap with healthy people in aspects of brain function. Nonetheless, despite important insights into brain function in schizophrenia from imaging studies, the origins of these problems, how they generate symptoms and the subjective experiences of the disorder, and how to treat them are far from clear. Therefore, as with genetics, the gap between our knowledge base and a comprehensive grasp of the nature of the disorder and how to treat it remains large.

In addition, in spite of major investments in the study of cognitive impairment—a factor thought to be closer to the basis of the condition than symptoms or behaviors—it remains difficult to isolate specific deficits from generalized cognitive impairments and motivational deficits, thus limiting our ability to understand the neural basis of the abnormalities. Behavioral studies of cognition generally have larger effect sizes than psychophysiological or neurobiological studies (Heinrichs 2001), which is the opposite of what was expected to occur with the application of techniques such as fMRI to studies of cognitive impairment in schizophrenia. Moreover, in both the behavioral and physiological domains, it is typical for an abnormal finding to be present in only 30–70% of patients, thus raising questions about the meaning of the deficit for the condition (Heinrichs 2001). Often, issues of diagnostic specificity are ignored, despite the fact that some of the most consistent findings from imaging studies (e.g., reduced hippocampal volumes) have been found in other populations (e.g., people who experienced childhood physical or sexual abuse; Bremner et al. 2003). This suggests that some findings may reflect nonspecific factors, such as chronic stress.

Unlike nonpsychiatric disorders (e.g., coronary artery disease), where the relationship between epidemiology and pathogenesis is generally understood, in schizophrenia, research on the interaction of these factors has, for the most part, remained separate (McGrath and Richards 2009). This has seriously limited the development of comprehensive theories of the disorder that integrate societal, environmental, biological, and developmental perspectives. Recent studies, however, indicate important roles for factors such as cannabis use, stress, negative family environments, physical and sexual abuse, viral exposure, and racial discrimination as well as other forms of chronic social defeat in increasing the risk for schizophrenia (e.g., González-Pinto et al. 2011; Kirkbride et al. 2008; Lysaker et al. 2007; Tienari et al. 2004). Therefore, frameworks that conceptualize the development of schizophrenia within a societal context need to be developed.

Progress in addressing these issues requires more than just incremental additions to the existing research base. We believe that new paradigms coupled with an integration of data from multiple levels of analysis (and new methods of doing this) are necessary. This Forum was viewed as a step forward in this larger process. Our expectation was that by the end of the Forum, progress would have been made in (a) identifying factors (e.g., paradigmatic, disorder-related, institutional, financial, societal) that are preventing breakthroughs and (b) exploring alternative and novel ways to conceptualize, model, diagnose, treat, and research the disorder. Below, we summarize the different themes of the Forum, the specific questions that served to spark each of the groups' discussions, and the outcomes of those discussions.

Group 1: Which Aspects of Heterogeneity Are Useful to Translational Success?

Issues

For many years, schizophrenia has been viewed as a single condition. However, there is no finding that is pathognomonic of schizophrenia, and the best available evidence indicates that specific abnormalities (e.g., in cognition, psychophysiology, neuroanatomy) are found in only 30–70% of patients (Heinrichs 2001). Genetic data increasingly indicate that schizophrenia is a heterogeneous disorder (Mitchell and Porteous 2011; Sebat et al. 2009). This suggests that what we now call schizophrenia may in actuality be a final common pathway of multiple etiologies, or a class of disorders that share some clinical similarities. This view is consistent with recent initiatives to redefine what we now call schizophrenia in terms of basic processes (Insel et al. 2010). The mission of the first discussion group (Corvin et al., Chapter 5, this volume) was to consider this and other evidence related to how schizophrenia is currently conceptualized. Guiding questions included:

- What are the core features of schizophrenia?
- Why has more progress not been made on the homogeneous–heterogeneous question, and what needs to occur to resolve this issue definitively?
- What are the most promising dimensions (e.g., genetic, cognitive, brain function) upon which efforts to clarify heterogeneity can be based?
- Within each dimension, to what extent do findings reflect basic widespread impairments (e.g., reduced cognitive coordination, reduced context-based modulation of neural processing due to NMDA receptor hypofunction, and reduced activity of parvalbumin-containing GABA interneurons) versus multiple independent abnormalities?
- In what ways do we need to revise our understanding of schizophrenia based on findings of genetic overlap with bipolar disorder and symptomatic overlap between childhood schizophrenia and autism spectrum disorders?
- How can we develop a theory of schizophrenia such that it is understood at multiple and interacting levels (e.g., biological, cognitive, phenomenological) in an integrated fashion?

Summary

In their deliberations, Corvin et al. (Chapter 5) began with the idea that schizophrenia is not a disease, because a disease is defined as a phenomenon with known etiology, pathophysiology, and course. Consensus emerged that schizophrenia is, at best, a syndrome, or a collection of signs and symptoms that statistically occur together. The group agreed that schizophrenia is an "open construct" in that its boundaries and many of its features overlap with other medical and psychiatric disorders. Corvin et al. also agreed that schizophrenia is best considered a category, such as dementia, epilepsy, or cancer. That is, what we now call schizophrenia is most likely a category of brain syndromes that bear some outward resemblance to each other, probably by virtue of sharing pathophysiological mechanisms. However, the number of individual syndromes that make up the category is unknown, as are the etiologies of the syndromes. With this in mind, a major agenda for research and treatment is to focus on identifying phenomena that go together, across multiple levels (e.g., biology, cognition, symptom, subjective experience), so as to better describe heterogeneity and move toward personalized treatment. Given that schizophrenia can be studied at so many levels, a key question is: Which levels of analysis are most important?

Consensus emerged that several levels are particularly important. The first level concerns etiological factors, such as genetics, and consequences of infection, such as inflammation, that affect brain function. A second level concerns pathophysiology, where cellular (e.g., neuropil loss), molecular (e.g., reduced GABA, excessive dopamine), and circuit (e.g., reward circuitry, effective connectivity) issues were all considered important. The third level can be broadly construed as the behavioral domain, including learning and other cognitive factors. The fourth, and most debated, level concerns observable or subjective phenomena, such as deficit symptoms (e.g., a loss of motivation) or an altered sense of self.

Because the biological bases of symptoms such as amotivation and hyperreflexivity (i.e., hyperawareness of normally tacit aspects of bodily or mental experience) are relatively unknown, skepticism was expressed as to how useful these constructs are at present for moving the field forward. However, there is a long tradition of a focus on symptoms, and research indicates that phenomena such as altered self-experience (Lysaker and Lysaker 2010; Nelson et al. 2013; Sass and Parnas 2009), despite its relatively unknown etiology, constitute some of the best predictors of schizophrenia; that is, who develops schizophrenia versus who develops bipolar disorder (Nelson et al. 2012). In addition, recent work suggests that disturbances in self-representation contribute to excessive inflammatory activity, thereby providing a potential link between psychological and biological abnormalities in schizophrenia (Barnsley et al. 2011; Corlett 2013). Therefore, a challenge to the field is to understand the psychological phenomena involved in schizophrenia and to advance integration across biological and psychological levels, in an effort to characterize heterogeneity. Methodological issues in studying covariation between phenomena at multiple levels were discussed, and the benefits of traditional linear model (e.g., correlational) approaches versus those that can model nonlinear relationships (e.g., coefficients of mutual information) were outlined. Finally, there was significant cross-fertilization with the discussions of other groups on

(a) the emerging view that schizophrenia is a lifetime disorder with evidence of impairment from birth, and the extent to which the dimension of "premorbid" developmental course can capture variance in heterogeneity relevant to current research and clinical efforts (see C. Morgan et al., Chapter 9, this volume); (b) the extent to which pathophysiological mechanisms can and should be studied individually without the need to model multiple clinical features, and how this can help us understand heterogeneity (see Mitchell et al., Chapter 13, this volume); and (c) which aspects of heterogeneity are most relevant for designing better treatments and treatment programs (see V. Morgan et al., Chapter 17, this volume).

Group 2: How Can Risk and Resilience Factors Be Leveraged to Optimize Discovery Pathways?

Issues

Much evidence indicates the presence of abnormalities that predate the diagnosis of schizophrenia. This includes enlarged ventricles in infants at genetic risk, "pandysmaturation" in infants at genetic risk, persistence of infantile motor activity into childhood, and poor motor, academic and social functioning in childhood and adolescence (Fish and Kendler 2005; Gilmore et al. 2010; Schenkel and Silverstein 2004; Schiffman et al. 2006; Walker et al. 1999). This evidence suggests that, for many people at least, schizophrenia involves a lifelong abnormality that may express itself differently over time, perhaps as a function of developmental changes in brain structure, regional activation level, and function. However, a simple unfolding of neuropathology is unlikely to account adequately for the life histories or clinical presentations of patients. For example, it is now known that environmental (e.g., toxic and psychosocial) factors affect whether schizophrenia develops and how it looks when it develops (for details, see C. Morgan et al., Chapter 9). In their discussions C. Morgan et al. aimed at integrating data across levels of analysis for the purpose of synthesizing a lifespan developmental perspective of schizophrenia, and, in doing so, addressed questions such as:

- How do environmental factors interact with genetic variables to increase or decrease the likelihood of first and later psychotic episodes?
- Do developmental data suggest a core dysfunction that accounts for multiple manifestations across the lifespan (e.g., motor, cognitive, phenomenological)?
- To what extent does abnormal subjective experience, and the concomitant distress associated with such changes, lead to further alterations in biological processes that increase the likelihood of psychosis emerging?

- Why has there been such a separation of pathophysiology from epidemiology research (e.g., on social defeat, poverty, physical and sexual abuse, drug abuse, exposure to specific viruses), and what can be done to change this?
- What do current results indicate about social and lifestyle factors and the development of schizophrenia, and what research needs to be done to understand this better?
- What needs to happen to improve our understanding of the genetic basis of schizophrenia?
- What is the role of epigenetic factors in schizophrenia?

Summary

Discussion in the group began with the recognition that if we want to prevent and treat disorders like schizophrenia, we must first understand the matrix of risk factors that underlies the etiology and pathogenesis of these syndromes. There has been considerable progress in our understanding of risk for schizophrenia and, more broadly, psychosis over the past 40 years, and this has spawned special clinics for young people considered to be at ultrahigh risk for psychosis. To date, however, most evidence suggests that although we can delay the onset of schizophrenia for one to two years in people in an at-risk mental state, we cannot prevent its eventual onset (Yung and Nelson 2011; Morrison et al. 2012). Exceptions to this include one relatively small study that used fish oil high in omega-3 fatty acids (and so, with anti-inflammatory properties) as the primary intervention (Amminger et al. 2010), and a study of a form of cognitive behavioral therapy specifically designed to address cognitive biases commonly found in people who develop schizophrenia (van der Gaag et al. 2012). In the latter study, however, the intention to treat analysis (i.e., including all subjects who entered the trial) did not reach statistical significance. Thus far, medication has not been shown to prevent schizophrenia. This discussion led to several insights and recommendations. One, agreed upon by all other groups, was that intervening at the point in time when a person begins to display prodromal indicators of schizophrenia is too late. Rather, recognizing that many psychiatric disorders share the same risk factors, an alternative—but largely untested approach—is to intervene much earlier (e.g., 9–13 years of age), when academic and behavioral difficulties typically emerge. The idea was that if we can prevent further deterioration of social and cognitive functioning during this "pluripotent risk state" (i.e., a phase during which a set of difficulties could develop into any of a number of later disorders), we are more likely to prevent schizophrenia, as well as several other conditions.

A second focus of discussion centered on the need to study the interaction of risk factors. Clearly, even the most promising risk factors only increase risk to a small degree. However, combinations and interactions of factors (e.g., genetic abnormalities, low intelligence, childhood abuse, stressful home environments, and drug use) are far more likely later to be associated with schizophrenia. Identifying protective and harmful interactions may also help us characterize heterogeneity and risk, as well as formulate rational public policies that are likely to reduce significant numbers of future cases of schizophrenia. Conversely, given the emerging recognition that psychotic phenomena in the general population are far more widespread than traditionally thought (e.g., Rössler et al. 2007), there needs to be an increased focus on factors, and their interaction, that promote resilience and reduce the likelihood of developing schizophrenia.

In addition to the group's call for greater study of the positive predictive value of *interactions* between risk factors, there needs to be greater integration of disparate fields of study. For example, integrating our understanding of genetic markers with their corresponding pathophysiological sequelae is in its infancy but has shown great promise. On a larger scale, C. Morgan et al. note that there has been insufficient exchange between fields such as epidemiology, sociology, and the neurosciences. This has led to a situation where we do not yet understand, for example, how, in terms of biology, certain risk factors (e.g., child abuse) increase the risk of developing schizophrenia later in life. Similarly, we do not yet fully understand the extent to which the incidence of alterations in specific mechanisms (e.g., hypothalamic-pituitary-adrenal axis dysfunction, viral infection) are made more likely by social and environmental factors (e.g., urban environments, poverty), although emerging evidence is beginning to reveal such relationships. There is also the longstanding issue of how the biological and cognitive factors associated with schizophrenia lead to the subjective experiences of psychotic symptoms and phenomena, such as disturbed experience of the self (Renes et al. 2013). To help resolve this issue, C. Morgan et al. stress the importance of and need for more cross-fertilization between scientists in fields such as computational modeling, neurobiology, and neurophenomenology. Greater clarity is also needed to distinguish better between concepts such as social adversity, social disadvantage, and social defeat.

Another important conclusion reached by C. Morgan et al. was that the traditional separation of child and adult psychiatric services negatively affects clinical care and research by forcing people to be seen in two different systems; it also minimizes exchange between researchers and clinicians in the different fields. They recommend that this separation be eliminated and envision a system wherein research on, and treatment of, mental and behavioral difficulties that emerge in childhood and adolescence would be informed by an understanding of factors that mediate and moderate the transition to adult forms of psychopathology. Specifically, they suggest that child, adolescent, and adult services be merged so that the population at greatest risk for psychosis can be better targeted and followed throughout the full developmental course. Finally, consensus emerged that it is not necessary for all preventive efforts to be carried out in psychiatric clinics. Already, school-based interventions have shown effectiveness for treating social and academic difficulties in young people. More efforts are needed, however, to examine the effects of these programs on young people in a pluripotent risk state, for improving cognitive, academic, and social functioning, and for reducing behavioral disturbance and later incidence of serious mental disorder (for elaboration on these issues, see Chapter 9).

Group 3: How Can Models Be Better Utilized to Enhance Outcome?

Issues

Much research on schizophrenia, especially in terms of neurophysiology and drug development, is done using rodents and, to a lesser extent, nonhuman primates. Indeed, as in other medical conditions, research on basic neurophysiology and drug development has required, and benefited from, decades of research on animals. However, schizophrenia has a number of features (e.g., language disturbances, altered sense of self) which suggest that it is a distinctly human condition. Therefore, the mission of this discussion group was to address the role of different types of modeling for furthering our understanding of, and ability to treat, the disorder. Their given questions were:

- To what extent is schizophrenia continuous or discontinuous with behavior disorders in animals?
- How can animal studies continue to enhance our understanding of the development and progression of schizophrenia and gene–environment interactions?
- Are there differences in the extent to which animal and neural network models can account for cognitive versus emotional abnormalities in schizophrenia?
- What kinds of animal research are necessary to develop new treatments?
- When treatment development is based on animal studies (i.e., when human subjective experience of the self and world is excluded), is there a limit to which treatments can be effective?
- To what extent do disturbances in the experience of the self lead to further abnormalities in biological processes, and how can interactions such as these be modeled in nonhuman systems?
- To what extent can nonhuman models account for the gene–environment interactions that are observed in schizophrenia?

Summary

Mitchell et al. (Chapter 13) affirmed the utility of animal models in investigating specific neurobiological underpinnings of schizophrenia. Importantly, however, this stance contrasts with the widely held idea that animal models can, or should be able to, recapitulate the disorder in its entirety or be used as a proxy for drug screening. Mitchell et al. note that it is obviously not possible to generate an animal model of the full syndrome of schizophrenia, given its etiological and phenomenological heterogeneity, and considering the uniquely human expression of many of its symptoms. Moreover, if schizophrenia is an open construct, the boundaries and features of which are difficult to delimit even in humans, then attempting to generate an animal model that recapitulates the disorder as a whole becomes entirely unrealistic. Furthermore, the expectation that a particular pathophysiological disturbance will manifest in an overtly similar behavioral impairment in animals and humans is not always justified. Manipulations that do not demonstrate face validity, in the sense of demonstrating an identical phenomenon in animals and humans (e.g., impaired prepulse inhibition), should thus not be rejected as irrelevant to understanding the condition, as long as it can be demonstrated that a biological process relevant to humans is being modeled. Based on these considerations, Mitchell et al. propose that the term "animal model" be used to refer to an animal that has been manipulated in a specific way that is either known to be of etiological relevance to schizophrenia or that is thought to recapitulate a phenotype of relevance to some aspect of schizophrenia phenomenology. In short, animal models can be useful to isolate and manipulate hypothesized etiological factors and their interactions, within and across levels of analysis. In this way, the group's discussions demonstrated how heterogeneity can be useful and lead to rapid advances in understanding the biology of schizophrenia, with obvious treatment implications.

Two recurring themes relevant to understanding and modeling heterogeneity are:

1. Many factors (including chance, intrauterine environment, social environment) determine how genes are expressed, and thus people with similar genetic factors may develop different clinical presentations.
2. Small changes at the micro level can interact and cascade to lead to macro-level changes in brain function, which are different from person to person.

An analogy was made to epilepsy, which is a heritable syndrome, but where the region of the epileptic focus can vary in people within the same family. Similarly, in schizophrenia, a genetic factor that leads to a neural circuit abnormality in one part of the brain might lead to one set of specific impairments (e.g., perceptual organization impairments resulting from occipital lobe abnormalities), whereas the same circuit dysfunction in another region (e.g., the frontal lobe) could lead to difficulties in organizing action plans, with a range of factors (including chance) determining in which region the abnormality is expressed. It is also possible that a single impairment (e.g., in dopamine signaling) could cause multiple problems (e.g., a reduced ability to learn from reward, working memory impairment). These types of complex relationships have not yet been modeled adequately. However, it is precisely these types of variations in expression of genetic factors, and the biological events to which they lead, as well as the interaction between these events that can be studied efficiently and effectively in animal models. Mitchell et al. state that only in this way is it likely that the heterogeneous nature of schizophrenia will be understood and that treatments truly tailored to the individual will be developed.

Mitchell et al. discuss how computational and human cellular (e.g., pluripotent stem cell) models could complement animal models. For example, within a computational framework it may be possible to predict the effect of a mutation in a specific gene on neural dynamics at various scales. It may also be possible to predict the behavioral or cognitive correlates of such alterations. It is important to note, however, that inferences in the reverse direction are much more difficult, since any phenomenon at a "higher" level can be the result of several causal pathways emerging from lower levels. For example, given a particular behavioral difference, it is usually not possible to infer what change in neural dynamics led to it. Similarly, an alteration in neural dynamics might have been caused by a change in any number of molecular components. Given this complexity, Mitchell et al. note that a major goal of experimental modeling of the effects of schizophrenia risk factors is to identify points and pathways of phenotypic convergence and possibly common pathophysiological states.

In addition to explaining pathophysiology, Mitchell et al. note that animal (and other) models can be used in longitudinal studies to clarify the development of prodromal features and the typical age of onset. The recent development of powerful small-animal neuroimaging methods offers the means to follow the same individual animal over time using a technique that provides data directly comparable to that from human patients. In short, they recommend that animal and other models not be used as proxys for the syndrome as a whole, but rather that these models are more likely to achieve advances by clarifying specific processes, their interactions, and their consequences. Because this work can be done much more quickly in animals than in humans, this new paradigm for modeling is critical for the development and targeting of treatment on an individualized basis (for elaboration on these issues, see Chapter 13).

Group 4: What Is Necessary to Enhance Development and Utilization of Treatment?

Issues

As discussed earlier, outcomes have arguably not improved significantly for people with schizophrenia over the last 100 years. However, it must be noted that only a small percentage of patients actually receive a full range of (and in some cases, any) evidence-based treatments: in the United States, for example, only 2–10% of patients who could benefit from assertive community treatment actually receive it (Lehman and Steinwachs 1998). In addition, the adoption of evidence-based practices into clinics is often slow. The mission of this discussion group was thus to address the following questions:

- To what extent are symptom severity and level of functioning driven by social factors (e.g., stigma, poor funding for mental health, unavailability of treatments, lack of evidence-based practices outside of academic medical centers)?
- Why has progress, in terms of developing new medications, apparently slowed?
- Is the continued predominance of the dopamine hypothesis based on science, inertia, and/or lack of evidence for other models?
- How can multidisciplinary work (e.g., genetics, imaging) accelerate progress?
- Why are effect sizes so small in well-designed studies of psychosocial interventions?
- Are our treatments simply not that good? Or are they good, but not acceptable to patients, many of whom may be unmotivated (e.g., due to negative symptoms, paranoia, poor insight, or severe side effects) to engage in them?
- Are the psychological models on which these are based outdated, and are there other conceptual bases upon which new behavior change methods can be based?
- To what extent could treatment outcomes be improved if there was a greater focus on social factors, in the form of, for example, widespread efforts at stigma and discrimination reduction, peer support, and education of family members, religious leaders, and other people in patients' lives?
- How do we integrate people with schizophrenia back into society in a manner amenable to both them and the community?
- To what extent are alterations of self and subjective experience primary phenomena in schizophrenia, and what are the implications of this for treatment development efforts?
- To what extent are discoveries regarding genetics informing treatment efforts? Can this happen to a greater extent than is now occurring?
- Do treatments need to be more tailored to specific symptoms or disability dimensions?
- Should drug development and clinical trials be left to the private sector and, if not, what should a government-run effort look like?

Summary

In their discussions, V. Morgan et al. (Chapter 17) considered the fact that there is still debate about what it is that we are trying to treat, due to all of the problems in defining the construct noted above. In particular, because of the heterogeneity in etiology and/or clinical features, treatments are less than maximally effective for most patients, and thus we need to develop a way of truly personalizing treatment. Other problems include deciding on what phenomena should be treated. There has been a relative separation between developing treatments that target pathophysiological processes thought to be involved in symptoms (i.e., the traditional focus of the pharmaceutical industry) and treatments which focus on reducing disability by improving cognitive and social functioning and promoting employment and independent living. In addition, there has been too little research on combinations of treatments.

V. Morgan et al. suggest that a more rational approach to treatment would begin by defining the problem space for intervention as involving primary, secondary, and tertiary levels, with the interventions and goals differing between levels. A radical proposition was that we might be able to prevent, rather than merely treat, schizophrenia if we were able to intervene early enough (i.e., primary prevention, little of which exists now for schizophrenia). For example, and as noted in several of the discussion groups, there is reason to believe— but no data yet to confirm—that an intervention to reduce cognitive decline (between 11 and 14 years of age) could reduce morbidity as well as prevent the onset of schizophrenia in late adolescence and early adulthood. However, given that schizophrenia involves multiple risk factors, important questions remain: How many different interventions would need to be developed to prevent the syndrome, and if such interventions were developed, where would they be delivered (e.g., school, after-school program, clinic)? How would such efforts be funded? Such questions speak to the need for involvement of policy makers and the larger society in efforts to prevent schizophrenia and other forms of serious mental disorders.

Even if, ideally, effective treatments were to be developed, a major problem at present is how to ensure that people who need the treatments actually receive them. For example, while there are many effective psychosocial treatments for schizophrenia, most are unavailable in typical mental health settings, even in developed countries. In addition, some countries, particularly the United States, have few mechanisms of payment for such effective treatments. Further, owing to factors such as poor insight, low motivation for treatment, and prior negative experiences with mental health professionals, many patients with schizophrenia choose not to adhere to treatment plans or attend clinics. Complicating this, many professionals are not trained in evidence-based practices for this population. Even when they are, decision-making processes engaged in by clinicians often lack sensitivity to contextual information and the patient's perspective, and thus often lead to less than optimal treatment or adherence with it. All of this speaks to the need to improve the education of people who work with schizophrenia patients and to address larger societal issues.

Recognizing the relative lack of technology used in the treatment of schizophrenia, compared to treatment of other chronic disorders, V. Morgan et al. recommend increasing the use of new momentary assessment technologies, such as handheld devices that can be used for experience sampling as well as to help monitor stress levels and the onset and offset of psychotic symptoms. Such technologies can augment interventions that have previously relied on cruder methods to assess these issues. In addition, virtual reality is a powerful tool for assessment and treatment that has been used successfully with disorders such as posttraumatic stress disorder. Thus far, this has not been used much for schizophrenia, and particularly to supplement or boost treatment effects. Other new technologies which show promise include real-time biofeedback via fMRI or variants of transcranial magnetic stimulation, to help patients reduce activity in areas related to symptoms or to increase activity in areas to enhance cognitive functioning. To date, however, the limited funding typically available for treatment of people with schizophrenia means that application of such new techniques is limited outside of clinical trials conducted in academic medical centers.

V. Morgan et al. emphasize that truly effective treatment of schizophrenia requires approaching each person with the condition as a unique person with biological vulnerabilities embedded within a matrix of environmental stressors; that is, these symptoms reflect this person with these genes and this brain in this environment with these stressors. Evidence for the necessity of this approach comes from many findings, including those on stress impact (Lincoln et al. 2009), or even walking through an urban environment (Ellett et al. 2008), on symptoms such as paranoia and anxiety, as well as the links between understimulating environments and negative symptoms (Oshima et al. 2003, 2005). In their report (see Chapter 17), V. Morgan et al. describe a treatment planning method (PROMIS) that—unlike typical approaches which focus primarily on symptoms—organizes treatment planning around disordered physiological processes, behavioral domains, and environmental stressors and other conditions. Although they recognize the utility of animal models, as discussed by Mitchell et al. (Chapter 13), V. Morgan et al. emphasize the importance of human models in driving systems neuroscience research, and the need to have these drive other scientific efforts as well (for further elaboration on these issues, see Chapter 17).

Finally, providing interventions external to the traditional medical or other treatment contexts may be useful, especially given the negative symptoms, poor insight, and other factors that reduce attendance at clinic-based treatments. For example, individual and family treatment has been provided in the home and has been effective in reducing relapse even when medication use is minimal (Lehtinen et al. 2000). In addition, cognitive behavioral therapy can be provided in patients' homes (Smith and Yanos 2009), as can cognitive remediation (Ventura et al. 2013). Although schizophrenia is typically seen as a poor outcome disorder, it remains to be seen what outcomes are possible if treatment is made more "user-friendly" in both type and location.

Further Synthesis and Final Thoughts

In our final plenary session we met to assess our overall progress and provide each group with feedback on their individual reports. In this section we wish to highlight the additional themes and ideas that emerged.

The Centrality of Cognition in an Understanding of Schizophrenia

Much evidence now suggests that schizophrenia is characterized by cognitive impairment and that cognitive impairment is an early aspect of the disorder, often predating the emergence of psychotic symptoms by more than ten years (see Kahn, Chapter 14, this volume). Alternately, it was suggested that since schizophrenia patients are impaired in all aspects of cognition, all cognitive impairments may reflect a generalized impairment, and thus these are not useful portals through which to search for clues about schizophrenia. Can these competing points of view be reconciled? What is the proper role and goal of cognition studies in schizophrenia?

First, we suggest that although cognition is definitely impaired in schizophrenia, the appearance of a generalized impairment is largely an artifact of the use of measures whose scores are confounded by multiple cognitive processes (especially attention lapses) and noncognitive factors (e.g., poor motivation or medication-related sedation). Strategies have been proposed to isolate specific impairments more effectively and to identify their neural correlates (e.g., Knight and Silverstein 2001; MacDonald and Carter 2002; Silverstein 2008), but these have rarely been used. In addition, some cognitive impairments are state-sensitive; thus, whether abnormal performance is observed can be a function of phase of the disorder (e.g., Keane et al. 2013; Silverstein and Keane 2009; Silverstein et al. 2013a). Better characterization of the covariation of specific impairments with state, as opposed to being trait (and perhaps endophenotype) factors, is an important but neglected area of research; attention to this could help us model how biology and cognition relate to symptoms, recovery, and functioning, thereby increasing the yield of cognitive treatment studies. Some of these insights have already been incorporated into clinical trials of cognitive remediation, where significant changes in performance have been found (Wykes et al. 2011). In addition to localized changes in brain activity (e.g., Wykes 1998; Wykes et al. 2002, 2011) and structure (Eack et al. 2010), recent studies are finding improvements in the functioning of neural networks in schizophrenia (Penadés et al. 2013). Such studies have the potential to improve our understanding of the effects of cognitive remediation, and of how these effects translate into normalized subjective experience, fewer symptoms, and improved functioning.

Second, we recommend that the view of what is cognitively impaired in schizophrenia should change. Thus far, this issue has been viewed in terms of the traditional categories of neuropsychology: perception, attention, memory, learning, reasoning, executive functioning, etc. When conceived this way, everything is seen as being impaired, to some degree, and therefore as evidence of a not particularly useful (for research purposes) generalized deficit (Dickinson et al. 2008). This habit of parsing cognition into pseudo-discrete functions may not, however, be the most appropriate strategy for maximally clarifying the pathophysiologies that underlie schizophrenia. Even less productive may be the strategy of identifying a single impairment, as is often done for working memory, as the basis from which all or most other cognitive impairments in schizophrenia emerge (e.g., Barch and Ceaser 2012; Wolf et al. 2006). Several reasons and examples demonstrate why this is unlikely to be a useful strategy:

1. It is clear that disorders of perception, long-term memory, and action are involved in schizophrenia in meaningful ways (e.g., Landgraf et al. 2012).
2. Some impairments in perception and attention, which do not appear to be secondary to disordered working memory (e.g., reduced visual acuity), can be demonstrated in children who later go on to develop schizophrenia, and it has been proposed that these play a causal role in abnormal neural development (e.g., Schiffman et al. 2006; Schubert et al. 2005).
3. Working memory impairment has been observed in relatives of people with schizophrenia (Conklin et al. 2005), which suggests that it is an endophenotype. Some perceptual impairments, however, have not been reported in this population or among people at risk and do not appear to be present even as late as the first episode of psychosis (Parnas et al. 2001; Silverstein et al. 2006b), thus suggesting that they are indices of syndrome progression, as well as state markers (given links with specific symptoms; Keane et al. 2013; Silverstein and Keane 2009).
4. Working memory impairments in schizophrenia are small, much smaller than in some neuropsychological patients with focal lesions whose symptoms have little overlap with schizophrenia.
5. The "work" that visuospatial working memory is assumed to do includes imagining transformations (e.g., mental rotation). We know of no evidence that such abilities are grossly impaired in schizophrenia (or present in animals used to model working memory deficits).
6. Cognitive impairment in schizophrenia typically reflects a process that is not working correctly, as opposed to a true deficit in function. Thus,

clarifying in which ways these systems are altered is a more valid perspective than generating a catalog of deficits.

Based on all of the above, we suggest that what is needed is not less concern for cognitive distinctions, but more concern for newer distinctions. One useful distinction that has already been applied to schizophrenia is that between coding and coordinating neuronal interactions (Engel et al. 2010; Phillips and Silverstein 2003; Silverstein 2010). However, recent work on canonical cortical computations—algorithms based in widespread circuitry that are used to solve a variety of problems (e.g., Carandini and Heeger 2012; Fuster 2003)—provides an emerging set of fundamental computational processes (e.g., gain control) which can be usefully applied to multiple impaired phenomena in schizophrenia (Butler et al. 2008; Phillips and Silverstein 2013).

As suggested by Mitchell et al. (Chapter 13), greater emphasis needs to be placed on the discovery of pathophysiological hubs through which etiology is channeled into behavioral and phenomenological symptoms. This approach has been useful in the study of epilepsy, and it can also be useful to study cognition in schizophrenia. Mitchell et al. agree that computational studies of neuronal dynamics can help reveal possible hubs at the level of pathophysiology and, as noted by Durstewitz and Seamans (Chapter 12), this is relevant for understanding cognition. Therefore, what is needed is continued development of modeling of causal links between brain dynamics, cognition, symptoms, phenomenology, and behavior. This will require novel ways of working between disciplines and funding agency incentives to do so.

Altering our view of how and why cognition is impaired in schizophrenia has obvious implications for how cognitive impairment should be treated and for the choice of outcome variables used in clinical trials. Importantly, however, we should not necessarily or blindly assume that treatments which target these cognitive difficulties will confer direct benefits to functioning, or that an absence of cognitive effects on these measures with treatment indicates a lack of improvement in real-world functioning. Often, as has been shown in both the traumatic brain injury and schizophrenia cognitive remediation literatures, test performance (i.e., impairment) and real-world functioning (i.e., disability) are independent of each other, and the extent of their change with treatment can vary independently of each other (e.g., Reeder et al. 2004; Silverstein et al. 2005; Wilson 1991, 1997; Whyte 1998; Wykes et al. 2012).

A Greater Number of Comparative Studies with Other Disorders Is Needed

Consensus emerged that schizophrenia is not a disease, but rather a syndrome that is best characterized as an open construct. In this way, it shares similarities with phenomena such as hypnosis: it can be characterized by alterations in consciousness, cognition, behavior, and physiology, but no one aspect of it is unique to the condition (e.g., Silverstein 1993). One implication is that it may be useful to further explore the similarities versus differences, or overlap versus nonoverlap, between schizophrenia and several other conditions, which thus far have been understudied in relationship to schizophrenia, as a means of clarifying the essential aspects of the syndrome(s).

One potential area of exploration involves the overlap between schizophrenia and other developmental disorders characterized by cognitive impairment. For example, many studies show an overlap between schizophrenia and both verbal and nonverbal learning disabilities. In terms of the former, there is an elevated rate of histories of dyslexia in people who grow up to have schizophrenia as well as in families of people with schizophrenia (Horrobin et al. 1995), and an elevated rate of schizophrenia and schizotypy in people diagnosed with dyslexia in childhood (Richardson 1994). In addition, anatomical abnormalities, as revealed by imaging, predict poor cognitive functioning in both disorders (Leonard et al. 2008), and both dyslexia and schizophrenia share specific visual processing impairments, such as in contour integration (Simmers and Bex 2001; Silverstein et al. 2009a) and magnocellular pathway processing (Revheim et al. 2006). Schizophrenia also shares social and cognitive abnormalities with nonverbal learning deficits (Silverstein and Palumbo 1995) as well as features of cognitive and social cognitive impairment (as well as genetics) with autism spectrum disorders (e.g., Lugnegård et al. 2013; Stone and Iguchi 2011). At the same time, schizophrenia and autism appear to represent opposite extremes on some dimensions (Crespi and Badcock 2008; Russell-Smith et al. 2010), and thus further investigation of the pattern of similarities and differences between these disorder classes may be quite revealing.

In addition to developmental cognitive disorders, schizophrenia is associated with a higher than normal rate of conduct disorder and antisocial personality disorder (Volavka and Citrome 2011), and these share aspects of reduced coherence in thinking and speech (Hare 1993) as well as biological abnormalities, such as reduced functional connectivity involving the frontal cortex (Motzkin et al. 2011) and cortical thinning (Ly et al. 2012). Physical and sexual abuse in childhood (Matheson et al. 2013) also increases risk for both antisocial personality disorder and schizophrenia, and its effects include violence and reduced thalamic volumes in both disorders (Kumari et al. 2013). Further investigations of these issues may sharpen our understanding of etiological and developmental pathways to schizophrenia syndromes. This would address similarities and etiological overlap between these conditions, which were proposed long ago (Bender 1959; Dunaif and Hoch 1955) but remain underexplored.

It may also be useful to study conditions which *reduce* risk for schizophrenia. Two notable examples of this are congenital blindness—where a case of schizophrenia has never been reported (Silverstein et al. 2012c, 2013b)—and rheumatoid arthritis, which occurs 70% less in people with schizophrenia than in other individuals (Mors et al. 1999). Data on congenital blindness has provided tantalizing clues regarding the role of crossmodal plasticity in reducing risk for cognitive and behavioral features associated with schizophrenia, and on the role that visual impairment may play in the development of schizophrenia. Data on rheumatoid arthritis may help clarify the role of lipid membranes, such as prostaglandin-2, platelet-activating factor, and the glutamatergic system in these two conditions (Oken and Schulzer 1999).

The Implications of Schizophrenia as a Disordered System

In addition to (stem) cellular, computational, and animal models of schizophrenia, we should not rule out the possibility of "macro" models. This suggestion is based on similarities which can be observed in complex systems, be they small or large, physical, biological, or social (Bar-Yam 1997, 2002; Csermely 2008; Freyer et al. 2012; Simon 1973). This includes characteristic dimensions such as sensitivity, stability, adaptability, and cooperation. In this view, not only biological but also social systems have the potential to inform us about processes involved in phenomena at other levels, such as brain function or behavior, in schizophrenia. To illustrate this, we suggest that examining social disorganization and its sequelae (including violence) may reveal insights about system-level disturbances associated with cognitive and behavioral disorganization in schizophrenia. For example, (a) both antisocial personality disorder and schizophrenia are associated with increased risk for violent behavior (Hodgins 2008) and reduced coherence in thinking and speech (Hare 1993); (b) increased rates of aggression in childhood are related to schizophrenia and in adulthood (Hodgins 2008); and (c) schizophrenia is associated with an increased rate of antisocial personality disorder (Jackson et al. 1991). It has also been suggested that paranoia is to thought, as aggression is to behavior (Gilligan 1996); both schizophrenia and violence reflect, in part, similar forms of breakdowns in adaptive response patterns (e.g., Broen and Storms 1966). Importantly, there are societal conditions associated with both violence and psychosis, and these conditions resemble, in terms of disruption of a system, what is found in schizophrenia. For example, it has been noted that both schizophrenia (Allardyce and Boydell 2006; Faris and Dunham 1939) and violence (Bouffard and Muftić 2006; Boyle and Hassett-Walker 2008; Sampson and Groves 1989) are more likely to occur in social systems where there is more disorganization—defined by residential instability, frequent vacant housing units, family disruption, reduced homogeneity in traditions and value systems among neighbors, less communication and cooperation between families in the same neighborhood, disrupted social closure or fewer interlocking ties or networks within communities and between families, and a general reduction in social capital (De Silva et al. 2005; Hagan et al. 1996; Sandefur and Laumann 1998). Are there ways in which symptom development in schizophrenia appears to parallel (in terms of system dysfunction) that which is found in disorganized social systems? Consider that hallucinations and delusions have been attributed to parasitic foci, where an attractor state forms and becomes isolated, and less influenced by surrounding cortical activity (Hoffman and McGlashan 1993). To the extent that this analogy is valid, what gains in our understanding of a system breakdown like schizophrenia might be won by better understanding disintegration of social systems and their sequelae? We believe that it is worth exploring whether these similarities represent more than an analogy and could even reflect causal relationships. For example, past theories and data have demonstrated excessive developmental neuroplasticity in schizophrenia and the related increased tendency for mental functioning to be molded by positive or negative features of the environment (Bender 1966; Reser 2007; Tienari et al. 2004). Rather than reflecting an isomorphism between social conditions and brain function in individuals vulnerable to such effects, it is also possible that aspects of social disadvantage may simply increase risk for outcomes such as violence and/or schizophrenia and reinforce other risk factors (Thornberry 1987; Toch and Adams 1989). We need to learn more about how this happens. In short, we suggest that the study of people with schizophrenia, or those at risk for it, could benefit from a greater understanding of brain dynamics within the context of, and in reaction to, the social environment.

At another level, future work should consider the role of the environment in planning treatment, beyond recognizing it as an etiological factor. Are there interactions between, for example, the level of social disorganization in a patient's past or current life and symptom expression or stress-sensitivity that may be relevant to treatment? Beyond this, can the dynamics of person–environment interactions form a dimension that can be used in characterizing heterogeneity? For example, is reactivity to the environment (e.g., Sturgeon et al. 1984) a variable upon which subtyping can be based? If so, what are the implications of this for diagnosis and treatment? One goal of these efforts would be to move beyond "personalized medicine" to "embedded medicine," in which treatment is based on person–environment interactions. The ultimate implication is that, as with other issues such as violence (Newman et al. 2004), intervention must be delivered at individual as well as community and national levels, in terms of public policy which affects social conditions that increase risk for schizophrenia. A novel paradigm that can express systems dynamics from molecular to social levels, model interactions between these levels, and characterize emergent phenomena such as schizophrenia appears necessary to move into the next phase of understanding and treatment. Finally, we also need to realize that it is unlikely that we will ever be able to predict completely who will develop schizophrenia. This is because all of the known risk factors are neither necessary nor sufficient—alone or in combination—for schizophrenia to occur. However, an increase in our understanding of the issues could be successful in lowering the risk for, rate of, or disability associated with the condition.

Finally, to study many of the issues described in this chapter, very large sample sizes will be necessary. This suggests the necessity of generating large databases and creating methods for investigators to contribute to and access data from them, as well as incentives for researchers to engage in this type of collaborative "cloud" research, as opposed to solely working on small datasets in individual laboratories. To study a condition as heterogeneous as schizophrenia, and to understand the relationships between multiple biological, psychological, and environmental variables and their covariation over time using mega-samples, strategies from informatics and novel data analysis techniques will have to be increasingly applied to schizophrenia research. Concurrently, there is also a role for largely forgotten idiographic methods (Allport 1962); that is, for more in-depth study of individual people as a way to understand and generate novel hypotheses about the development of schizophrenia and the factors that protect against, cause, and modify expression of the condition(s).

Heterogeneity

2

What Kind of a Thing Is Schizophrenia?

Specific Causation and General Failure Modes

Angus W. MacDonald III

Abstract

The status of schizophrenia as a disorder has been controversial since its original description by Kraepelin and Bleuler. This chapter critiques a prominent theory of schizophrenia espoused by Meehl in 1962 that spurred a great deal of research into its genetic origins and subthreshold manifestations. In particular, a decade of findings on the meta-structure of mental disorders, the development and course of at-risk youth, and genetic epidemiology can be understood as direct challenges to the idea of a specific etiology for the disorder. Instead of a well-mannered diagnostic entity, schizophrenia and thought disorder more generally delineate a psychosis spectrum linked to a number of other psychiatric outcomes, including, but not limited to, bipolar affective disorder. In addition, studies of the cognitive impairments associated with the disorder show that a generalized deficit is a prominent behavioral feature of the disorder. This chapter concludes by noting that spectrum constructs do not preclude generating and testing falsifiable hypotheses. The use of a fault tree analysis, as employed in reliability engineering, may be helpful in delineating such hypotheses explicitly. This perspective gives rise to a new set of priority questions.

Introduction

Marilyn Monroe had just died under mysterious circumstances. John F. Kennedy announced that within ten years the United States would put a man on the moon. Back in the Kremlin, Nikita Khrushchev decided that within ten days the Soviet Union would position missiles in Cuba. In the midst of such tumult, the histories written about September 2, 1962, tend to overlook the fact that a generation of genetic epidemiology, experimental psychopathology, and nosology was being inspired. A generally anodyne affair, the American Psychological Association's presidential address would be transformed this evening by Paul Meehl (1962). In his customary fashion, he managed to be both flamboyant and tightly argued. Neologisms—schizotaxia, schizotypy, and hypocrisia—were introduced. Rather than symptoms of thought disorder, they would enter the language with which we would think about schizophrenia and risk for schizophrenia. There is no hyperbole in claiming the speaker that night was one of the greatest thinkers about psychology and psychopathology of his generation, and even his century. He was also wrong.

Examining the way in which Meehl was wrong in his presidential address on schizotaxia, schizotypy, and schizophrenia allows us to approach a more intransigent question: What kind of a thing is schizophrenia? What kind of a thing is schizophrenia that it should afflict so many, across the world, irrespective of potential or position? What kind of a thing is schizophrenia that families are sundered and lives brought to a standstill or even ended? And, from a scientific perspective, what kind of a thing is schizophrenia that it might so deceive the insights of our clearest thinkers and scatter our efforts to understand it across all corners of the brain?

I begin by reviewing the theory espoused by Meehl to elucidate its dependence upon a specific etiological mechanism. Thereafter I draw together evidence from four domains, including work on the meta-structure of psychopathology, work on the development of the disorder, its course over time, and its genetic epidemiology. These data challenge Meehl's emphasis on a specific etiology by showing that schizophrenia may be best thought of as a syndrome that can be described as an open concept that is linked by correlational, rather than necessary or sufficient, relationships with various symptoms and with other forms of psychopathology. These findings are consistent with additional work on the cognitive neuroscience of schizophrenia which I will review. Finally, I suggest some ways to make progress when hypothesizing about syndromes.

A Legacy of Specificity and Falsifiability

A Falsifiable Hypothesis of Schizophrenia

From the podium that night, Meehl hypothesized that "the statistical relation between schizotaxia, schizotypy, and schizophrenia is class inclusion: All schizotaxics become, on all actually existing social learning regimes, schizotypic in personality organization; but most of these remain compensated" (Meehl 1962:832). Importantly, for this philosopher of psychology, the theory made predictions that were falsifiable. He hypothesized that few individuals with schizotaxia (carriers of a dominant schizogene) decompensated to such a degree that they were diagnosed with schizophrenia. Schizophrenia, therefore, waited at the end of a probabilistic chain of events. The chain began with the inheritance of a dominant schizogene whose proximal effect was synaptic slippage, or hypokrisia. Synaptic slippage affected the organism in a number of different ways, including "soft" neurological and psychological signs, associative thought disorder, and an exaggerated experience of negative feedback. These dysfunctions were what then led to the schizotypal personality, as distinguished by loose associations, anhedonia, ambivalence, and interpersonal aversiveness.

While Meehl broadened the phenotypes that he saw as relevant to schizophrenia, he narrowed what he called the "etiological specificity" of schizophrenia. In this regard he followed closely on Bleuler's 1911 theory that schizophrenia was fundamentally a disorder of disconnectivity (Bleuler 1911/1950), which in its turn was informed by the Kraepelinian dichotomy between schizophrenia and affective psychosis (Kraepelin 1919/1971). Meehl stated that "what makes schizotaxia etiologically specific is its role as a necessary condition" (Meehl 1962:831). In the parlance of modern developmental psychopathology, Meehl's theory stressed the *equipotentiality* of the dominant schizogene, which is to say this single cause might be manifest in a number of different ways (Cicchetti and Cannon 1999). Anyone without schizotaxia who had the same experiences might develop some other disorder, but they would not be schizotypal and they could not go on to develop schizophrenia. Therefore, it was not mothering that caused schizophrenia (a popular theory among many in Meehl's audience). Some mothering styles (or other environmental factors) might more readily potentiate a psychotic episode, but these could not be considered the ultimate cause.

This theory came at a crucial time, when psychodynamic explanations were waning and learning theories were joining with a newer kind of psychology that emphasized cognition and affect. Meehl's became a theory of central importance to schizophrenia, and psychopathology more generally. The early work of Irving Gottesman and James Shields (1967, 1972) grew from the perspective of a latent biological risk factor that may be expressed in a manner other than manifest psychosis. This developed in time into our current conception of an endophenotype, or intermediate risk indicator (Gottesman and Gould 2003). In conjunction with a modern neuroscience of brain mechanisms, this concept is central to the U.S. National Institute of Mental Health's emphasis on Research Domain Criteria (Insel and Cuthbert 2011). Taxometrics (i.e., the search for statistically distinct groups within a continuous distribution of such indicators) was developed in large part to test Meehl's schizotaxia hypothesis (Faraone et al. 2001). Meehl's work also resonated in the Danish Adoption Study conducted by Seymour Kety, David Rosenthal and colleagues, where disease manifestation was found to trace through genes more than environments, and the first signs of the unexpressed genetic liability to schizophrenia would later be described and labeled schizotypy (Kety et al. 1971). Meehl's ideas inspired the scales developed by Loren and Jeanne Chapman, who began a cottage industry of self-report scales for identifying people (generally undergraduate students) at risk for schizophrenia, and later psychosis more broadly (e.g., Chapman et al. 1994; Eckblad and Chapman 1983). In a related development, the emphasis on a specific biological etiology led them to emphasize that indicators of that biological etiology need also be specific, rather than a general risk factor (Chapman and Chapman 1973a). One consequence of this elegant idea has been a four-decade long obsession in the experimental psychopathology of schizophrenia with the psychometric properties of experimental tasks and how well or poorly they match the properties of control tasks (Chapman and Chapman 1973b, 1978; MacDonald 2009; Strauss 2001). In the ensuing 50 years, the impact of the ideas of Meehl's presidential address has been profound and far-reaching.

The bone of contention for this chapter is not that schizophrenia does not have important biological antecedents, or that Meehl had those wrong. Indeed, much of Meehl's theory, including much of his self-deprecating physiologizing, seems remarkably prescient: description of the nature of failures of synapses would be recognizable as an early formulation of a glutamate or even a GABA hypothesis. The concept of a schizophrenogenic mother is dead and this chapter will not resurrect her. Nor will this chapter fault Meehl's contention, even as late as 1990, that a single major locus underlies schizophrenia (Meehl 1990). This is known to be patently wrong. What is most interesting about the way in which the theory is wrong is that schizophrenia is not caused by schizophrenia genes per se. This assertion lies at the heart of Meehl's hypothesized specific etiology. What is interesting about it is that schizophrenia is caused by genes that encode many proteins, across many brain and nonbrain systems. These genes appear to raise the risk for many kinds of psychiatric and perhaps other kinds of disorders. Therefore, what is most striking about how incorrect Meehl was is how *nonspecific* the biological etiology of schizophrenia now appears to be.

Conceptualizing Concepts

What makes schizophrenia troublesome is that it inhabits two worlds at once. It is, on the one hand, a *diagnosis*, and thus, as a diagnosis, we can determine who has it and who does not have it. From there one can go on to ask how common it is, which treatments do and do not ameliorate its symptoms, and what pathophysiological features people with the diagnosis share. As a diagnosis, schizophrenia is a closed concept. Closed concepts are constructs to which one can provide a definition stating what is necessary and sufficient for membership. For example, one can state the necessary and sufficient conditions for a fuel-efficient vehicle or a positive urine drug screen. Just so, diagnoses according to the modern Diagnostic Statistical Manual (DSM) or the International Classification of Diseases (ICD) are closed concepts about symptoms, the impact of which will be covered at greater length in the next section. Meehl's theory also very consciously closes the concept of schizophrenia. The diagnosis of schizophrenia applies only to people with a dominant schizogene who decompensate. Anyone else who appears to be psychotic (of whom there should be very few) does not have schizophrenia but rather a "phenocopy." Unfortunately, this does not comport well with the way the world works.

The other world schizophrenia occupies is the one in which we live, wherein the disorder is part of a *syndrome*. Syndromes are different from diagnoses in that they reflect symptoms and other measurable signs that often occur together. Such descriptions are helpful in that the presence of one or more features of a syndrome alerts the caregiver to probe for the presence of its other aspects. As a syndrome, schizophrenia is an open concept. The definition of an open concept cannot be precisely specified; there may be no necessary or sufficient conditions for membership. Instances of an open concept have a "family resemblance" to one another, and we recognize members of the class by their similarity to exemplars of the concept. Most concepts are open concepts, acquired through experience rather than definitions. One could hazard a closed concept for a chair, perhaps defined as "a piece of furniture designed for a single individual to sit upon." Such definitions are beset by boundary conditions where reasonable people will disagree. Is a beanbag chair a chair? How about a comfortable rock? Of course schizophrenia is not a chair, but like a chair, a rigid definition can have unintended consequences. It can limit thinking and progress by suggesting an artificial homogeneity among the members of the class and an artificial boundary between members of the class and other informative conditions.

Schizophrenia is a kind of failure of mental functioning, but it is not a particular kind of failure. A laundry list of the symptoms people with schizophrenia or psychosis share with other diagnoses is wide and varied. As with bipolar disorder, dementia, or delusional infestations, schizophrenia can give people mistaken ideas about the motivations of others. As with Parkinson's disease, schizophrenia can cause people to perceive things that are not present. As with overmodulated posttraumatic stress disorder, schizophrenia can blunt peoples' emotional reactions to those perceptions, or like undermodulated posttraumatic stress disorder, schizophrenia can enhance those reactions. Like Alzheimer's, schizophrenia can result in problems in encoding and recalling information. Phenomenological similarities occur all across medicine and are in this sense nothing special. However, they are far from trivial insofar as they are often used to guide treatment. People with delusions are often prescribed antipsychotic medications irrespective of whether their diagnosis is bipolar disorder, dementia, delusional infestations, or schizophrenia. People with cognitive deficits have been found to benefit from computer-guided cognitive remediation regardless of whether their diagnosis is stroke, Alzheimer's disease, or schizophrenia.

The syndrome of schizophrenia is, therefore, not a particular kind of failure of mental functioning, it is a constellation of failures that tend to co-occur, and this co-occurrence can begin to take on a coherent shape when examined across a sufficiently large group of people. Although much of the remainder of this chapter will examine studies that have delineated schizophrenia as a closed concept, one with necessary and sufficient conditions designed for the purpose of reliable diagnosis, our purpose is to examine the broader features of the syndrome as an open concept. That is, although much of the data available treats schizophrenia the way we wished it would exist in the world for scientific purposes, with a specific manifestation derived from a specific etiology, we will avoid this temptation. Instead, we will try to discern schizophrenia as an open concept to see if we have the means to examine the spectrum in the world that we are given.

Finding Schizophrenia in the World We Are Given

Finding Schizophrenia in the Meta-Structure of Psychopathology

A number of analytic techniques have been devised over the years to allow researchers to ask questions about open concepts, including exploratory and confirmatory factor analysis. One advantage to such methods is that, using the data from the world we are given rather than the world we desire, they can help determine what an open concept refers to and does *not* refer to within a single framework. Such models thereby specify, within the limits of the granularity of the data, what kinds of symptoms group together and, just as importantly, what symptoms do not. Although there is a very large factor analytic literature, only recently have these tools been aimed at understanding how psychosis fits within the meta-structure of other symptoms.

Using a combination of exploratory and confirmatory factor analysis, Markon (2010) examined the structure of psychopathology using a British psychiatric epidemiological survey of over 7,000 adults. Here, psychosis symptoms were quantified using items from the Psychosis Screening Questionnaire and the SCID-II personality disorders screening questionnaire, and analyses were performed on the covariation between all the various DSM-IV criteria. In this analysis, a single factor labeled "thought disorder" was the label used for the latent symptom factor most closely associated with the syndrome of schizophrenia, or the schizophrenia spectrum. Thought disorder was identified as a higher-order factor defined by loadings from paranoia (.95), disorganized attachment (.70), inflexibility (.63), schizoid characteristics (.61), and eccentricity (.57) (see Figure 2.1a). In addition to their loadings on thought disorder, hostility (.60) also loaded on externalizing (.30), and hallucinations and delusions (.54) also loaded on internalizing (.30). Unfortunately, mania was poorly represented among the items included. In addition, the higher-order

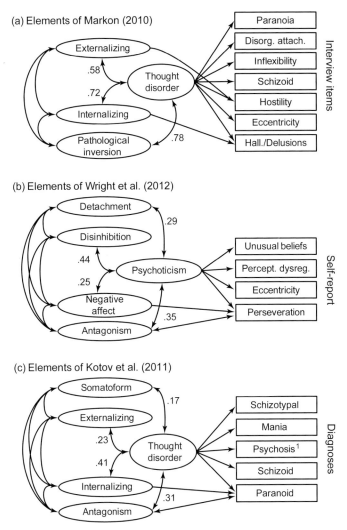

Figure 2.1 Three models of the meta-structure of psychopathology derived from (a) Markon (2010), (b) Wright et al. (2012), and (c) Kotov et al. (2011). Latent symptom dimensions are labeled within ovals. Only correlations with thought disorder/psychoticism are reported. Rectangles show indicators of thought disorder/psychoticism and are listed in descending order of loading strength. Indicators for other latent symptom dimensions are not reported. [1]Because Axis I psychotic diagnosis imposes a series of hierarchies, psychotic disorders in Kotov et al. (2011) were grouped into a single category and were not further distinguished.

thought disorder factor was closely associated with the three other factors in the model: pathological inversion, internalizing, and externalizing. A second study that used interview data from a smaller sample of schizophrenia patients, their first-degree relatives, and controls also discerned closely overlapping factors (Tackett et al. 2008). The term "thought disorder" comes from this labeling convention and is not meant to convey the more specific meanings and distinctions associated with thought disorder that occur in the schizophrenia literature.

Apparently this structure is also evident in self-report questionnaires, assuming those measures include adequate pathology across the factors. In this case, Wright et al. (2012) examined responses from 2,900 undergraduates who had completed the 220-item Personality Inventory for DSM-5. This measure was designed to be a self-report measure of the 25 facets underlying the proposed personality disorders for DSM-5. As illustrated in Figure 2.1b, the five-factor solution reported was similar to that found in previous work. There were two factors of internalizing disorders (detachment and negative affect) and two factors of externalizing (antagonism and disinhibition). Finally, there was a factor labeled psychoticism, which was somewhat related to all of the other factors (correlations from .25–.44). Psychoticism was most closely related to the facets of perseveration, eccentricity, unusual perceptual experiences, and unusual beliefs.

In both Markon's (2010) and Wright et al.'s (2012) models, there is a high to very high rate of covariation between the latent factors. One possibility is that the relationships among these higher-order factors reflect common mechanisms that fail across the two disorders that covary. For example, keeping distance from one neighbor is rated as an instance of eccentricity and as an example of an indicator variable for antagonism. However, it is useful to note that this covariation may also have been modeled by a general factor. (For a formal model of a general psychopathology factor in a study that did not evaluate thought disorder, see Lahey et al. 2012.) If such a model is appropriate, what might this general factor represent? One possibility is that it represents a general psychopathology vulnerability factor. A second possibility is that it reflects the *result* of psychopathology; that is, the impact of a dysfunction in the world that has broad implications for mood and cognition. A third possibility it that it is spurious, simply representing a response bias. This third possibility could perhaps be addressed to some degree by evaluating this covariation at a diagnostic level rather than at an interview or self-report level.

Another study to have looked at the meta-structure of diagnoses was Kotov et al. (2011) who examined co-occurrence patterns from diagnostic interviews among 2,900 adults seeking outpatient treatment. Because only diagnosis and the number of mood episodes were available for analysis, and because Axis I psychotic diagnosis imposes a series of hierarchies, psychotic disorders were grouped into a single category and could not be distinguished with finer grain resolution. Even so the best-fitting model demonstrated five higher-order factors quite similar to those of Markon (2010): internalizing, externalizing, somatoform, antagonism, and thought disorder (see Figure 2.1c). In this case, thought disorder also showed the highest loadings for schizotypal personality disorder (.91), mania (.72), psychosis (including schizophrenia and a number of other hierarchical categories of psychoses, .70) and schizoid (.51) personality disorder. Paranoid personality disorder was significantly related to thought disorder (.21), but was also related to internalizing (.36) and antagonism (.41), thereby demonstrating one of the most ill-mannered patterns of covariation of any of the 28 disorders measures. Therefore, this approach found nothing like a Kraepelinian dichotomy between schizophrenia and bipolar affective disorder.

In addition to relationships between diagnoses and higher-order factors, the Kotov model (Figure 2.1c) demonstrated relationships between the higher-order factors themselves. In this case, thought disorder showed strong relationships with internalizing and antagonism and somewhat lower, but significant, relationships with externalizing and even somatoform symptoms. These relationships, though, were notably lower than when using item-level data. This may reflect the desirable result of removing response biases which could have moderately inflated reports of symptoms across disorders. Alternatively, this covariance may mean that information about psychopathology and distress which did not meet the cut-off for diagnoses was thrown away and not further modeled. Despite using diagnosis-level data, there continues to be evidence for a general psychopathology factor. Nested within that general psychopathology factor is a thought disorder factor that does not conveniently split schizophrenia from any of these other forms of psychosis.

This growing literature approaches psychosis as an open concept, guided by the covariance structure of symptoms or items as they coalesce into facets and syndrome and then into factors. It is reminiscent of the discussion *within* the schizophrenia literature that has percolated since the 1980s, focusing on the appropriate factor structure for schizophrenia (for review, see Peralta and Cuesta 2001). Note that in contrast to the meta-structural approach highlighted herein, the within-schizophrenia factor structure work has been predicated on a diagnostic boundary around the schizophrenia construct (see Corvin et al., this volume; Figure 2.1b vs. 2.1c). It does, however, provide a level of granularity for resolving heterogeneity that the meta-structural approach has not yet addressed. The extent to which these two approaches can mutually inform each other requires further exploration.

From this approach, thought disorder emerges as a dimension of psychiatric symptoms regardless of whether it is measured using traditional psychiatric interviews or self-report scales. The thought disorder factor is most easily observed in samples with high rates of psychopathology, but it can also be observed in healthy samples if the samples are both large enough and the number of items with psychotic content is sufficiently large. In previous work, such as in the model of Eysenck (Eysenck et al. 1985), thought disorder was likely invisible because as Eysenck's scales developed, psychoticism items took on more of the content of antagonism or psychopathy, perhaps due to the relative prevalence in the population of individual differences in antagonism relative to thought disorder. Importantly from the perspective of thinking about schizophrenia as a specific, taxonic entity as Meehl would suggest, thought disorder here is distinct from, but correlated with, other psychiatric symptom dimensions, such as internalizing and externalizing, even when looking at diagnoses rather than self-report as a means to control for response bias. Still, this is only a partial control. Perhaps specificity for the diagnosis would be more evident when examining children at risk for developing the disorder.

Finding Schizophrenia in the Development of Psychopathology

Studies of children at risk for developing schizophrenia have a long tradition dating back to the early 1970s (Erlenmeyer-Kimling and Cornblatt 1987; Erlenmeyer-Kimling et al. 1984). Initially, these projects targeted only the offspring of schizophrenia patients as a means to enrich the sample that would eventually convert to psychosis. As the number of risk indicators increased, family history of schizophrenia became only one of several criteria for being considered at high risk for converting to psychosis. A number of studies have now used multivariate combinations of signs and symptoms to predict onset of a psychotic syndrome, defined as a more-or-less open concept with various criteria depending on the needs and data available to the investigators (for a review, see Goldstein et al. 2010). In one study, genetic risk for schizophrenia with recent functional deterioration, unusual thought content, suspiciousness, social impairment, and a history of drug abuse were the strongest predictors of decompensation to psychosis within 2 1/2 years (Cannon et al. 2008). A number of such schemes now exist that allow us to speak with varying levels of precision about *ultra high-risk status* and *at-risk mental states* (Goldstein et al. 2010; Yung et al. 2007, 2008).

While much of this work has focused on prediction of psychosis, there is a growing sense that such states do not *specifically* predict schizophrenia, or even a broader vulnerability to the schizophrenia spectrum. Reports focusing on the offspring of patients as a single risk factor report significant increases in any kind of psychosis with and without concurrent affective disorders and some cases show nonsignificant increases in affective disorders without psychosis (Goldstein et al. 2010). This is consistent with more recent work, mostly discussed at conferences but not yet published, where there is growing evidence that at-risk mental state criteria capture a population of youth who are vulnerable to a much wider variety of psychiatric conditions. One report on nonconverters selected for being at clinical high risk for schizophrenia reported high levels of anxiety and depression at both baseline and follow-up (Addington et al. 2011). It was noted that these levels declined from baseline but remained elevated.

Longitudinal studies of this nature are critically important if we are to realize the goal of preventing schizophrenia. Interestingly they provide some insight into the broader swathe of mental disorders for which these young people are at risk and which, in the face of effective prevention programs, might be ameliorated. However, given samples that number in the hundreds, they may be of limited power to reflect the equipotential of genes that represent the schizophrenia spectrum. In addition, the direction of inference is upside down if our effort is to challenge the specificity of schizophrenia as defined by Meehl. An ardent defender of a schizogene taxon might say that the cases which did not go on to develop schizophrenia still had schizotypy or were only mistakenly believed to be at risk for schizophrenia initially, due to imperfect inclusion criteria of the at-risk mental state. It may therefore be particularly useful to examine even larger, epidemiological cohorts to determine how the presentation of the disorder changes over time.

Finding Schizophrenia in the Course of the Disorder

Another way of asking the developmental question is to determine what, if any, changes in diagnosis occur over the course of the disorder. This question has frequently been asked in the context of whether subtypes of schizophrenia, such as catatonic or paranoid, are consistent over time. This approach is useful for the purpose of determining whether these subtypes are natural kinds. In fact, the subtypes have proved to be of such little use that they will not continue to be used in DSM-5 (Tandon and Carpenter 2012). Similar strategies can determine how the diagnosis of one disorder affects risk for other disorders later in life. One study examined more than 16,000 Danes born between 1955 and 1991 and admitted to the nation's psychiatric clinics or hospitals with a diagnosis of schizophrenia, schizoaffective, or bipolar affective disorder (Laursen et al. 2009). In relying on chart diagnoses, one must assume subsequent diagnosticians would rely to a great degree on previous psychiatric diagnoses, thereby lending an artificial high consistency to diagnoses that truly independent raters would not experience. Indeed, in smaller studies a much higher rate of diagnostic hopping has been reported using independent research diagnoses (Bromet et al. 2011). Still, given the rarity of these disorders it does not take a large number of shifts to new diagnoses to discern significant levels of comorbidity, based on the likelihood of switching from one diagnosis to another irrespective of which diagnosis was made first. Here, risk for bipolar affective disorder among patients diagnosed with schizophrenia and vice versa was twenty times higher by age 45 than the risk for either disorder in the general population. Risk of schizoaffective disorder converting to schizophrenia or vice versa was sixty times higher, whereas risk for schizoaffective disorder converting to bipolar affective disorder was over one hundred times higher. These findings are largely consistent with analyses of the meta-structure of psychopathology, insofar as the manifestation of one of these disorders markedly increases the likelihood of manifesting another subsequently. These data are somewhat difficult to interpret, however, as they may be influenced by help-seeking behaviors or reflect comorbidity only in those most liable for psychiatric conditions. One way to improve on this perspective might be to examine comorbidity within families to examine the etiologic specificity of liability genes.

Finding Schizophrenia in the Genes of Families

To test whether the relationships among latent symptom dimensions are spurious is to examine whether genetic risk is transmitted in a manner consistent with a general thought disorder factor that is also significantly related to other symptoms factors. Two early studies that asked whether genetic risk for schizophrenia was specific or shared with other psychiatric disorders suggested some level of specificity (Kety et al. 1971; Onstad et al. 1991). However, genetic epidemiology for rare disorders requires large sample sizes, and these studies were of a more modest scale. Two other recent studies have found that, in the same populations from which the earlier samples were drawn, genetic risk does indeed appear to be shared with other psychiatric disorders. One study, which looked at the families of 35,000 people with schizophrenia and another 40,000 with bipolar disorder, reported that risk for bipolar disorder among the parents and offspring of schizophrenia patients (relative risk 5.2%; 95% CI 4.4–6.2) was similar to risk for bipolar disorder among the parents and offspring of bipolar patients (relative risk 6.4%; 95% CI 5.9–7.1) (Lichtenstein et al. 2009). The relatives of bipolar patients were also at a significantly increased risk of schizophrenia, although this was somewhat lower than the risk among the relatives of schizophrenia patients. A second study went beyond bipolar disorder to examine the genetic epidemiology of psychiatric disorders more generally, including developmental disorders, in the entire population of Denmark (Mortensen et al. 2010). In addition to schizophrenia, the relatives of schizophrenia patients were at increased risk for eight other types of psychiatric disorders, including bipolar and other affective disorders, substance use, as well as personality and "other" mental disorders. The exceptions were nonsignificant increases in risk for Alzheimer's disease only in the offspring of schizophrenia patients, and for bipolar affective disorder only in the siblings, but not the parents, of schizophrenia patients. Though beyond the scope of this review, there is a growing catalog of the specific genetic polymorphisms and mutations that show up as risk factors for multiple psychiatric disorders (e.g., Fanous et al. 2012).

Taking a long view, schizophrenia, or dementia praecox, was originally conceived of as an open concept in terms of its signs and symptoms. Over time, the openness of the original diagnostic concept became reified or treated as an object for study and treatment. The fact that this may have occurred for good reasons (e.g., to generate falsifiable hypotheses or reliable diagnostic categories) may not justify clinging to the nosology of a closed concept if it detracts from scientific and clinical progress, for example by artificially polytomizing psychopathology into bins to be studied in isolation. The desire for having a closed concept of schizophrenia serves as the proverbial lamppost under which we are looking for keys we dropped; despite all the evidence to suggest that we dropped our keys somewhere in the shadows where concepts are not so clear cut, we keep hunting within this little pool of light because that is where we are comfortable looking.

I have argued here that the statistical and methodological tools developed to address open concepts show that schizophrenia shares features and risk factors with a number of other diagnoses characterized by thought disorder. Thought disorder is, in turn, related in some way to other disorders on the internalizing and externalizing spectra. Meehl and other theorists would have predicted that there would be evidence for a specific etiology of schizophrenia. Instead, not only is genetic risk for schizophrenia shared with related forms of thought disorder, there is also evidence that some aspects of genetic risk are quite general. That is, risk is conferred across psychiatric symptom factors broadly. Given that we have failed to find specificity at the level of signs and symptoms, let us now examine whether cognitive or affective mechanisms have been useful in characterizing specific aspects of thought disorder.

Pinpointing the Cognitive Mechanisms That Underlie Schizophrenia

In arguing above for schizophrenia as an open concept, I made the points that schizophrenia shares symptoms, a developmental course, and etiological factors with a number of schizophrenia spectrum disorders. It also shares etiological factors and symptoms with other serious and persistent mental disorders, such as bipolar affective disorder and, to a lesser degree, unipolar depression and the internalizing spectrum, and perhaps also attention deficit disorder and the externalizing spectrum more broadly. In addition, it certainly shares symptoms, and may share some etiological factors, with disorders of aging such as Alzheimer's disease. Still, this may not be considered particularly strong evidence against a closed concept of schizophrenia. Could there not still be a core cognitive dysfunction from which thought disorder cascades? Such a cognitive process might reflect Meehl's hypothesized development of schizotypy, such as a loosening of associations, a lack of pleasurable responses, indecisiveness, or negative responses to interpersonal interactions. Alternatively, failure might be more evident as the direct effect of schizotaxia reflected in "soft" neurological and psychological signs, associative thought disorder, or abnormalities in processing negative feedback. A failure of working memory functioning has also been suggested as an alternative failure from which other psychotic symptoms follow (Goldman-Rakic 1991). Therefore, one thing that would be useful to find would be a cognitive process—and by this I mean to include affective and interpersonal processes—that was awry in schizophrenia.

There has been evidence at one point or another for failures in all of these processes in schizophrenia patients. The problem is an abundance of cognitive impairments, because the literature is rife with such cognitive impairments. This widespread reduction in patient performance has been spoken of as a *generalized deficit*. To date, the largest and most consistent signal associated with patient performance is the generalized deficit. Across a broad variety of tasks, the generalized deficit is about 1.0 standard deviation (Dickinson et al. 2007). In comparison to effect sizes in other domains, such as pathophysiology, this generalized performance deficit remains the most reliable way to distinguish patients from controls (Heinrichs 2005). Performance deficits of this magnitude might reflect medication side effects or some other impact of the condition, such as demoralization. Medications generally do little to ameliorate this deficit in performance (Green et al. 2004a), but there is little evidence to suggest that medication or other treatment factors are primarily to blame. Prospective studies show that in many cases these deficits appear to be present well before the disorder is diagnosed (Brewer et al. 2006). The deficit may also reflect part of the genetic liability of schizophrenia. A deficit with an average magnitude of about .34 standard deviations is also found across many neuropsychological tasks in unaffected first-degree relatives of patients with schizophrenia (Snitz et al. 2006).

This evidence of a generalized deficit among patients with schizophrenia, somehow related to an unexpressed genetic liability, has been largely unsatisfactory. Some scholars have responded by arguing that the appearance of a generalized deficit, like thought disorder itself, could result from a failure of a specific process that just happens to be required for all tasks. For example, the observations of Kraepelin, Bleuler, and others inspired the notion that attentional processes may be particularly disturbed in schizophrenia patients. Intelligence and neuropsychological tests require attention to the test administrator and the stimuli to perform accurately. Therefore, a specific deficit in attention might impair performance on tests of many different abilities. Numerous studies support this notion; patients are impaired on tasks thought to tap attention, generally studied as selective attention (Luck and Gold 2008). One supportive piece of evidence is that an impairment on a putative attentional task, the AX continuous performance task (CPT), was more predictive of conversion to psychosis than any other test in the classic New York High-Risk Project (Cornblatt and Erlenmeyer-Kimling 1985). However, the case based on this evidence is problematic for the following reason: in the context of a disorder with a large *generalized* deficit, one needs to demonstrate a *differential* deficit; that is, a deficit over and above the impaired performance on other tasks (Chapman and Chapman 1973b). This is often accomplished in terms of a group by task interaction. However, this approach is only valid if the tasks being compared are psychometrically matched. This means that the tasks must be at least equally sensitive to a generalized deficit, a criterion that rarely occurs by chance. (Additional approaches to this problem have been discussed by Knight and Silverstein 2001.) Due to this tangle we do not know, for example, whether the children who went on to become schizophrenia patients performed worse on the AX-CPT because they had a specific deficit in selective attention or whether the AX-CPT was simply more sensitive to their generalized deficit.

Cornblatt and Erlenmeyer's finding is in no way unique in this regard; another large-scale prospective study found verbal memory performance discriminated prodromal individuals from controls, and among prodromes it predicted a faster conversion to psychosis (Seidman et al. 2010). Since this was not a differential deficit, it is unclear whether there was anything special about this cognitive domain, or simply something special about the test; that is, it was the most sensitive instrument to variation in a generalized deficit that was what really predicted the outcome.

The desire to understand schizophrenia by determining what specific and differential cognitive deficit underlies patients' generalized deficit has motivated some investigators to adapt experimental cognitive tasks for studying individual differences in clinical traits such as psychosis (Carter and Barch 2007; MacDonald and Carter 2002). These efforts have still to yield a definitive account of how any specific deficit may underlie the generalized deficit, much less how any specific deficit may lead to the symptoms of schizophrenia. An example of the problem comes from two research programs, which have been mindful of the interpretive snarls in patients' performance, that have examined two very different functions of the brain: visual integration and context processing. First, visual integration is the capacity to extract larger percepts from a field of stimuli. Because this is generally quite easy, Silverstein and colleagues (Silverstein et al. 2000, 2012b; Uhlhaas et al. 2004) developed a task which places a larger circle among a field of irrelevant stimuli to make it more difficult to integrate the pieces of the circle. By manipulating the strength of the signal that unifies the edges of the circle, it can be made to disappear into the background in the manner that traces each participant's psychometric function suggestive of a specific deficit. The psychometric functions of patients with schizophrenia show much lower levels of visual integration than controls. At the other end of the brain, namely the prefrontal cortex, investigators have been studying context processing. Context processing refers to the aspect of cognitive control that represents and actively maintains task-relevant information despite subsequent noise, and is in this way related to both selective attention and some aspects of working memory. Using a variant of the AX-CPT, Cohen and colleagues have demonstrated a differential deficit in one condition of the task sensitive to context processing relative to another condition that does not measure context processing but is of similar difficulty and therefore, perhaps, as sensitive to a generalized deficit (Cohen et al. 1999; Jones et al. 2010a; MacDonald et al. 2005b). While it would be simplest if patients had only one specific deficit, evidence for two or more specific deficits might be useful if they reflected different aspects of the condition. Indeed, the two tasks draw upon two very different networks: visual integration relies largely on visual cortices (Silverstein et al. 2009a) whereas context processing relies primarily on prefrontal-parietal networks (e.g., MacDonald et al. 2005a). Unfortunately, the two tasks appear to reflect on the same aspect of psychotic heterogeneity.

Performance on both tasks was related to disorganization symptoms to about the same degree: $r = .47$ for visual integration (Silverstein et al. 2000) and $r = .41$ for context processing (Cohen et al. 1999; see also Gold et al. 2012). One could be forgiven for seeing these efforts as again identifying factors related to general disease severity rather than lighting upon a special key for understanding schizophrenia.

The prominence of the generalized deficit in schizophrenia gives rise to another line of thinking. There has long been a notion of "g," known also as generalized intelligence or positive manifold (Spearman 1904). The broad nature of the generalized cognitive deficit in schizophrenia raises the question as to whether thought disorder in broader population studies and disorganization in patient studies reflects a "deficit g" or a negative manifold. Recent work addressing this question in an epidemiological sample examined the siblings and twins of people who went on to develop schizophrenia (Fowler et al. 2012). This study found that the genetic correlation between schizophrenia liability and intelligence was modest but significant: -0.26. The fact that this relationship is not stronger may reflect limitations of the assessment battery. Standardized tests for military service from which these data were drawn may not be optimized for probing the relevant portion of the distribution. Alternatively, they may reflect the fact that psychosis, while significantly related to cognitive ability, also stands apart from this additional generator of individual differences.

In this section we sought to determine whether specific cognitive processes, to include both affective and interpersonal processes, could help us determine whether there was a key cognitive mechanism whose failure led to schizophrenia. This is an ill-posed question because a definitive answer requires a thorough search of all possible cognitive mechanisms. Even so, to date the literature shows that the most prominent aspect of schizophrenia-related cognition is the generalized deficit. Many have argued for a number of specific deficits (Cohen and Servan-Schreiber 1992; Grace 2000; Hall et al. 2009; Howes and Kapur 2009; Phillips and Silverstein 2003), and it still may be the case that a single deficit, or a canonical cortical dysfunction (Carandini and Heeger 2012), rooted in a basic and widespread aspect of cortical circuitry, accounts for a wide range of observed cognitive impairments in schizophrenia. For our current purpose, suffice it to say that a cognitive perspective has not yet provided a key for unlocking the nature of schizophrenia and, given the effort and intellect thus far expended, we must consider the possibility that it cannot. It appears to have struck upon some of the same, poorly bounded psychopathological severity that we observed when considering symptoms, development, course, and genetics. Thus, let us move on to the question of how to make scientific progress with this ill-defined construct as we find it in the world. To do this, I will suggest that we turn for inspiration from experimental psychology to the engineering sciences.

Falsification and Failure Modes

For the purposes of record keeping and billing, there is nothing as reassuring as a nicely delineated category. This intuition has infused our science with the ideal of rigorous definitions and of necessary and sufficient conditions. We are seduced into thinking that we need closed concepts to formulate strong, falsifiable hypotheses. In this final section, let us consider whether science and treatment are hindered by open concepts and, if so, whether this cost can be reduced in any way.

Clinically, psychiatrists and other clinicians treat disorders one individual, even one story, at a time. In much of practice, diagnostic criteria are used to generate questions about symptoms that may not be proffered. The treatment itself, however, is more often focused on symptoms, or even anecdotes. Thus, whether you come to a psychiatrist with schizophrenia or Alzheimer's disease, if the presenting symptom is persecutory ideation you are likely to be prescribed an antipsychotic medication and it will be a D_2 dopamine antagonist. Unfortunately, if you present with prominent negative symptoms, whether from schizophrenia or posttraumatic stress disorder, the psychiatrist will not have a particularly rich armamentarium. So, a more open conception of psychosis may not imply major changes in clinical practice. In science, though, one of the reasons we are reluctant to move toward a more open conceptualization of thought disorder and the psychosis spectrum is a concern that we lack the tools for thinking about how insufficient and unnecessary factors can influence each other in a manner that results in disordered thought. Even where we are comfortable *thinking* about such things, it is a challenge to build falsifiable hypotheses about causes that are difficult to characterize. This reluctance may give way to enthusiasm, or even obviousness, should the proper tools be made available for understanding what kind of a thing schizophrenia is.

We have already alluded to some of the statistical tools that can be used to gain an understanding of the type of a thing that comprises schizophrenia. Confirmatory and exploratory factor analyses and model fitting cope with open concepts through covariance structures. Also in the domain of modeling, but of a different sort, artificial neural networks work from exemplars of a concept rather than explicit definitions. This provides neural networks with flexibility to integrate more information as a means to, for example, discriminate between two groups or predict symptoms from brain data. A branch of the engineering sciences known as reliability engineering may provide another complementary tool to allow us to integrate studies and make explicit hypotheses about open concepts.

Reliability engineering, a subdiscipline of systems engineering, addresses the capacity for a system (in this case the brain) to perform its required functions within specified parameters over the course of its lifetime. The psychiatric equivalent of nosology and pathophysiology is a failure modes and effects analysis. In reliability engineering, a failure is defined as "the termination of the ability of an item to perform a required function" (International Electrotechnical Commission 1990). It may be odd to think of the brain as an item, but it is quite common to think of it as a system "designed" as it were by evolution, with many components that work in concert. It is certainly not a stretch to imagine those components having a range of operations, and for any one of them to fail to perform its required function within that system. A failure mode is the effect, or the symptoms, by which such a failure is observed for that item.

I submit for consideration that what we call thought disorder broadly presents several failure modes of the brain. This is perhaps most easily thought of as one failure mode for the moment. This failure may represent a small and limited failure event, which simply reflects a failure in that particular state. A failure may also be more extensive, called a fault, which implies the failure is a trait and that in most cases the system will not be able to perform a required function. Such a trait could be thought of as what we call "schizophrenia." The idea of failures and faults would certainly be recognizable to Meehl, as they are reflected in the theory of schizotaxia as subthreshold signs of vulnerability or schizotypy, and only in some cases would this lead to a fault: fully decompensated schizophrenia. However, the perspective of failure modes and effects analysis would seem to open up a number of additional ways of thinking about the problem of schizophrenia, thereby providing access to formal thinking and tools for examining the brain system and its schizophrenic failure modes.

Consider Figure 2.2a, which illustrates a conceptual framework introduced by Cannon and Keller (2006) for thinking about the cascading and cumulative effects of different genes on the manifestation of the disorder. This framework was introduced to provide a unifying model for the many diverse gene systems implicated in the schizophrenia diatheses and to illustrate a systematic set of hypotheses about how endophenotypes up and down the watershed might be more or less related to their sources (genes and perhaps environmental or stochastic factors) and more or less related to their outcome (schizophrenia, or system failure). For example, Cannon and Keller use a working memory deficit as an example of an endophenotype that may be the result of many genes (not) working together. This, in turn, may work with other endophenotypes to increase risk for symptoms, which are then very likely to be manifest as a disorder. Along the way, none of the contributors to working memory deficits are necessary or sufficient. Similarly, no endophenotype is necessary or sufficient for the expression of symptoms. The particular mix of tributaries, however, will contribute to the heterogeneity of symptoms observed in the symptoms and the disease presentation, as suggested by the width of the watershed at the terminus. By providing a framework for conceptualizing how diverse factors might summate, the watershed model approaches the failure mode idea of reliability engineering. It may have a number of additional virtues, but one thing it does not do is make explicit predictions. To the contrary (correctly, I believe), it tells the scientist what kind of prediction *not* to make about schizophrenia (one gene or neurotransmitter system → one mechanism → one disorder). It

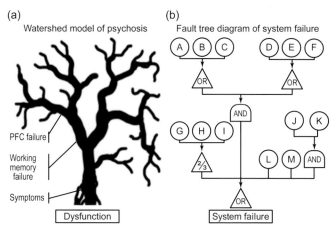

Figure 2.2 Two frameworks for understanding complex etiology: (a) Watershed model after Cannon and Keller (2006). PFC = prefrontal cortex. (b) Fanciful fault tree diagram illustrating the Boolean logic gates that connect contributory factors or events (A–M). For an AND gate, all factors must be present; for an OR gate, any factor must be present; for a "Voting OR" gate (e.g., 2/3), a minimum number of factors must be present (after Cannon and Keller 2006).

would be more useful to have a framework that allowed us to combine these tributary factors within a more explicit framework.

Figure 2.2b illustrates a way in which the watershed framework might be reconceptualized as a fault tree analysis. Fault tree analysis is an approach developed by Bell Laboratories in the 1960s for the U.S. Defense Department using a reliability engineering approach. The tree is a top-down structural model that shows the logical paths connecting various contributing causes and specifying the manner in which these can lead to a system failure (ReliaSoft 2012; Ericson 2011). Among other virtues, for example, it allows one to calculate the likelihood of a system failure if the likelihoods of the constituent events are known. For example, the reliability of the first OR gate (the likelihood of it *not* contributing to a system failure) is equal to the product of the reliabilities of all its constituent events (e.g., $R_A \times R_B \times R_C$). In turn, the reliability of the top AND gate is equal to the sum of the reliabilities of its constituent events minus the likelihood of both failures occurring (e.g., $R_A \times R_B \times R_C + R_D \times R_E \times R_F - [R_A \times R_B \times R_C \times R_D \times R_E \times R_F]$). Such calculations could therefore cascade through the diagram. Conversely, the likelihood of a system failure can be taken into account to push the algebra backward to identify failure rates of constituent events needed.

The scheme may provide a number of advantages for psychopathologists, and the study of schizophrenia in particular, insofar as it suggests ways in which we might pull apart and systematize the complexity of these disorders. The diagram and these operators just scratch the surface of the kinds of causal relationships available for consideration in such an analysis. To whet the appetite, other such gates include exclusive OR (XOR), priority AND, load sharing, standby, sequence enforcing, inhibiting, and transfer gates—each with its own computational characteristics. An additional role for a formal fault tree analysis is that it allows one to make more objective calculations about where an intervention is likely to be most effective in improving the overall reliability of the system.

From Equipotentiality to Multipotentiality

To anyone who already recoils at how people are treated like machines in modern medicine, the reliability engineering approach will only confirm their worst suspicions. However, for those of us who believe the choice between open concepts and falsifiable hypotheses is a false choice, reliability engineering, generally, and fault tree analysis, more specifically, may be particularly appealing. This is because it provides a means to model explicitly how multiple miniscule factors, none of which are necessary or sufficient, can summate into a disorder like schizophrenia with its personal tragedies, family crises, and large societal costs. In the parlance of development psychopathology, this is *equipotentiality* (Cicchetti and Cannon 1999), and it suggests that there are multiple pathway models of disease development.

Whereas one of the shortcomings of the watershed analogy is that the upstream factors might be thought to contribute inevitably to downstream manifestation, like water flowing downhill, the fault tree is better able to capture basic facts about the etiology of schizophrenia. For example, a fault tree could be used to explain why heritability of liability to schizophrenia might be 80%, but MZ twin concordance could be only 50%. That is, most of the ultimate factors are genetic, but those genes must combine with nongenetic (stochastic or environmental) factors that now serve as rate-limiting factors and reduce the genes' penetrance. It may also be relevant to heterogeneity in treatment response, and could be applied to understanding how premorbid functioning affects clinical presentation and outcomes. Most importantly, such explicit models allow for testable, albeit more complicated, hypotheses.

The careful reader may not yet be convinced of the usefulness of a fault tree analysis for understanding schizophrenia as a failure mode of the brain. The central pillars of this chapter have been the lack of a specific etiology for schizophrenia and the lack of a consistent presentation of the disorder, across people and within the same person over time. If there is one thing the fault tree analysis clearly does, it is that it uses a diverse set of risk factors to predict a *specific* failure. Figure 2.3 illustrates another fanciful diagram, insofar as the causes and relationships refer to no particular factors or disorders in particular. It does, however, suggest two ways in which fault tree diagrams may be superimposed or combined to generate comorbidity. The comorbidity between Disorder 1 and Disorder 2 is driven by the factors A–F that can lead people, depending on the status of conditions G–M, to have just Disorder 1 or both

Schizophrenia: Specific Causation and General Failure Modes 45

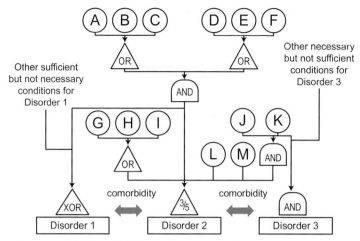

Figure 2.3 Fanciful layered fault tree diagram allowing for hypotheses about sources of comorbidity. XOR = exclusive OR; 3/5 = "3 out of 5 Voting OR" gate (see Figure 2.2). Note that here disorder represents any particular failure mode of the brain, including a failed mechanism, decreased ability, or psychiatric symptom.

Disorder 1 and Disorder 2. Disorder 1 can also be caused by other, completely unrelated, factors. Comorbidity between Disorder 2 and Disorder 3 is driven by J and K. Together, J and K are one of several potential risk nodes needed for Disorder 2. Together with another set of distinct risk factors, J and K contribute to Disorder 3. In this manner, the kinds of patterns of comorbidity observed in Figure 2.1 can be hypothesized and evaluated.

A reliability engineering research program may proceed from several angles. Currently, we examine whether people with schizophrenia are more likely to have a particular etiological factor. Fault tree diagrams could allow one to test specific predictions about rates of psychiatric morbidity in people with a particular etiological factor, or better yet across multiple etiological factors. Alternatively, one could test predictions about rates of various etiological factors in people with a particular condition. Such hypotheses could be bootstrapped using data mining techniques to identify the key etiological factors and their interactions to quantify their impact on schizophrenia or related conditions; analyses could then be extended to examine the impact of those etiological factors on "near-by" disorders.

Working With the Heterogeneity of Thought Disorder

Conceptualizing schizophrenia as being the final manifestation of a single cause propagated through a series of further probabilistic conditions and events, as proposed by Meehl (1962, 1990), was a productive spur for innovative research about the etiology of schizophrenia. Many other theories have followed that likewise propose a specific etiology. Unfortunately, theories predicated on specific etiologies have also led us astray. For example, we were overly optimistic about the probability of finding schizophrenia genes and the specificity inherent in such theories fails to predict the blurring of diagnostic lines. This blurring of lines is reflected in the comorbidity of psychotic symptoms with other affective and interpersonal symptoms, the nonspecific risk associated with prodromal states, and patients' movement across mutually exclusive diagnostic boundaries over the course of a lifetime.

In this chapter I have grappled with the question of what kind of a thing is schizophrenia, given a broader view of symptom co-occurrence, development, and genetic epidemiology. The perspective that I have adopted here still pushes us toward the conclusion that schizophrenia is part of a broader, open concept of a thought disorder syndrome. In the space available, I have not unpacked and addressed all of the data that defenders of a specific etiology and pathophysiology might bring to bear. Thus, an important discussion to have is whether there is remaining evidence of such specificity that provides an intellectual redoubt of specificity for the disorder. Such a foundation may be built of pharmacological, neuroanatomical, or cognitive evidence; in its current state, molecular genetic evidence would appear to be an unlikely source of such findings.

Thought disorder syndrome may represent a failure mode of the brain, of the kind that can be quantified and illustrated using a fault tree diagram. Such a perspective may be useful for reconciling and organizing diverse findings across the study of schizophrenia. This possibility opens a number of questions, some of which focus on the idea of a fault tree analysis itself:

1. If the brain is a graded system, its performance is more akin to small differences contributing to *variability* rather than *failures*. Is the brain really even amenable to fault tree analysis, which focuses on a dichotomous outcome?
2. If so and the brain is amenable, what are the main branches of a fault tree analysis? What are the contributory components to those branches?
3. Are our measurement tools and hypotheses of a sufficiently precise nature to test specific branches of a fault tree for psychosis?
4. Many deficits associated with thought disorder appear to propagate into diverse domains of cognition and behavior. Is a fault "web" of causality a more appropriate representation of events rather than a top-down "tree"?

Another set of questions relates to nosology. For example, how does a failure mode perspective of psychopathology reflect on new nosological systems, such as DSM-5 put out by the American Psychiatric Association or the Research Domain Criteria (RDoC) defined by the U.S. National Institute of Mental Health? DSM-5 is built around a system of categories that largely ignores these sources of comorbidity. RDoC is built around cognitive neuroscience mechanisms. As RDoC is conceptualized, mental disorders manifest as a failure of one or several of these mechanisms. Can the tools of reliability engineering be integrated within these frameworks to account for this evidence of nonspecificity of thought disorder, and its relationships with other forms of psychopathology?

Finally, is there a category of risk factor that is general across the thought disorder spectrum, or across psychopathology even more generally? For example, the established risk factors for schizophrenia include a number of general stressors such as age, season of birth, prenatal factors, substance abuse, urbanicity, minority or migrant status, autoimmune disease, and socioeconomic development. Are these also risk factors for other psychiatric and neurological disorders?

Our understanding of schizophrenia, psychosis, and thought disorder has been guided by the work of many great thinkers. Findings from the last fifty years have pushed us toward an ever more inclusive view of the causes and effects of the constellation risk factors and symptoms related to schizophrenia. If we are to systematize and build upon this literature, the next fifty years will require better use of our tools to cope with open concepts.

3

How the Diagnosis of Schizophrenia Impeded the Advance of Knowledge (and What to Do About It)

William T. Carpenter Jr.

Abstract

Schizophrenia is sometimes conceptualized as a disease entity where all patients share the same fundamental causal mechanism and core brain pathophysiology. Alternatively, it is viewed as a clinical syndrome comprising several different causal mechanisms and pathophysiologies. In the latter concept, differences between individuals may be substantial and this heterogeneity reduces the power of most study designs. Currently schizophrenia is viewed as a mental disorder with implications of a clinical syndrome and without compelling evidence of a homogeneous disease. Most investigations over the past century, however, have been designed without addressing heterogeneity. Acquisition of knowledge has thus been impeded.

Recent paradigm shifts in the schizophrenia construct are intended to provide more valid and more robust approaches to new knowledge. These include:

1. Identifying patient subgroups to enrich study cohort homogeneity on causal pathway and pathophysiology.
2. Deconstructing schizophrenia from the top down by identifying key domains of psychopathology using each domain as the pathology of interest.
3. Approaching the deconstruction from the level of the neural circuit or behavioral construct to investigate molecules, genes, and pathways related to known neural circuits and behavioral constructs which, in turn, are related to psychopathology domains.
4. Using stages of vulnerability development prior to fully manifest schizophrenia as study targets, to conceptualize causal pathways to early vulnerability that are not specific to schizophrenia as well as later stages associated with pathological variables which have greater disorder-outcome specificity.

The first paradigm shift can be informative for a form of schizophrenia that may not generalize to all forms of the disorder. The last three provide for more specific study targets but address pathologies that will cut across current disorder boundaries. The fourth paradigm, in particular, calls attention to preventive and resiliency factors as well as causal factors.

Introduction

Schizophrenia is a mental disorder with the status of a clinical syndrome rather than a specific disease entity. The central thesis presented here is that equating a heterogeneous clinical syndrome with a disease entity has impeded the acquisition of knowledge. For over 100 years, the paradigm of schizophrenia as a disease entity has been dominant, and the implications of this have been profound. Before discussing the limitations of the disease model and alternatives with greater heuristic value, a definition of key terms may be helpful.

- *Nosology* refers to the classification of medical diseases. A disease class has greatest clarity when based on known etiology/cause and/or specific pathophysiology. Diagnostic classes are also necessary in the absence of etiology/pathophysiology knowledge and may be better considered disorders or clinical syndromes.
- *Disease entity* is a disease based on etiology/pathophysiology, presumed or proven, that distinguishes it from other diseases. It is the knowledge of cause and mechanism that distinguishes a disease entity from a disorder, and uniformity of cause and mechanism that distinguishes it from a clinical syndrome.
- A *syndrome* is the association of several clinically recognizable features such as symptoms and signs that often occur together in patients.
- A *mental disorder* or *mental illness* is a psychological or behavioral pattern which deviates from normal and is generally associated with distress, dysfunction, and/or disability. Mental disorders are generally defined by a combination of how a person feels, acts, thinks, and/or perceives.
- A *domain of psychopathology* comprises signs and symptoms conceptualized as relating to a single construct. In schizophrenia, diagnosis is based on a combination of signs and symptoms. Domains attempt to reduce heterogeneity by defining unified symptom/sign constructs such as hallucination or avolition.
- *Dimensions of psychopathology* place domains of psychopathology on a severity continuum in terms of specific observable variables or hypothesized underlying processes.
- *Deconstruction* means identifying domains of psychopathology within a syndrome, recognizing that persons classified within the syndrome

- will vary as to which domains are actually present. In schizophrenia, no domain is unique to the disorder.
- For present purposes, let us consider a *behavioral construct* to be a behavior that can be specified as a phenotype whose physiology involves a known neuroanatomic framework or neural circuit (e.g., fear, working memory).
- *DSM* and *ICD* are diagnostic manuals published by the American Psychiatric Association (APA) and the World Health Organization to provide a nosology with information and criteria for diagnosis of each class. In DSM, "A criteria" define the symptoms required for diagnosis of a case.
- A *schizophrenia construct* is an organized view of the concept of schizophrenia including principle defining features. Diagnostic prototypes and criteria relate to the construct and may change as the construct is revised over time.
- *Heterogeneity of schizophrenia* poses the central problem for a clinical syndrome where individual cases vary substantially on key features. One case may have disorganized thought and behavior and negative symptoms, another may have hallucinations and delusions without negative symptoms or disorganized behavior, while still another may have disorganization with psychomotor abnormalities but without negative symptoms. A study design that includes such diverse cases while testing for a neural circuit for hallucinations or a gene associated with negative symptoms or a treatment for disorganization is weakened to the extent that some or many subjects do not actually have the phenomena of interest.
- *Medical model*: schizophrenia has traditionally been considered a medical disorder. A medical model implies disease pathophysiology, but an understanding of pathways to the pathophysiology as well as an understanding of the consequences of this pathology require integrating information across levels of human functioning. A broad medical model encompassing social, psychological, and biological data and concepts is essential to achieve an integrated view that relates to individual cases as well as to an overarching construct. Use of a medical model, however, does not imply biological reductionism, since causes of psychopathology can come from any number of levels, including interpersonal and environmental ones. Experimental designs are reductionistic by necessity (at any level of a functioning organism). Biomedical reductionism is adequate for certain study designs, but not for a construct of schizophrenia.

Schizophrenia has traditionally been conceptualized, via the medical model, as a disease and until recently, this concept has driven research efforts and understanding of what appears instead to be a clinical syndrome or a class of disorders. Much has been accomplished, but most study designs treat schizophrenia as a disease entity, in part because methods to reduce heterogeneity decisively have only recently emerged. The proposition addressed here is that schizophrenia, when treated as a disease rather than a clinical syndrome, has *limited the acquisition* of knowledge.

The history of the dominant construct can be briefly summarized as follows. In the late nineteenth century, the disease entity approach was validated through the identification of various infectious diseases, including tertiary syphilis—a disease entity associated with psychosis. Mental disorders had putative disease entities such as hebephrenia, catatonia, and paranoia, each associated with psychotic symptoms. In the late nineteenth century, Kraepelin (1919/1971) postulated a unifying pathological process involving the unique combination of avolition (e.g., weakening of the will such that initiation of action and thought are impaired) and dissociative psychopathology. He also proposed dementia praecox as a disease separate from manic depressive disease. Bleuler (1911/1950) provided strong support for the disease entity concept by viewing dissociative pathology as primary and fundamental in all cases. However, he also raised the issue of syndrome, referring to the group of schizophrenias. The behavioral manifestations of dissociative pathology were broad and often subtle (e.g., separation within thought, loss of intimate connection between thought and action and thought and feeling, fragmented or vague speech). When schizophrenia was diagnosed with subtle abnormalities of thought, the boundary of the disorder became inflated and the link with avolition was weakened.

The concept of schizophrenia changed again in the middle third of the twentieth century. Schneider (1959) attempted to clarify and narrow the concept by emphasizing understandability of special experiences as being the symptom pathology of first importance in identifying cases. Experiences such as hearing a voice with a running commentary, referring to the patient in third person, or bizarre forms of delusions such as thoughts being inserted by alien forces shifted the concept toward reality distortion pathology and away from avolition and disorganized thought and experience. Langfeldt (1939) explicitly addressed the perceived problem of diagnosing schizophrenia in cases that did not have the same affliction as found in true schizophrenia. Using Schneiderian first-rank symptoms and other reality distortion phenomena (e.g., massive derealization) he separated true schizophrenia from pseudo-schizophrenia in an attempt to define the core or nuclear aspects of the construct.

It is important to note that with dementia praecox, Kraepelin established a putative disease entity based on avolition/dissociative pathology. Bleuler accepted this view but considered dissociative pathology the fundamental and primary pathology. Bleuler made clear that reality distortion symptoms were secondary phenomena and not fundamental to the construct. However, applying the Bleulerian concept also involved inward withdrawal (autism), affect pathology (e.g., restricted experience and expression of emotion), ambivalence in thought and action, as well as associative pathology—often referred to as Bleuler's four As. These manifestations may be subtle in many cases, and the movement from the Kraepelinian emphasis on observable signs and course criteria to less observable psychological constructs had the unintended consequence of broadening the concept and raising doubts as to the validity of the diagnosis. The construct based on work by Schneider and Langfeldt was viewed as addressing this problem by separating true or nuclear schizophrenia from pseudo-schizophrenia. This was accomplished, however, through a major shift in the construct away from avolition and dissociative pathology and toward reality distortion without validating the true versus pseudo distinction. Nonetheless, this latter approach was very influential as DSM-III was prepared and published in 1980.

A final note before proceeding to DSM-III: Jaspers (1963) viewed impaired empathy as fundamental to schizophrenia. Empathy here refers to the sense that one appreciates the mind and feelings of another through automatic processing not dependent on complex language communication. Impaired empathy was considered to be in the same class of special schizophrenia experiences described by Schneider. Schneider (1959) identified a set of "first-rank" symptoms and made a clear distinction between understandable delusions (e.g., delusions of poverty in a depressed person) and bizarre delusions (e.g., believing an unknown external source is responsible for one's thoughts). First-rank symptoms represented a pathology of ego boundary or reality distortion quite distant from the avolition/dissociative pathology described by Kraepelin (1919/1971). Jaspers's concept was, perhaps, misconstrued and poor rapport may better represent impaired empathy. Note that the most discriminating features in differential diagnosis are omitted from the A criteria in DSM-III and are only partly in place in DSM-IV. These are: poor rapport, lack of insight, and restricted affect (Carpenter et al. 1973).

With its publication in 1980, DSM-III put a new paradigm for diagnosis in place. Previously developed for research, explicit criteria on which to make a diagnosis were formulated for each disorder. Clinical, research, and epidemiologic diagnoses were to be based on ascertainment of the specific criteria. Previously, clinicians would rely on training and experience, an understanding of prototypes for various disorders, and a general description of each disorder. The approach now included the explicit determination of criteria that needed to be met in each case. With the broad international acceptance of DSM-III, the schizophrenia concept at the symptomatic level was explicitly related to delusions, hallucinations, disorganized thought, and psychomotor abnormalities and required the presence of at least two of these four psychopathology domains. With DSM-III the field had operationalized criteria with documented reliability. Little noticed was the remarkable shift in concept in the direction of reality distortion and away from avolition. Negative symptoms, characterized by experience and expression of emotion and avolition/anhedonia/asociality, were not included in the DSM-III criteria. Cases of schizophrenia could now be defined by the presence of just hallucinations and delusions. This is remarkable considering two empirical findings in the 1970s. First, symptoms of first rank had been documented in other mental disorders. Separating broadly defined schizophrenia into true and pseudo-schizophrenia with Schneider's or Langfeldt's criteria failed to support validity based on disorder development, course, outcome, or functional status. Second, the most discriminating features between different psychotic disorders were restricted affect, poor rapport, and poor insight, none of which were included in the A criteria. Parenthetically, negative symptoms were added to the A criteria in DSM-IV, but so was the criterion that a single bizarre delusion or hallucination could fulfill A criteria.

DSM-IV, published in 1994, contained two significant changes related to the avolition/reality distortion dialectic. First, negative symptoms were added to the A criteria and now two of the five were required: delusion, hallucination, disorganization, psychomotor, and negative symptoms. Second, an exception was made to allow A criteria to be met by a single hallucination or delusion if considered bizarre. Bizarre, for practical purposes, can be considered a first-rank symptom of Schneider. Parenthetically, schizoaffective disorder was introduced in an attempt to address cases where schizophrenia criteria are met in the context of extensively overlapping major mood episodes.

By viewing schizophrenia as a disease entity based on Kraepelin's dementia praecox, reinforced through Bleuler's view of the primary and fundamental pathology being found in all cases, the disease entity concept was expanded to a construct of "schizophrenia as a brain disease"—a construct which was used to endorse a medical model and, in theory, to reduce stigma. The concept has been further reinforced as the neurodevelopmental hypotheses gained traction. Most research data is generated in study designs which compare people diagnosed with schizophrenia to psychiatrically healthy subjects or schizophrenia as a disorder compared to other disorders. Only a fraction of reported studies attempt to reduce heterogeneity and relate study findings to a specific pathology. The impediment to acquisition of knowledge can be seen in genetic studies where the design accepts a diagnosis of schizophrenia as the phenotype despite the broadly held view that multiple phenotypes exist and vary from case to case. Another telling example is the equating of schizophrenia with psychosis and sixty years of developing dopamine antagonists for psychosis and viewing them as anti-schizophrenia drugs. This resulted in sixty years of "me-too" drug development for one aspect of the construct while the therapeutic needs in other critical pathologies (e.g., impaired cognition and negative symptoms) remained unmet (Buchanan et al. 2005; Kirkpatrick et al. 2006).

The remaining discussion assumes that the proper construct at this point in time for schizophrenia is that of a clinical syndrome with heterogeneity of manifestation across individual cases (widely documented) and presumed heterogeneity at the level of etiology and pathophysiology. This heterogeneity, if not addressed in study designs, weakens the opportunity for discovery. To illustrate the problem, imagine an imaging, genetic, postmortem study of dementia where a relatively small number of cases are compared to controls without a brain disorder. If subjects are selected based on impaired short-term memory, the study cohort may include cases of Alzheimer's disease, multi-infarct dementia, Pick's disease, normal aging, and pernicious anemia. This mixture will reduce the chances of discovering pathology associated with each specific form of dementia. Fortunately, for many forms of dementia there is sufficient knowledge to reduce the heterogeneity with diagnosis. As a clinical syndrome, schizophrenia presents, however, some of the problems associated with dementia before separate disease entities could be defined.

Why Classification Failed

Classification is, of course, essential for many valid purposes. To advance knowledge of disease etiology, pathophysiology, treatment, prevention, and cure, the dominant paradigm is quite limited and represents a flaw, often fatal, in many research designs. Failure to address heterogeneity in schizophrenia has resulted in the following:

- Biomarkers or endophenotypes are not established to validate the diagnosis in the individual case.
- Drug discovery cannot be rationally based on known molecular pathophysiology.
- Risk factors have not led to effective prevention.
- Psychopharmacology has made very limited progress since chlorpromazine was introduced sixty years ago. Scores of "me-too" antipsychotic drugs have been approved but only clozapine is recognized for its superior effectiveness.
- Many genes with small effects have not yet been linked effectively to meaningful phenotypes (as suggested by genome-wide APA studies).
- Group findings with many variables that distinguish a schizophrenia cohort from a non-ill cohort may have little discriminating power between schizophrenia and "near-by" psychotic disorders.
- Power in any research design is presently reduced by the incorrect expectation that all subjects with schizophrenia actually have the pathology related to the variable of interest.
- When a variable hypothesized to be related to schizophrenia is observed in all subjects with the diagnosis, likely explanations include antipsychotic drug effect, shared lifestyles, or that the variable is related to psychosis in general and is thus not specific for schizophrenia.

Genome-wide APA studies generally treat a clinical syndrome as though it were a disease entity. Classification, however, has failed because of heterogeneity across diagnosed individuals on variables such as risk factors, etiology pathways, developmental pathways, endophenotypes, onset, manifest symptoms, course, treatment response, and associated features, such as neurological soft signs and cognition. Currently, there are no pathognomonic manifestations of schizophrenia and no biomarker with the sensitivity and specificity required for diagnosis in the individual case.

Overcoming Classification: How to Accelerate the Acquisition of Knowledge

The field of research has generally accepted schizophrenia as a disease entity paradigm for the past hundred years or so. To shift from a dominant paradigm is usually difficult, although alternative paradigms, new and old, are available:

1. Reduce syndrome heterogeneity by identifying subgroups, each representing a putative disease entity. Examples: traditional subtypes such as hebephrenia or paranoid, good versus poor premorbid development, and deficit versus nondeficit based on presence or absence of primary negative symptoms.
2. Deconstruct the syndrome construct into psychopathology domains. Examples: Strauss et al. (1974) stipulates six domains, Cuesta and Peralta (1995) eight domains, and eight domains have been placed in Section 3 of DSM-5 (American Psychiatric Association 2013).
3. Establish separate developmental pathways, each being an independent variable in the study design. Example: the interactive developmental model represented in Figure 3.1.
4. Establish stages of psychopathology development with each stage being an independent variable. Examples: stages in the neurodevelopmental model (Weinberger 1987; Murray and Lewis 1987) or clinical staging (Hogarty et al. 1995; McGorry et al. 2010).
5. Address psychopathology from a behavioral construct representing a phenotype closely related to clinical manifestations on the one hand and to brain anatomy on the other. Example: the NIMH Research Domain Criteria initiative (RDoC 2011).
6. Hypothesize a biomarker as the point of entry and select subject according to the presence of the biomarker at a level thought to represent pathophysiology (e.g., predictive pursuit in eye movements). Example: select subjects according to deviation from norm on a biochemical, imaging, or psychophysiological phenotype (Braff et al. 2007; Turetsky et al. 2007; Schork et al. 2007; Gur et al. 2007; Thaker 2008).

Earlier efforts to reduce heterogeneity involved identifying subgroups within the schizophrenia syndrome. Traditional subtypes of schizophrenia reflect some important subgroup differences (e.g., genetics of hebephrenia vs. paranoid subtypes), but have not proven to be strong heuristics for investigative purposes. This, in part, is because subtypes are not stable within the individual

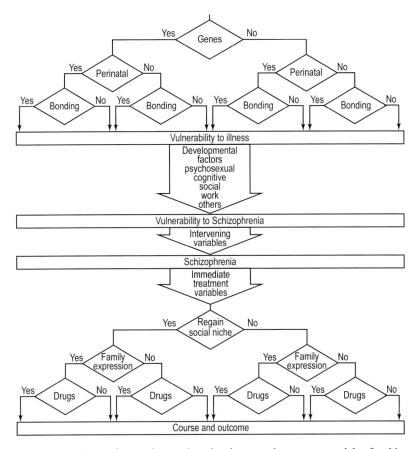

Figure 3.1 Schema for an interactive developmental systems model of schizophrenia; reprinted from Strauss and Carpenter (1981) with permission from Springer Science+Business Media B.V.

and many features are shared across the subtypes. A more robust approach has been to subdivide schizophrenia according to dichotomies, for example, reactive versus process schizophrenia, acute versus insidious onset schizophrenia, and good versus poor prognostic schizophrenia. These subdivisions were robust from a premorbid and course of illness perspective. Shortcomings, however, include (a) the failure to validate the etiological implications of the process and reactive dichotomy, (b) presuming that insidious and acute referred to the nature of onset of the same disease rather than distinguishing separate etiopathological pathways, and (c) conceptualizing a poor developmental pattern as prognostic rather than as an early manifestation of a syndrome subgroup.

These earlier attempts failed to establish a strong candidate disease entity within the schizophrenia syndrome. More recently, investigators at the Maryland Psychiatric Research Center segregated schizophrenia into a subgroup with and a subgroup without primary negative symptoms. Hypothesized as a disease entity, several lines of evidence support this proposition (Messias et al. 2004; Kirkpatrick et al. 2001).

Deficit schizophrenia systematically differed from schizophrenia without primary negative symptoms on a range of variables, including epidemiological risk factors and aspects of neuroimaging and postmortem tissue analysis. However, while using a single domain of pathology to reduce heterogeneity in the deficit subgroup, it fails to address heterogeneity in the larger nondeficit subgroup. The hypothesis that deficit schizophrenia has a distinct etiopathophysiology that leads to psychosis (i.e., different from the pathway to psychosis in nondeficit cases) is interesting, but it may not be as robust as a construct that views negative symptoms as, or along a continuum of, a domain of psychopathology. This alternative interpretation—that primary negative symptoms are a domain of pathophysiology rather than necessarily marking a different pathway to psychosis—may offer a stronger heuristic approach. In this context, negative symptoms represent an independent variable in study designs. Domains of psychopathology form the primary target for etiological, pathophysiological, and therapeutic discovery (Carpenter et al. 1988; Carpenter and Buchanan 1989).

This explanation was advocated in 1974 with negative symptoms, positive symptoms, and pathology observed in the interpersonal sphere as three candidate domains (Strauss et al. 1974). This framework was prompted by Strauss, who considered dimensions as alternatives to categorical classification (Strauss and Carpenter 1975; Strauss 1969).

In recent years, this approach has gained traction with domains of cognition and negative symptoms being identified as critical unmet therapeutic needs. The DSM-5 Psychosis Work Group has pursued the deconstruction paradigm in parallel with categorical classification. Based on prior evidence for the utility of a dimensional approach to schizophrenia (Strauss et al. 1974; Peralta and Cuesta 2001; Cuesta and Peralta 1995), DSM-5 describes eight dimensions in Section 3, each conceptualized as a domain of psychopathology important to psychotic illnesses, but varying in manifestation in subjects within each of the relevant syndromes. The psychopathology domains, rated from 0–4 for severity, are: delusions, hallucinations, disorganized thought, psychomotor abnormalities, restricted affect, avolition, depression, mania, and cognition impairment. This paradigm raises the possibility of identifying a range of pathologies as the target of investigation with a diagnostic class being involved for general relevance; however, the domain of pathology has specific relevance. Here one seeks genes for, say, depression across diagnostic classes rather than genes for major depression disorder, or for reality distortion across syndromes rather than for schizophrenia as defined by reality distortion. There is potential here for a dramatic reorganization of scientific enquiry.

Four Paradigm Shifts to Consider

Heterogeneity characterization is involved in the first three paradigms that will be discussed whereas the fourth involves staging.

The first paradigm views disease entities within the syndrome as independent variables rather than schizophrenia at the syndrome level. In a sense, this is an old and frequently tried paradigm that has not been very productive. The traditional subtypes have not provided a strong heuristic approach. Patients often have features of several subtypes, and symptoms and subtype presentation can vary significantly across psychotic episodes in the same person. Other approaches have been more closely associated with course and prognostic variables (e.g., good vs. poor prognosis, process vs. reactive). These dichotomies have lost traction as subgroups for two related reasons. First, the tautology between predictor and predicted became evident: asociality prior to psychosis predicts asociality after psychosis, prior occupational function predicts future function, etc. Second, many factors associated with prognosis are now conceptualized as early morbid manifestations of the disorder, not independent moderator factors. Prominent in more recent work within this paradigm is the development of deficit schizophrenia as a putative disease entity. This subgroup appears validated by factors associated with etiopathophysiology. However, separating deficit schizophrenia from other schizophrenia leaves the larger cohort without precise defining features, and the number of disease entities that remain is not known. Despite modest progress to date, future research within this paradigm is expected to be more robust when the syndrome can be subdivided based on biomarkers. The term "biomarker" refers to variables robustly associated with a pathology that results in a more valid grouping of cases. Increased homogeneity at the biological level is assumed, but biomarkers may be derived from any level of functioning (e.g., genetics, physiology, cognition, behavior).

Consider now a second paradigm based on deconstructing the syndrome into psychopathology domains. The fundamental assumption here is that symptom/sign complexes can be defined with greater homogeneity, and dependent variables may relate to a domain rather than to all subjects in a syndrome cohort. From a clinical perspective, psychopathology domains represent the evaluation and treatment targets that clinicians address. This paradigm calls for ascertaining the domain of interest in each subject and shifts the discovery process in the direction of psychopathology across diagnostic boundaries. From a drug discovery perspective, each domain can now be viewed as a potential indication for regulatory approval. In this framework, investigators will seek gene associations for depression, reality distortion, or psychomotor abnormalities either within the syndrome, by identifying specific cases, or across disorder boundaries. DSM-5 introduces in Section III a series of psychopathology domains to be used as dimensions across psychotic disorders.

The third paradigm calls for a more aggressive integration of neuroscience and behavioral science to identify behavioral constructs that have specific relationships to neural substrates; as such, they constitute the independent variables needed to address fundamental mechanisms as they relate to pathophysiology at the neural circuit level. Here it is assumed that several behavioral constructs are related to psychopathology (e.g., impaired positive valence in anhedonia related to depression). NIMH has prioritized moving research in this direction, cutting across diagnostic boundaries and levels of severity.

The fourth paradigm, which relates to staging, assumes that several pathways are involved in the formation of a general vulnerability toward mental disorders (see Cadenhead and de la Fuente-Sandoval as well as C. Morgan et al., this volume). During development, other factors may determine the direction taken in progression toward a diagnosable disorder. At the first stage, there may be very broad sharing of risk factors. Moving toward a particular disorder in the second stage may involve a more discrete set of variables. A third stage relates to the onset of a disorder or of psychopathology domains. Finally, still other factors may be involved in altering the course once a disorder is present. In this paradigm, investigators need to determine the stage of the independent variable and create evidence that bears on the development of features at that stage. The first two stages are particularly relevant to primary and secondary prevention as well as to the study of resiliency.

Two projects are nearing completion and will be influential in shifting research focus to domains of pathology or behavioral constructs:

1. DSM-5 and dimensional ratings of symptom domains across psychotic disorders. Hallucinations, delusions, disorganization of thought, restricted affect, avolition, psychomotor abnormalities, cognition, mania, and depression are specified in Section 3. These domains require clinical evaluation and treatment, but can also impact on discovery by orienting science away from syndrome and toward individual domains of pathology (e.g., cognition, mood, arousal, motor functioning). For instance, we think that the Food and Drug Administration in the United States and other regulatory bodies will recognize the domains as a consensus in the field and consider them as indications for drug approval. In time this may extend to include several disorders in clinical trials based on sharing the domain of interest. Already in place are methods for addressing cognition and negative symptoms in the context of schizophrenia (Buchanan et al. 2005; Kirkpatrick et al. 2006).

2. RDoC, with its elaboration of five behavioral constructs and related neural circuit substrates, represents NIMH's tactical approach, consistent with their strategic plan to develop information on pathophysiology at the neural circuit level for mental disorders. It is a direct repudiation of discovery based on clinical syndrome classification. Nonetheless, it will be essential to relate the behavioral constructs to specific clinical

manifestations of disorders. Guidelines for this translation are currently being developed.

For purpose of illustration, five behavioral constructs with demonstrated links to psychopathology are shown in Table 3.1. Methods are currently being developed to integrate the two approaches discussed above.

The deconstruction of schizophrenia according to the DSM-5 domains of pathology framework and RDoC behavioral construct/neural circuit framework and integration into clinical and preclinical study designs will change the acquisition of knowledge in the near future. The first paradigm shift, which identifies putative disease entities within the schizophrenia syndrome, will be available as biomarkers gain traction in separating a subgroup from the whole. Recent illustrations involve latent class analysis in gene association studies, where candidate genes appear to separate a deficit form of schizophrenia from other subgroups (Fanous et al. 2008; Holliday et al. 2009).

The fourth paradigm shift has been introduced for clinical therapeutics, where the nature of interventions are different for various stages, for example, prodromal, first psychosis, impaired cognition addressed during clinical stability, rehabilitation of functioning in chronic stages, etc. (Hogarty et al. 1995; McGorry et al. 2010). Moving this paradigm for discovery involves reconceptualizing the schizophrenia psychopathology. Rather than a clinical syndrome that evolves over time, the paradigm suggests that early risk factors may produce a general vulnerability for mental dysfunction. Later risk factors may shape the development of disorders where schizophrenia is only one of perhaps many disorder outcomes. Once a particular disorder is present, the focus of study may evolve from primary etiological and preventive factors to secondary

Table 3.1 Examples of behavioral constructs with demonstrated links to psychopathology (courtesy of Bruce Cuthbert, NIMH).

1. Negative Valence Systems • Acute threat ("fear") • Potential threat ("anxiety") • Sustained threat • Loss • Frustrative nonreward	3. Cognitive Systems • Attention • Perception • Working memory • Declarative memory • Language behavior • Cognitive (effortful) control
2. Positive Valence Systems • Approach motivation • Initial responsiveness to reward • Sustained responsiveness to reward • Reward learning • Habit	4. Systems for Social Processes • Imitation, theory of mind • Social dominance • Facial expression identification • Attachment/separation fear • Self-representation areas
5. Arousal/Regulatory Systems • Arousal and regulation, multiple • Resting state activity	

prevention. At each stage of disorder development, resiliency factors as well as causative/promotional factors are relevant (see Figure 3.1 for a representation of staging; Strauss and Carpenter 1981). Current studies that report, for example, overlap in candidate genes between schizophrenia and bipolar disorder would be reconceptualized in this staging paradigm as genes contributing to vulnerability to mental disorders, whereas candidate genes unique to each disorder would be conceptualized at the stage of a vulnerable individual developing a specific disorder.

The above paradigms relate to methods for obtaining knowledge on etiology, pathophysiology, prevention, treatment, and cure. Another shift in concept, perhaps paradigm, may be additive or synergistic with the above. This relates to concepts of resiliency and compensatory processes. As the disorder develops in any individual, a series of adaptive challenges unfolds. Attempts to prevent or repair dysfunctional mechanisms are central to prevention and therapeutics. An alternative view relates to determining how individuals successfully cope with impairment and reinforce natural strengths and/or determine how compensatory mechanisms can be enhanced.

Conclusion

Conceptualizing schizophrenia as a disease has impeded the acquisition of knowledge because of the heterogeneity of individuals with the diagnosis and the clinical syndrome status of the disorder. Four paradigms are currently available and may accelerate discovery in the near future by addressing heterogeneity. These paradigms identify subgroups as putative disease entities using psychopathology or biomarkers, deconstruct the syndrome into psychopathology domains, use behavioral or neural circuits as independent variables, and reconceptualize the development of mental disorders in stages, progressing from general vulnerability to more specific psychopathology outcomes. An additional consideration addresses personal characteristics and compensatory mechanisms that enable an individual to minimize illness effects and progression. It is expected that advancing knowledge on etiology and pathophysiology will provide a basis in the future for substantial reconsideration of the classification of mental disorders and the schizophrenia construct.

4

What Dimensions of Heterogeneity Are Relevant for Treatment Outcome?

Leanne M. Williams and Chloe Gott

Abstract

Schizophrenia is a disorder, or a class of disorders, of cognition. Defining features include a loss of coordination in core perception, attention, memory, and executive functions together with the dysregulation of emotion. These features are the strongest contributors to burden of illness. Diagnostic criteria, clinical trials, and popular conceptions typically focus, however, on the more florid positive symptoms of psychosis, such as hallucinations. As a result, impairments in cognitive–emotional function remain largely undiagnosed and untreated, with no current treatments in routine use that target these impairments. The evidence base for developing new treatments requires cognitive–emotional measures that link to functional capacity as well as to brain changes involved in schizophrenia pathophysiology.

This chapter looks at five aspects of cognitive–emotional function in schizophrenia: Which cognitive–emotional impairments characterize schizophrenia patients at first onset? Are functional capacities predicted by these impairments at first onset? What brain systems are involved? How do cognitive–emotional impairments, and their relationships with functional capacity and brain function, progress over time? What are the implications for treatment outcomes? Focus is on the first episode of schizophrenia, since early intervention is likely to have the best impact for improving outcomes.

Which Cognitive–Emotional Impairments Characterize Schizophrenia Patients at the First Psychotic Episode?

Even though schizophrenia is increasingly conceptualized as a cognitive disorder (Moran 2006; Nuechterlein et al. 2004), this has not been adequately reflected in clinical trials or routine clinical care. This translational gap has been exacerbated by the lack of effective measures of cognition that link to functional outcomes as well as to direct measures of brain function relevant to the pathophysiology of schizophrenia (Hyman and Fenton 2003; Moran 2006).

Many efforts have attempted to close this translational gap (summarized in Table 4.1), as was highlighted in a recent review of completed and ongoing clinical trials that included cognitive assessments in the protocol (Keefe et al. 2013). As identified by Keefe et al. (2013), current evidence from these trials is limited: the low rate of publication in peer-reviewed journals (only 19 of the 61 completed trials) has impeded access to the information by clinicians; many trials utilized small sample sizes and/or short follow-up periods (less than eight weeks); and objective endpoints are lacking to assess specific domains of cognitive function. In addition, to date trials have tended to assess participants with chronic schizophrenia. The two exceptions completed so far have focused on first-episode schizophrenia (Hill et al. 2008; Levkovitz et al. 2010). One advantage of assessing first-episode samples is that cognitive impairments may be identified in the absence of potential confounds from the effects of chronicity and long-term medication use (Keefe et al. 2013).

A major initiative in helping to close the translational gap has been MATRICS (Measurement and Treatment Research to Improve Cognition in Schizophrenia). Set up by the National Institute of Mental Health (NIMH), its domains have been recognized by the U.S. Food and Drug Administration for treatment trials. MATRICS has published consensus guidelines for cognitive domains derived from previous literature (Bilder et al. 1992; Riley et al. 2000) to assess schizophrenia. These domains include speed of processing, attention/vigilance, verbal learning, visual learning, working memory, reasoning/problem solving (also known as executive function), and social cognition (management and identification of emotion) (Green et al. 2004a, b; Nuechterlein et al. 2004):

- *Speed of processing*: Tasks that are utilized in the measurement of this domain emphasize speed of performance. They target aspects of cognition that are relatively basic, involving perceptual and motor components, and include both verbal and nonverbal processing (Nuechterlein et al. 2004). Speed of processing ability has been consistently found to be impaired in chronic patients with large effect sizes ranging from −1.57 to −0.88 (Dickinson et al. 2004, 2007; Gladsjo et al. 2004). These deficits are present even at first episode (Lucas et al. 2009); a meta-analysis of 2,204 first-episode patients reported a large effect size for this domain of −0.96 (Karaka et al. 2003).
- *Attention/vigilance*: In normal populations, a combined domain of working memory and attention has been implicated (Tulsky and Price 2003). However, in schizophrenia samples, tasks which specifically target attention and vigilance, such as continuous performance tests (CPT), load onto a separate factor from tasks based on working memory span. This implies that an attention domain which includes vigilance,

and is separate from working memory, is appropriate in schizophrenia research (Nuechterlein et al. 2004). Current research indicates that there are considerably large deficits in attentional abilities in chronic patients, with effect sizes ranging from –0.86 to –1.16 (Dickinson et al. 2007; Heinrichs and Zakanis 1998). These deficits are mirrored in first-episode samples (Lucas et al. 2009), though possibly to a slightly lesser extent, where a more moderate effect size of –0.71 has been reported (Karaka et al. 2003).

- *Working memory*: Tests which target this domain involve the temporary online storage and mental manipulation of information (Nuechterlein et al. 2004). This domain includes both verbal and nonverbal processes, and although verbal measures have traditionally been used more in schizophrenia samples, visual working memory tasks have better animal model analogs, which are useful in drug development (Nuechterlein et al. 2004). Working memory processes have consistently been shown to be impaired in chronic patients, with moderate to large effect sizes ranging from –0.61 to –1.01 (Dickinson et al. 2004, 2007; Gladsjo et al. 2004). Studies examining first-episode samples have found similar deficits in working memory capacity (Lucas et al. 2009); a meta-analysis reported an effect size of –0.79, well within the range reported for chronic samples (Karaka et al. 2003).
- *Verbal learning* is associated with the immediate or delayed recall of verbal material that exceeds working memory span. Tasks include word lists, paired associates, or more narrative style information. While there is some overlap between this domain and the learning of visual material, factor analyses in schizophrenia samples tend to separate verbal learning tasks from visual learning ones (Nuechterlein et al. 2004). Furthermore, schizophrenia patients tend to have deficits in either verbal and/or visual episodic memory, further indicating that these two cognitive domains seem to be separable within schizophrenia research (Nuechterlein et al. 2004; see also Aleman et al. 1999). Verbal learning tends to have considerable deficits for chronic (Dickinson et al. 2007; Gladsjo et al. 2004; Keefe et al. 2004) as well as first-episode samples (Lucas et al. 2009), with large effect sizes for both groups ranging from –0.90 to –1.41 (Dickinson et al. 2007; Gladsjo et al. 2004; Keefe et al. 2004) and –1.2 (Karaka et al. 2003), respectively.
- *Visual learning and memory*: Visual learning is associated with the long-term or immediate recall of visuospatial material, including the recognition of faces, the immediate or delayed recall of family scenes, reproduction of line drawings, and memory of nonfamiliar figures (Nuechterlein et al. 2004). There is less evidence for impairment in this domain in the literature compared to the other cognitive dimensions; however, new visual learning has still been shown to be impaired

Table 4.1 Clinical trials in the public domain as reported by Keefe et al. (2013).

Peer-Reviewed Trials: Chronic schizophrenia samples

Source	Sample Characteristics	Cognitive Measure
Buchanan et al. (2011)	Clinically stable, nonacute (age: M = 42.7)	MCCB
Kane et al. (2010a)	Acute exacerbation of psychotic symptoms, (age: M = 43.2)	MCCB
Javitt et al. (2012)	Clinically stable (age: M = 43.3)	MCCB
Lieberman et al. (2009)	Clinically stable, nonacute (age: M = 40.5)	MCCB
Marx et al. (2009)	Clinically stable (illness duration greater than 1 year) (age: M = 51.1)	MCCB, BACS
Friedman et al. (2008)	Not reported	BACS
Ritsner et al. (2010)	Clinically stable (illness duration greater than 2 years) (age: M = 38.5)	CANTAB
Buchanan et al. (2007)	Not specified (age: M = 43.5)	Tests for processing speed, verbal fluency, processing speed, attention, auditory memory, visual spatial memory, auditory working memory, visual spatial working memory, and executive function
Buchanan et al. (2008)	Chronic (age: M = 49.7)	WAIS-III letter-number sequencing, BACS number sequencing, CVLT, BVMT, grooved pegboard, WAIS-III digit symbol and symbol search, GDS-CPT
Freudenreich et al. (2009)	Clinically stable (age: M = 45.3)	NAART, TMT, DS-CPT, HVLT, WMS-III, WCST, WAIS-III letter-number sequencing, LCF, grooved pegboard
Goff et al. (2008b)	Clinically stable (age: M = 42.9)	NAART, TMT, DS-CPT, CVLT, WMS-III Faces and Family Pictures, WCST, LCF, WAIS-III letter-number sequencing, grooved pegboard
Goff et al. (2008a)	Not specified (age: M = 49.1)	NAART, WMS-III, HVLT, WCST, TMT, LCF, WAIS-III letter-number sequencing, grooved pegboard, LMT WMS-R
Goff et al. (2009)	Chronic (age: M = 49.7)	HVLT, WAIS-III letter-number sequencing WAIS-III digit symbol test, WAIS-III category fluency, CPT-XX, WAIS-III-spatial scan, LMT-WMS-R

Dimensions of Heterogeneity Relevant for Treatment

Table 4.1 (continued)

Source	Sample Characteristics	Cognitive Measure
Honer et al. (2006)	Not specified (age: M = 37.2)	WAIS-III letter-number sequencing, Brown Peterson procedure
Kelly et al. (2009)	Not specified (age: M = 49.0)	WAIS-III letter–number sequencing, BACS number sequencing, WAIS-III digit symbol search, grooved pegboard, WAIS-III letter fluency, Woodcock Johnson Planning Test, CVLT, BVMT, GDS-CPT
Kinon et al. (2011)	Symptomatic (age: M = 38.8)	BACS Symbol Coding Task

Peer-Reviewed Trials: First-episode schizophrenia samples

Source	Sample Characteristics	Cognitive Measure
Levkovitz et al. (2010)	Early phase (within 5 years of exposure to treatment, aged 18-35 years) (age: M = 24.9)	CANTAB
Hill et al. (2008)	Drug naïve, first-episode patients (age: M=25.97)	Stroop Color Word Naming Test, COWAT, TMT, CVLT, WMS-R Visual Production, WAIS-R digit span, WAIS-R digit symbol search, grooved pegboard

Non-Peer-Reviewed Trials

Source	Sample Characteristics	Cognitive Measure
Memory Pharmaceuticals Corp (2008)	Not reported	MCCB
Allon Therapeutics Inc (2009)	Stable (age range 18–65)	MCCB
AstraZeneca and Targacept (2008)	Not reported	IntegNeuro
Merck & Co (2011)	Stable (age range 21–55)	BACS
Cephalon Inc (2010)	Stable (age range 21–55)	BACS

Abbreviations:
BACS: Brief Assessment of Cognition in Schizophrenia
BVMT: Brief Visual Memory Test
CANTAB: Cambridge Neuropsychological Test Automated Battery
COWAT: Controlled Oral Word Association Test
CPT-XX: Continuous Performance Test–Identical Pairs
CVLT: California Verbal Learning Test
DS-CPT: Degraded Stimulus-Continuous Performance Test
GDS–CPT: Gordon Diagnostic System–Continuous Performance Test
HVLT: Hopkins Verbal Learning Test
LCF: Letter and Category Fluency
LMT WMS-R: Logical Memory Test Revised Weschler Memory Scale
MCCB: MATRICS Consensus Cognitions Battery
NAART: North American Adult Reading Test
TMT = Trail Making Test
WAIS-III: Weschler Adult Intelligence Scale, 3rd ed.
WCST: Wisconsisn Card Sorting Test
WMS-III: Weschler Memory Scale, 3rd ed.

in chronic samples with effect sizes ranging from −0.43 to −1.03 (Dickinson et al. 2007; Heinrichs and Zakanis 1998). This has also been observed in first-episode patients (Lucas et al. 2009), where a meta-analysis has reported an effect size of −0.79 (Karaka et al. 2003). Smaller differences from controls in visual compared to verbal learning, particularly at first episode, imply differential patterns of impairment between these two domains in schizophrenia. This may give additional weight to the separation of these two memory domains.

- *Reasoning and problem solving* involves higher-level cognitive processes that require complex reasoning or utilization of strategies. While problem-solving tasks can be either verbal or nonverbal, and often involve relatively basic motor or perceptual abilities, they all require additional higher-order skills in decision making and planning (Nuechterlein et al. 2004). These skills have been found to be impaired in long-term chronic schizophrenia patients, where effect sizes ranging from moderate to large (−0.68 to −1.11) have been reported (Dickinson et al. 2007; Gladsjo et al. 2004; Heinrichs and Zakanis 1998). First-episode patients show similar difficulties (Lucas et al. 2009), where a large effect size (−0.83) fits within the range of those described for chronic samples (Karaka et al. 2003).

- *Social cognition* refers to the mental processes required to understand and participate in social interactions. These include the ability to perceive and interpret one's own and others' emotions accurately, as well as the capacity to generate appropriate responses in accordance with these interpretations (Green et al. 2005). Because social cognition is a relatively recent area, comparatively few studies have included tests of emotional and social cognition. Substantial impairments in emotion identification have, however, been observed (Edwards et al. 2002; Sachs et al. 2004; Williams et al. 2007b). Meta-analysis reports a large average effect size of −0.91 across 86 of these studies, and some debate remains as to whether the impairment is specific to certain emotions, and whether or not deficits in emotion recognition are independent of problems with perceiving the face as a whole (Bryson et al. 1997; Edwards et al. 2002; Johnston et al. 2006; Kohler et al. 2000, 2003, 2009; Kosmidis et al. 2007). Schizophrenia patients also show impairments on emotional intelligence measures, which contribute uniquely to separating them from healthy controls (Brune 2005; Penn et al. 1997). This finding was subsequently replicated in a first-episode sample (Lucas et al. 2009; Symond et al. 2005). At first episode, patients also show impairment in emotion recognition, which demonstrates an increased degree of negativity bias (Symond et al. 2005). Meta-analysis of the five studies assessing social cognition in first-episode samples report a moderate overall effect size of −0.77 (Karaka et al. 2003).

Impairments in MATRICS domains have been found to be trait-like, persisting with remission of psychotic symptoms, and occurring in first-degree relatives (Friedman et al. 2001b; Harvey et al. 1996; Heaton et al. 2001; Snitz et al. 2006). Using traditional paper and pencil batteries, a meta-analysis of 53 longitudinal studies of cognition in chronic schizophrenia suggests that impairment is not progressive (Szoke et al. 2008) and that improvements may reflect practice effects, corresponding in size to those in healthy controls. Longitudinal studies of first-episode patients are comparatively patchy, and none have examined MATRICS domains. The focus has been on traditional measures of intelligence. At a group level, these studies show relative consistency of poor intelligence from one to five years after the first episode (Gold et al. 1999; Leeson et al. 2011). When examining individual patients, a distinction can be made between patients with clear impairments and patients with relative preservation (Leeson et al. 2011). In the "impaired" subgroup, poorer IQ at first episode of schizophrenia was the strongest predictor of poor functional outcomes (in occupation and hospital readmission) at one- and three-year follow-ups (Leeson et al. 2011). Studies of individual cognitive tasks also suggest differential trajectories according to cognitive domain; for instance, more than other domains, verbal learning may decline after first episode (Hoff et al. 1999; Townsend and Norman 2004). There may also be deterioration on tasks which show preservation at baseline, such as visual learning; areas of decline implicate frontotemporal brain systems (Stirling et al. 2003).

To provide a starting point for treatment targets, these domains need to be assessed in more systematic longitudinal studies so that the heterogeneity of schizophrenia can be explicated. For instance, such studies would determine if subgroups of "impaired" versus "preserved" cognition are present at first episode, and whether these groups show deterioration versus preservation over time. With this type of evidence, candidate new treatments could target those patients with impaired status at first episode who show a profile of cognitive–emotional function, which predicts deterioration over time, with the goal of halting or potentially reversing this deteriorating course.

Assessing cognitive–emotional functions for treatment targets is likely to require standardized testing that is consistent across site and time as well as cost-effective to implement. Computerized tests, validated against consensus domains and previously established manually administered tests, offer one solution. By computerizing tests, and standardizing their administration and scoring, tests might be comparatively easier to translate into treatment trials, cognitive remediation, and clinical practice. These advantages are similar to a traditional paper and pencil battery test, which typically takes longer to administer and relies on specialist neuropsychologically trained personnel to administer and score responses manually.

Are Functional Capacities Predicted by Cognitive Impairments in Schizophrenia Patients at First Episode?

Poor functional capacity refers to the reduced ability to function in the real world (e.g., in school, at work, in relationships, with regard to self-care) and in schizophrenia, it produces the burden of disease. Individuals with schizophrenia consistently identify meaningful relationships and the capacity to perform at school (or work) as their main goals.

Cognitive impairments are the strongest predictor of poor functional outcomes in social, occupational, and independent living capacity in schizophrenia. This relationship suggests that impairments in cognition are a key factor in the loss of functional capacity (Green 1996; Green et al. 2004a, b; Koren et al. 2006; Nuechterlein et al. 2004).

At baseline, distinct aspects of functional capacity have been linked to impairments in specific domains of cognition (Friedman et al. 2001a, 2002; Heaton et al. 2001):

- attention/vigilance with poor social functioning,
- verbal learning and memory with poor social, occupational, and independent living capacity,
- reasoning/problem solving with poor independent living, and
- processing speed with poor employment capacity.

More recent findings demonstrate a generalized relationship between multiple domains of cognition and each aspect of these functional skills, at least in older schizophrenia patients (Bellack et al. 2004b).

Longitudinally, a review of 18 studies has shown that cognitive impairments prospectively predict functional outcomes in the community over periods of at least six months (Green et al. 2004a). Of these studies, 12 had medium to large effect sizes. In chronic patients, improvements in cognitive domains have been shown to predict improvements in social skills performance with effect sizes of up to 0.92 (Nuechterlein et al. 2004). In first-episode patients, impairments in attention/vigilance were found to predict work/school resumption, accounting for 52% of variance in this outcome, even after controlling for clinical symptoms (Nuechterlein et al. 1999; Nuechterlein et al., pers. comm.).

What Brain Systems Are Involved in Cognitive–Emotional Impairments at First Episode?

The brain basis of cognitive impairments in schizophrenia has been elucidated using direct measures of brain function relevant to its temporospatial pathophysiology (Gallinat et al. 2004; Harrison et al. 2007; Lawrie et al. 2002). Functional magnetic resonance imaging (fMRI) provides a high spatial resolution measure of neural connectivity and electroencephalogram (EEG) recordings provide a high temporal resolution measure of neural connectivity (known as gamma synchrony) (Basar-Eroglu et al. 2007; Light et al. 2006; Spencer et al. 2004; Whitford et al. 2006). These two fields of schizophrenia research—brain imaging and EEG gamma synchrony—have progressed by using their own sets of activation tasks, which are typically different from those used in neuropsychological and psychophysical research, including research using the MATRICS battery. New initiatives—CNTRICS (Cognitive Neuroscience Treatment Research to Improve Cognition in Schizophrenia; http://cntrics.ucdavis.edu/) and its successor "CNTRACS" (Cognitive Neuroscience Test Reliability and Clinical Applications for Schizophrenia consortium; http://cntracs.ucdavis.edu/)—have advanced this goal by developing perceptual and cognitive tasks solidly grounded in cognitive neuroscience, including well-understood brain circuitry. These and future studies that delineate the relationships between MATRICS domains and fMRI and EEG brain measures should enable greater use of perceptual and cognitive measures with demonstrated neural construct validity to be included as treatment targets and outcome predictors in clinical trials and in clinical practice.

Functional Neuroimaging

The field of brain imaging in schizophrenia, using cognitive activation tasks, is growing rapidly. Imaging findings highlight a loss of activation and connectivity in frontotemporal networks. Also apparent are alterations in the functional connectivity of these temporolimbic and frontal brain systems in fMRI and positron emission tomography (PET) data (Engel et al. 1991; Engel and Singer 2001; Goldman et al. 1992). Those relevant to MATRICS[1] domains that are most impaired in schizophrenia are as follows:

- *Attention/vigilance domain*: CPT tasks are the most commonly used in schizophrenia brain imaging studies (relevant to the attention/vigilance domain of MATRICS). They show a consistent hypoactivation of the frontal (particularly dorsolateral) cortex (Jansma et al. 2004) and fronto-temporal-parietal networks (Meyer-Lindenberg et al. 2001).
- *Verbal learning domain*: Verbal learning tasks, such as the California Verbal Learning Test (CVLT), engage frontal and temporal (especially hippocampal) activity in controls (Johnson et al. 2001). On these tests, chronic schizophrenia patients show reduced activation in these regions (Heinze et al. 2006).
- *Social cognition domain*: Within this domain, emotion identification tasks are used to engage frontal and temporolimbic networks (Williams et al. 2006). First-episode schizophrenia patients show abnormal

---

[1] In brain imaging publications, the CPT (n-back) task is typically referred to as one of working memory (in contrast to the MATRICS recommendation that this task defines attention/vigilance).

connectivity for this task (Das et al. 2007; Williams et al. 2007a). One observation is a reversal of normal connectivity between temporolimbic regions, such as the amygdala and frontal areas (Das et al. 2007).

EEG Gamma Synchrony

One of the most significant recent breakthroughs in understanding the brain basis of "real-time" cognition is a candidate marker of neural synchrony in the 40 Hz gamma band of the EEG. Gamma synchrony is purported to underlie the binding of information and associated neural networks for coherent perception and cognition (Engel et al. 1991; Engel and Singer 2001). Initially synchronous gamma activity was demonstrated in response to coherent perception in cat and primate cortices (Engel et al. 1991; Engel and Singer 2001). In humans, it is quantified from EEG scalp recordings.

Current thought in schizophrenia research indicates that patients have an overall absolute increase in background gamma synchrony, yet relatively smaller increases in synchrony, compared to healthy adults, during perceptual and cognitive activity (Uhlhaas and Singer 2010; Williams et al. 2009b), and this is observed as early as the first episode (Silverstein et al. 2012a). This pattern of results suggests that schizophrenia patients have difficulty in separating relevant information from background noise (Uhlhaas and Singer 2010; Williams et al. 2009b). These cognitive findings and their relevance to MATRICS domains are as follows:

- *Attention/vigilance*: Chronic schizophrenia patients show a loss of normal gamma synchrony on CPT tasks, particularly over the frontal brain (Basar-Eroglu et al. 2007). Chronic patients also show a loss of gamma synchrony over frontal and temporal regions in response to a simple auditory attention task (Light et al. 2006). In the same auditory attention task, a corresponding loss of frontal gamma synchrony has been observed in first-episode patients (Symond et al. 2005).
- *Social cognition domain:* Abnormal gamma synchrony has been observed in first-episode patients during an emotion identification task relevant to the MATRICS domain of social cognition (Williams et al. 2009b). These abnormalities were particularly pronounced over temporal brain regions.

How Do Cognitive–Emotional Impairments and Their Relationships with Functional Capacity and Brain Function Progress over Time?

To date, we do not know if patients with the greatest deterioration in cognition show the poorest functional outcomes, compared to patients with comparative preservation. We also do not know whether cognitive impairment versus preservation relates to corresponding changes in brain function. Addressing these gaps in knowledge would contribute a valuable evidence base upon which new pharmacological and cognitive treatments could be developed and evaluated for early intervention in schizophrenia to limit, and ultimately prevent, the debilitating burden of illness. New treatments that target cognition and provide solid cognitive endpoints are needed on which to base evaluations.

What Are the Implications for Treatment Outcomes?

Antipsychotic Medication

To date, there is no consistent evidence for the effects of medication on cognition in schizophrenia (Goff et al. 2011). As noted previously, effective intervention early on is likely to improve outcomes (Wyatt 1991). To date, however, very few studies have analyzed treatment outcomes for early onset schizophrenia, and even fewer treatment studies have been conducted with cognitive–emotional test predictors or endpoints. In the small number (ten) of blinded randomized controlled trials that have examined antipsychotics in early episode schizophrenia (Kumra et al. 1996, 2008; Pool et al. 1976; Realmuto et al. 1984; Shaw et al. 2006b; Sikich et al. 2004; Spencer et al. 1992), half have included a first-generation antipsychotic (FGA) agent (such as haloperidol) as well as second-generation antipsychotic (SGA) agents. Across these trials, the results indicate a similar response rate and side-effect profile to those found in adults, although in clinical practice the traditional FGA agents are usually relegated to second-line treatments. A meta-analysis of 15 treatment trials which did not necessarily have strict controls indicated that SGAs may be more efficacious in early onset patients (Kryzhanovskaya et al. 2009). However, given the small number of trials, sufficient evidence is lacking to determine whether SGAs with different mechanisms of action produce different profiles of clinical efficacy in this group. The recently completed TEOSS (Treatment of Early-Onset Schizophrenia Spectrum disorders) trial (in patients aged 8–19 years) indicates that risperidone (SGA) was not more efficacious than molindone (FGA) over eight weeks (Green et al. 2008). Higher self-reports of akathisia with molindone suggest, however, that it may produce more extrapyramidal side effects over longer time periods. SGA medications, such as risperidone, might also be relatively more efficacious over longer periods. In the TEOSS study, the capacity to compare between SGAs (risperidone to olanzapine) was removed since the olanzapine condition was discontinued following adverse weight gain (Green et al. 2008; Patterson et al. 2001).

Cognitive Remediation

Similarly, there is a dearth of research on cognitive treatments for schizophrenia, especially at early and first onset. Cognitive remediation (also known as "rehabilitation") is an approach that has been found to improve systematically the everyday functioning of people with schizophrenia, yet it is not the focus of research funding (Wykes 2010). A meta-analysis of methods and effect sizes for this therapy (Marder 2006; 104 patients) has shown durable effects on global cognition and functioning. Stronger effects are found when cognitive therapy is combined with psychiatric rehabilitation and treatment in a strategic and adjunctive approach (Wykes et al. 2011). User-friendly methods for cognitive remediation include computerized programs integrated into usual care (Hodge et al. 2010); tailored computerized "brain games" have been shown to be effective in randomized controlled trials (McGurk et al. 2007).

There is a need for studies that examine cognitive and emotional predictors of response to cognitive treatments, and the most appropriate outcome measures for these treatments both alone or in combination (e.g., combined use of precognitive medication and cognitive remediation). Such studies could focus on first-episode patients to limit effects of previous treatment history and include brain imaging and/or EEG measures to test the mechanistic biological basis of treatment–outcome relationships. Overall, a focus on "personalization" is likely to be one way to use the heterogeneity of schizophrenia to determine which subcohorts of patients show particular profiles of brain–cognition–emotion functioning, and how these vary in a coherent way with treatment.

First column (top to bottom): Aiden Corvin, Will Carpenter, Angus MacDonald III, Matcheri Keshavan, Angus MacDonald III, Louis Sass, and Michèle Wessa
Second column: Robert Buchanan, Louis Sass, Robert Buchanan, Michèle Wessa, James Kennedy, Will Carpenter, and Angus MacDonald III
Third column: Michèle Wessa, Will Carpenter, Matcheri Keshavan, Robert Buchanan, Aiden Corvin, James Kennedy, and Aiden Corvin

5

Which Aspects of Heterogeneity Are Useful to Translational Success?

*Aiden Corvin, Robert W. Buchanan,
William T. Carpenter Jr., James L. Kennedy,
Matcheri S. Keshavan, Angus W. MacDonald III,
Louis Sass, and Michèle Wessa*

Abstract

Schizophrenia brings challenges of heterogeneity at multiple levels related to symptomatology, behavior, outcome, genetics, and pathophysiology. The clinical disorder may capture more than one disease mechanism, which has certainly been an impediment to research progress. This chapter summarizes discussions on the utility and problems of the current, syndromal diagnosis. Three potential conceptual paradigms for addressing the heterogeneity problem in schizophrenia are identified and discussed, as are the potential opportunities and challenges for future research using these conceptual frameworks.

What Do We Mean by the Schizophrenia Construct?

Schizophrenia is a common disorder with significant personal, medical, and societal implications. Diagnosis is based on observed behavior, the duration of symptoms, and impaired functional outcomes. Many clinical symptoms are described in schizophrenia but the core clinical symptom domains are delusions, hallucinations, disorganized speech, disorganized psychomotor behavior, and negative symptoms (e.g., avolition). This categorical diagnosis is operationalized in current classification systems by the World Health Organization (ICD-10) and the American Psychiatric Association (DSM-5).

The schizophrenia construct, originally termed dementia praecox, emerged from the work of Kraepelin in the late nineteenth century (see Carpenter, this

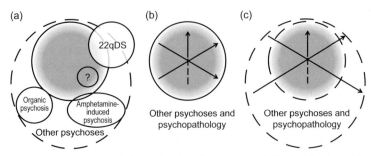

Figure 5.1 Three models of the schizophrenia construct, delimited in gray: (a) categorical heterogeneity, (b) internal dimensional heterogeneity, and (c) broadly dimensional heterogeneity. Each model suggests a different research agenda to make progress in understanding the disorder.

volume). Kraepelin proposed a unifying pathological process for a putative disease state that involved avolitional and dissociative psychopathology and poor clinical outcome. Modern classification was influenced by the subsequent work of Bleuler, Schneider, and others with some shifting from the original concept. A reliable approach to the definition of schizophrenia has emerged. However, the boundaries of this construct are not the same as those of the original construct, where there is a shift in emphasis away from the combination of negative symptoms and thought disorder toward reality distortion symptoms (i.e., hallucinations and delusions).

A diagnosis based on specific etiology and pathophysiology defines a disease entity and is a closed construct (Figure 5.1). In a closed construct, membership is defined by necessary and sufficient conditions. In contrast, a diagnosis in the absence of this knowledge is more appropriately described as a disorder or syndrome and, as such, constitutes an open construct that may comprise a number of diseases not yet specified by etiopathophysiology, with potential overlap between signs and symptoms of these and other diseases. In either case, clinical diagnosis can be based on specific criteria predicated on symptoms, onset, and course through which clinicians understand disorder prototypes. Matching an individual patient to the most likely prototype provides a differential diagnosis. Within the open, syndromal construct of schizophrenia, there are at least some identifiable closed constructs or specific disease entities, such as a small subgroup of patients defined by the presence of genetic etiology based on 22q11.2 deletions (Murphy 2002).

What Have We Learned from the Schizophrenia Construct?

An experiment is a question which science poses to Nature, and a measurement is the recording of Nature's answer.—Max Planck (1949)

For a century, there has been an ongoing debate about the utility of the schizophrenia construct. It is well to bear in mind the assessment of this long-running debate made by Karl Jaspers (1946/1997:567): "For many years the border between manic-depressive insanity and dementia praecox has vacillated considerably in a kind of pendulum movement without anything new emerging." Jaspers recognized the difficulty of precisely defining the border between these conditions, and he acknowledged the near-impossibility of deciding on a diagnosis in certain cases. He did not, however, doubt that there is something valid about this distinction to which we seem always to return, writing that "there must be some kernel of lasting truth not present with previous groupings" (Jaspers 1946/1997:568).

Diagnostic approaches for schizophrenia in DSM and ICD provide a basis for reliable classification of cases. The current, open construct, syndromal definition of the disorder has been validated by evidence at the genetic, physiological, brain imaging, psychological, social, and epidemiological levels (Keshavan et al. 2008; Tandon et al. 2008). Using these classification systems, groups of cases can be distinguished from comparison groups (usually non-ill controls) on variables such as:

- a family history of schizophrenia and other mental disorders,
- genetic risk factors,
- paternal age,
- a history of prenatal insult,
- perinatal complications,
- neurodevelopmental deficits,
- childhood abuse,
- cognitive deficits,
- structural or functional brain differences, and
- physical health issues (e.g., reduced insulin sensitivity).

Despite robust group differences, diagnostic biomarkers with sensitivity and specificity for classification of individual patients have not yet been developed.

The schizophrenia construct also defines a patient group which has particular needs. This has been important in the development of treatments, in particular in the development of antipsychotic medication for the treatment of positive symptoms (Leucht et al. 2011) as well as in the development of a number of psychological (Wykes et al. 2008, 2011) and psychosocial interventions (see Mueser, this volume) that have focused on improving functionality and the promotion of recovery. In addition, the significant personal, family, public health, and societal consequences of schizophrenia are enormous and have implications from the level of individual patient care to health policy. People with schizophrenia have reduced life expectancy (17–25 years across available international studies; Hennekens et al. 2005; Tiihonen et al. 2009; Kilbourne et al. 2009; Chang et al. 2011), substantial comorbidities (Carrà et al. 2012), increased rates of homelessness and incarceration (Foster et al. 2012), and high levels of unemployment and impoverishment of social roles (Salkever et al. 2007). Further clarification of the disabilities produced by schizophrenia is necessary, as, to date, this information has had only a minimal impact on advocacy, peer support, and healthcare planning and provision.

Despite more than a century of research, our understanding of the biological basis for schizophrenia is limited. If there is a core disease entity, this is unlikely to map neatly to clinical boundaries based on behavior, symptom duration, and impaired functioning. Furthermore, there is marked variability in clinical presentation within the schizophrenia syndrome. Any two given patients may share no symptoms at all, at least as defined according to standards, and thus have very different courses of illness. Therefore, it is reasonable to ask whether the construct of schizophrenia has been useful or an impediment to research and treatment. Next we raise challenges to the use of the schizophrenia construct and propose ways in which the construct continues to be useful.

Is the Construct Useful or an Impediment to Translation?

Schizophrenia poses challenges of heterogeneity at multiple levels (Figure 5.2). No two patients with this diagnosis present similarly to the clinician: symptom constellations and behavioral abnormalities vary between individuals, and often even within individuals across time points (clinical and behavioral heterogeneity). The pathophysiological substrate of the illness varies between individuals: although the various biomarkers differ between patients with schizophrenia and healthy subjects, considerable overlap exists and none of them is present in the vast majority (pathophysiological heterogeneity). Finally, etiological factors have been proposed (most notably genetic factors), but none is necessary and sufficient for disease causation. The limited progress in developing pathophysiological and etiological understanding of schizophrenia is often attributed to such heterogeneity.

This issue squarely raises the question as to whether, or to what extent, the construct of schizophrenia is useful or constrains further progress in understanding the diagnosis, etiology, treatment, or even prevention of the syndrome. The fact that schizophrenia does not have a set of necessary and sufficient criteria leads to considerable differences of opinion as to its origins, and disagreement regarding proper treatment. To some degree this may lead to confusion and misperceptions by others outside the field. However, this must be viewed in the context of treatment gains that have been made based on pharmacological, psychological, and social interventions. An argument has been made that the schizophrenia construct is stigmatizing. In some countries, such as Japan, an attempt has even been made to change the name of the disorder (Sato 2006). Society rather than biology determines how we respond to people with mental illness. Based on experience from other areas (e.g., mental retardation or learning disabilities), where the establishment of a new nomenclature

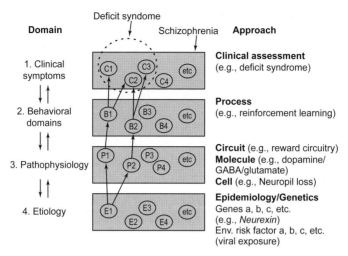

Figure 5.2 Modeling schizophrenia at the level of clinical symptoms (C), behavior (B), pathophysiology (P), and etiology (E). The figure indicates how future research could be integrated across different levels of analysis.

did not result in treatment improvements or outcome, we feel that changing the name is unlikely to resolve the problem.

The current definition, which is based on Kraepelin's proposition for a unifying pathological process, assumes that schizophrenia is a single brain disease. However, there is much evidence to suggest that schizophrenia is not a single disease entity. What is far from clear is the extent to which the current definition of the syndrome captures all relevant disease entities, how many diseases there might be, and, if there are many, whether these represent one or more distinct pathological processes.

Diagnosis has been useful, as it allowed specific treatments to be applied to groups of people who share (at least some) similar features and permitted these interventions to be evaluated. However, it is increasingly recognized that diagnosis has limited utility, that new ways of conceptualizing the disorder(s) are needed, not least to aid the development of more effective treatments. For example, there is substantial variability between patients with schizophrenia for measures of cognitive impairment, positive symptoms, negative symptoms, disorganization, and insight into the nature of their condition. In addition, many people with schizophrenia have comorbid conditions, such as mood disorders, anxiety disorders, and substance abuse, that are often not assessed in clinical practice but can significantly affect outcome. The overemphasis on the development of dopamine antagonists/antipsychotics for the treatment of positive symptoms has resulted in important unmet therapeutic needs (i.e., cognitive impairments, negative and anxiety symptoms) for which these agents have limited, if any, benefits. The use of dimensional measures of psychopathology has been helpful, for example, in fostering the development of psychological interventions that target specific symptoms (e.g., cognitive behavioral therapy for persistent positive and negative symptoms, cognitive remediation for cognitive impairments). Given the evidence for certain subgroups within the disorder of schizophrenia (e.g., deficit syndrome), there are likely to be better ways to engage with and address treatment heterogeneity; however, the etiological bases for this remain to be determined.

A further criticism of the syndromal diagnosis is that it is potentially too simplistic in operationalizing complex phenomenology. The experience of psychosis needs to be seen in the context of the person and his/her environment, as this may define the experience of symptoms. The converse is also true in that at a symptom level, a delusion involving perceived harm by others may have quite a different quality and treatment implications than a delusion involving the control of movement by an external entity. For example, the biases in attribution and jumping to conclusion style that is characteristic of paranoia (Lyon et al. 1994; Moritz et al. 2012) have proven amenable to cognitive behavioral therapy (Chadwick et al. 1996), with its emphasis on behavioral experiments and examination of evidence for and against different ideas. On the other hand, it has been proposed that delusions involving control of movement and self-representation in space, including passivity phenomena, may result from visual processing disturbances (Landgraf et al. 2012), suggesting that an improvement in visual functioning might prove beneficial in addressing these delusions. Dimensional measurement may help in this regard and has been recognized in the diagnostic revisions for schizophrenia made in DSM-5.

The dimensional approach also comes with the potential challenge of pseudospecificity. For example, a number of studies have demonstrated that negative symptoms significantly improve when antipsychotic medications are administered to people in the midst of an acute episode of the illness. However, these beneficial negative symptom effects have been shown to be secondary to the primary effect of antipsychotic medications on positive symptoms. In follow-up studies, when the same agents have been used to treat clinically stable people with schizophrenia with persistent negative symptoms, only limited benefits for this aspect of the illness have been observed.

For the clinician, the syndrome is useful in identifying the wider treatment challenges, where a simpler psychosis construct would not. The emphasis on functional impairment and duration of symptoms, although arbitrary and even if they are perhaps the most relevant dimensions from a clinical perspective, brings into focus a group of patients who are likely to have enduring mental health problems and require planning for future care delivery.

Despite some progress, and as noted by Carpenter (this volume), progress in finding the causes of schizophrenia has been slowed by including what appear to be several heterogeneous syndromes in the same group in most of schizophrenia research. The problem is becoming more apparent as the armamentarium of investigative methods expands. Schizophrenia is a "black box" with a fuzzy boundary definition. We do not have sufficient evidence to replace the definition with constructs incorporating newer neuroscience or cognitive-based theories, such as "failure of neuroconnectivity syndrome" or "social and cognitive deficit disorder." We can certainly alter the dimensions of the "box," but in the absence of empirical evidence, we cannot know whether it is more useful to widen or narrow the diagnostic criteria. Findings from schizophrenia genetics illustrate this point.

From genetic epidemiology, we can define a family of schizophrenia spectrum disorders which share many of the group features of schizophrenia previously described (Kendler et al. 1993). It is well established that many of these group features (e.g., imaging, electrophysiological, and cognitive variables) extend to family members with no psychiatric diagnosis. More recently, large register-based population cohorts indicate clustering of other psychiatric disorders, including bipolar disorder, depression, and autism in the families of schizophrenia patients (Lichtenstein et al. 2009; Mortensen et al. 2010; Sullivan et al. 2012b). These data suggest significant fusion or overlap of the boundaries across a number of psychiatric diagnostic categories. This sharing could relate to a shared core pathology, a shared noncore pathology (e.g., anxiety), and/or a pre-disorder vulnerability platform shared by both disorders or nonpathological confounds (e.g., lifestyle factors more common to these groups). Evidence from molecular studies (e.g., genome-wide association studies) provides empirical support, at least for the overlap of common, small (odds ratio <1.3) genetic risk factors between schizophrenia and bipolar disorder (International Schizophrenia Consortium 2009). Molecular studies have also revealed a series of rare structural genomic variants (copy number variants, CNVs) of larger effect on risk (odds ratio 3–20) in a subgroup of patients (for a review, see Malhotra and Sebat 2011). As was the case with 22q11.2 deletion syndrome, the newer risk CNVs (e.g., deletions at 1q21.1, 3q29, 15q11.2, and duplications at 16p11.2 and 16p13) all show evidence of pleiotropy. This means that each CNV has variable influences on multiple phenotypic outcomes (e.g., schizophrenia, autism, intellectual disability, epilepsy, and obesity).

The critical question in genetics is how to translate these findings into an etiological understanding of schizophrenia. The heterogeneous population identified by the DSM-IV and ICD-10 criteria represents potentially many different pathophysiological or etiological processes. With no clear strategy to reduce heterogeneity, disambiguating the statistical association with small effects in large numbers of subjects or large effects in very small groups is challenging. Investigating genetic findings requires an ability to validate these findings using other paradigms (e.g., animal models). To do this requires some understanding of which heuristic framework to use to account for heterogeneity optimally (for further discussion, see Williams and Gott, this volume). The current genetic data is insufficient to support widening the schizophrenia construct into a broader "neurodevelopmental disorder" construct. However, CNV findings challenge us to ask if having a pathogenic mutation of, for example, a gene like *Neurexin-1* (Rujescu et al. 2008; Kirov et al. 2008) or *VIPR2* (Vacic et al. 2011) independently affects risk of autism and schizophrenia, or whether the mutation impacts a neurodevelopmental mechanism which can have as its consequence either autism or schizophrenia, or some features of both. These rare forms may represent discrete disease entities within the syndrome or diseases that overlap with the syndrome. Under the current criteria, schizophrenia is diagnosed by exclusion of cases with known medical etiology. How will these new genomic disorders shape diagnostic practice? In our view, to restrict the syndrome of schizophrenia by excluding novel genomic disorders, which may be etiologically informative, appears unhelpful.

Paradigms for Addressing the Heterogeneity Problem in Schizophrenia

> Measure what is measurable and make measurable what is not. —attributed to Galileo (1564–1642)

The challenge facing our field is to develop paradigms in addition to a diagnosis that address the significant heterogeneity and provide more robust approaches to etiology, pathophysiology, prevention, and therapeutic research. Heterogeneity is not a problem unique to schizophrenia. Cancer serves as a useful analogy. In the case of breast cancer, diagnosis was initially made according to phenotypic features (e.g., tumor node, metastases); later classifications included histological grading (based on tumor cell differentiation) and even more recently include underlying causes and mechanisms (e.g., those with known genetic causation, those with altered estrogen receptor sensitivity). Throughout all of this, the concept of cancer still remained and the field happily accepted diagnostic pluralism, with all three approaches to classifying the disease being used as needed for the appropriate purposes. At the molecular level, it is interesting to note that some cancer centers conduct DNA sequencing on the tumor tissue of every individual patient and use this to inform treatment; this suggests a unique disease entity for each person. Here, heterogeneity alone was not a barrier to improving nosology and treatment.

Investigating disorders of the brain naturally presents additional challenges. However, progress has been made for brain disorder syndromes, including epilepsy and intellectual disability. In the case of schizophrenia, we are hampered by the lack of an objective measure or biomarker for defining individual cases of the disorder. In the absence of specific markers or pathology, what may be most relevant at this stage are paradigms with a strong evidence base that optimize discovery, treatment, and patient care by reducing heterogeneity. Meeting the standard principles of scientific measurement and addressing differences in study design, research measures, and research settings are general challenges to any new paradigm. We recognize that discovery is an iterative process and that useful paradigms must be sufficiently flexible to allow new hypotheses to be challenged and incorporated or rejected as we try to understand the syndrome. This will require integration across levels of analysis and disciplines. In a wider sense, to bring about advances in neuroscience and other disciplines to bear on the study of schizophrenia, we need to have greater interaction across research disciplines to improve discovery.

We have identified three potential conceptual paradigms for considering schizophrenia (see Figure 5.1). The first model (Figure 5.1a) represents categorical heterogeneity. In this case the open construct of the syndrome can be parcellated into different, identified closed constructs. These can stay within the syndrome for further research (e.g., novel 1q21 deletions or future disease entities) or can be "carved out" from the construct once a medical cause is elucidated (e.g., syphilis as a cause of psychosis). Currently, closed constructs would be removed from the syndrome in this model if they have the symptom criteria but fail the etiological rule-out or symptom duration specifier for schizophrenia (e.g., amphetamine-induced psychosis). Whether this approach is helpful for studying etiology or pathophysiology may require further consideration, although it is a clinically useful distinction. To take another example, many neurodevelopmental syndromes (e.g., verbal and nonverbal learning disabilities) can include features found in schizophrenia such as negative symptoms, disorganization, poor social cognition, and visual processing impairments (see the discussion on comparative studies in Silverstein et al., this volume), features that are not generally considered in the diagnostic criteria for these syndromes. This has led to a relative lack of potentially useful studies comparing, for example, schizophrenia and neurodevelopmental syndromes on their similarities and differences in etiology, course, and phenomenologies. Here, clinical but not research utility has been served. As more is understood about the genetic etiology of intellectual disability syndromes (many are associated with well-characterized genetic mechanisms), these could be conceptualized as potentially useful schizophrenia models. The defining characteristic of this model is that heterogeneity is reduced by identifying and separating different categories within the schizophrenia construct.

In the second model (Figure 5.1b), we conceptualize dimensions within the schizophrenia construct. The internal dimensional heterogeneity model could include the dimensional constructs in, for example, hallucinations, delusions, depression, mania, disorganization of thought, restricted affect, psychomotor abnormalities, avolition, and cognition impairment. However, it could also accommodate other dimensions related to how the disorder presents (e.g., temperament or other personality dimensions). In contrast to the first model, heterogeneity within the construct derives from continuous factors, and identifying the heterogeneity is addressed by understanding these factors and accounting for them in treatment.

The third model (Figure 5.1c) is similar to the second, but allows for extension of the dimensions beyond the core syndrome and could include traits (e.g., in related psychotic disorders) that vary on a continuum in the general population (e.g., anhedonia, paranoia, psychosis) (van Os et al. 2009). This model emphasizes broadly dimensional heterogeneity and is similar to the Research Domain Criteria (RDoC) approach, which includes dimensional measures of behavioral and neural circuit response (e.g., negative valence systems or arousal/regulatory systems). An extension would be to see this model within a broader platform of risk states (e.g., high-risk studies for schizophrenia or the concept of a pluripotent risk state). Again this could include aspects of normal variation (e.g., lability, introversion, alienation). In contrast to the second model, heterogeneity within the construct of schizophrenia is thought to be related to continuous factors that are also operative in other conditions. From this perspective, heterogeneity within schizophrenia informs and is informed by symptoms and individual differences found in other psychiatric disorders and in the general population.

Working with These Paradigms

A key purpose of developing new paradigms is to facilitate analysis across disciplines and at multiple levels to maximize discovery. There are examples from the literature where including this second dimensional level of measurement has proved helpful for analysis across disciplines (e.g., with the deficit syndrome concept). We propose a framework that involves four levels of analysis: clinical symptoms, behavior/cognitive domains, pathophysiology, and etiology (see Figure 5.2).

This is a multidirectional framework, in which there is integration across analyses. We suggest that this framework could be helpful in allowing more collaboration between clinical research in patient populations and basic scientists working with model systems. This framework is similar conceptually to the RDoC but is less constrained.

Such an approach generally emphasizes the building of associations between two levels of analyses (e.g., a pathophysiological process and a behavioral measure). There are additional examples where research reaches across three levels of analysis; for example, transgenic mouse models which examine both the pathophysiological consequence of a mutation and alterations in behavior, or human neuroimaging studies which link activation abnormalities to impaired cognitive processes and symptoms. These links are generally tested using as few variables as possible to avoid multiple comparisons.

In addition to collecting data across multiple levels of analyses, a newer generation of statistical and data-mining algorithms have opened up new ways to examine these data. There are, for instance, increasingly viable methods to examine relationships across multiple levels of analysis using many variables at once. Procedures such as *independent components analysis* can be used to sort large matrices of data into simpler covariance structures. Such correlational links can provide important targets for understanding causality, particularly in cases where multiple pathways can be examined and compared. In addition, a new approach to data-mining algorithms, known as *frequent pattern mining* (Ceglar and Roddick 2006; Han et al. 2007; Tan et al. 2005), has been developed for just these kinds of problems across a number of areas in industry and science. Frequent pattern mining has a solid theoretical foundation derived in part from formal concept analysis (Ganter et al. 1999), and it now has a number of efficient algorithms which seamlessly incorporate inferential statistics, intelligent algorithms, data reduction, and pattern-pruning strategies to maintain statistical power and increase computational efficiency, even when the number of variables and patterns considered is large (Fang et al. 2010, 2012). Such data-mining approaches provide another viable means for drawing together data across heterogenous data sets and extracting method-related variance to observe more clearly the relationships associated with potentially causal pathways.

Measured parameters may be defined, but these need to be validated and reliably measured using available technology. This will change, in some cases rapidly. For example, there are opportunities to develop level 1 assessment tools for symptoms, psychological or phenomenological states through the Internet or mobile phone interfaces. These could provide unprecedented access to "real-world" subjective phenomena in large patient samples.

At level 1 in the model, the investigator would define symptom domains of interest such as obsessive symptoms, avolition, or thought disorder. This would enable more specific and robust investigation of the phenomena at levels B (behavioral), P (pathophysiological), and E (etiological). A symptom domain may identify a subgroup of interest as illustrated by using primary negative symptoms to separate deficit schizophrenia from nondeficit schizophrenia with substantial differences between the two groups at levels B, P, and E. Based on phenomenological exploration, there is also the potential for discovery of more subtle but perhaps more decisive dimensions or matrices of subjective experience. For example, this could be based on disturbance of the "minimal self" or "basic self" sometimes termed "ipseity" (the basic sense, usually implicit, of existing as a subject of experience or as an agent). It might also target the closely related issue of fundamental temporal structuring of experiences. In turn, this could generate hypotheses for testing at level 2 or 3, by examining psychological processes or normal circuits subserving self-experience temporality or interconnected processes of motivation and emotion.

At level 2, there is the possibility of investigating how process relates to clinical symptoms (level 1) as well as to circuitry (level 3). Level 2 is an important intermediate level which might also be understood as a relay between pathophysiological processes and clinical symptoms. Taking reinforcement learning as one example, a level 2 behavioral process might translate into different clinical symptoms, such as depressive mood, perseveration, or risk-taking behavior. On the neural level, disturbances in an orbitofrontal-limbic-striatal circuit are supposed to mediate reinforcement learning (Cools et al.

2002; Remijnse et al. 2005). However, differential activation (hypo- or hyperactivation) has been associated with the anticipation of and response to reward or punishment, or the switch of one's own behavior according to the feedback (e.g., Linke et al. 2012; O'Doherty et al. 2001). Therefore, a neural systems perspective which spares the currently ongoing behavioral/cognitive process would not solve heterogeneity in observed neural activation patterns in, for example, schizophrenic patients and related psychopathologies. We would hence argue that experimental investigations should not focus on one single level of analysis but acquire data on clinical symptoms (C) as well as on psychological (B) and pathophysiological (P) processes. The accurate measurement of the different domains, which implies thorough operationalization according to principles of test theory (i.e., validity and reliability), is indispensable for such an approach. With fast developing utilities of, for example, machine learning algorithms and clustering methods, we should be able to challenge an integrative analysis of these different levels in the future. With common correlational approaches, multimodal data can be related even to date.

An example at Level 3 is the recent implication of alterations in gamma oscillations, which reflect aberrant synchronization of neural activity in parvalbumin-positive cortical GABA neurons as underlying executive function and working memory impairments in schizophrenia (Gonzalez-Burgos et al. 2011).

At level 4, we could identify a novel genetic risk factor and use this to identify molecular subgroups within the schizophrenia syndrome to be examined multifactorially at levels 1–3. At a physiological level, we could investigate at a molecular or circuit level using neuroimaging approaches, but also at a cellular level using animal systems or human-induced pluripotent stem cells from patients who carry a particular risk factor. The same types of human and animal experimental work can be applied to investigate at a behavioral domain and clinical symptom level.

To be effective, this approach relies on information sharing across groups and the development of large patient cohorts with information available on many parameters. This would allow for hypothesis generation using correlation analysis across levels of analysis in human subjects, which could then be validated in model systems. The reverse approach could also be applied, with wider availability of information on models from levels 2 and 3. The development of data sharing could be open source or through summaries of available measures by interested researchers. This could be provided at an "exchange" for potential collaborators.

Into the Future

The framework that we have suggested may be helpful in reducing heterogeneity in studies of schizophrenia. In this we call for multidimensional analysis. We wish to emphasize that this is not a standard call for large databases and even larger international consortia. The potential application of this approach for standard statistical approaches, data mining, machine learning, or crowd sourcing may be apparent. We believe that this framework has the flexibility to allow individuals with creative ideas to examine novel hypotheses by bringing together unusual clusters of symptoms, risk factors, or other measures. These, in turn, can be tested across levels and validated to confirm or be discarded by a researcher so as to prioritize a more heuristically informative hypothesis.

The neurodevelopmental etiology of schizophrenia needs to be seen in the context of limited (current) understanding of the normal trajectory for neurodevelopment. Schizophrenia needs to be conceptualized in a developmental or staging context (Hickie et al. 2013; Wood et al. 2011). There is evidence for people at perceived high risk of schizophrenia, based on risk factors such as family history and being within the age of greatest risk. A subset of people will develop subsyndromal symptoms (e.g., anxiety or prodromal symptoms) while another will develop schizophrenia. These stages of the evolving illness may be determined in a temporal framework. For example, genetic and early environmental factors, such as viral exposure and periadolescent psychosocial stress, can serve as subsequent etiological "hits"; these etiological events may interact to produce the sequential evolution of pathophysiology and clinical features of the premorbid, prodromal, and psychotic phases of the schizophrenic illness (Figure 5.3). Epidemiological evidence suggests that a pluripotent risk state can be identified with a range of potential outcomes from a return to normal function to the development of the schizophrenia syndrome. However, better understanding of risk and resilience factors is needed to develop effective primary and secondary prevention strategies. Also required are prospective studies that examine interaction of risk and protective factors over time (for a discussion, see C. Morgan et al., this volume).

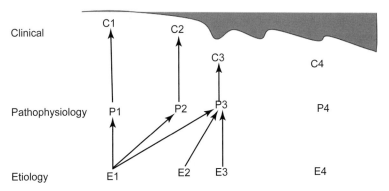

Figure 5.3 Representation of how different etiologies (E1, E2, etc.) could potentially contribute to different pathophysiology processes (P1, P2, etc.) with different clinical presentations across life span. This shows how etiological mechanisms may interact with each other and with brain development to influence a disease process.

Risk and Resilience

6

How Should Resilience Factors Be Incorporated in Treatment Development?

Peter B. Jones

Abstract

This chapter reviews the definitions of resilience with respect to psychological disorder and to schizophrenia, in particular. Alternative meanings of resilience emphasize innate characteristics and the steeling effect of experiences; these are not mutually exclusive and both could be harnessed in terms of treatments for the disorder. The implications of resilience are already well known in the sphere of psychosocial interventions and recent developments in cognitive therapies. The notion of building structural or physical resilience of the brain to prevent the onset of schizophrenia is not new: Kraepelin discussed such an approach in the conclusion of his most definitive description of dementia praecox a century ago. To do this successfully, however, remains a challenge, but much could be done if studies on risk modifiers and causes were reformulated toward public health intervention. Finally, new domains for inquiry into developmental resilience are explored, with a focus on neural connectivity and healthy brain growth.

Definitions of Resilience

What does not kill us makes us *stronger*…—Friedrich Nietzsche (1887, in Bittner 2003:188)

Like many seemingly precise terms, resilience can mean substantially different things to different people and in different contexts. In the physical sciences, resilience refers to a characteristic of materials and summarizes the extent to which they return to their prior form following some form of deformation. This meaning does not infer that something is rigid and unaffected by stress, but measures the extent to which it returns to the status quo. Such resilience is an innate characteristic but, even so, may be subject to environmental influences; for instance, rubber may be more or less resilient at different temperatures.

This meaning is readily translated into psychology as an individual's tendency or ability to recover from the effects of adverse events is well known to lead to disorder. Thus, it captures a degree of adaptability that may arise from innate as well as acquired factors: the ability to bounce back. Resilience is something revealed only in the face of adversity (Rutter 1985) and is distinct from the concept of a protective (or risk) factor.

Resilience can have another related meaning and involve cases where adversity has no effect, whatsoever, on an individual: resilience so strong that no deformation occurs, and physical or psychological adversity has no effect. This is relevant to schizophrenia and other psychiatric research, especially in the realm of risk research. Risk factors are simply measurable features that are statistically associated with an outcome when groups which present these features are compared to groups that do not. Such risk factors can usefully be distinguished between those that, themselves, modify the probability of disorder (i.e., risk modifiers) and risk indicators that are simply pointers toward the former.

Genetic risk for schizophrenia, manifest as family history, is conventionally accepted as a risk modifier, indicating the presence of disease-related genetic factors and a manyfold increase in risk of the disorder (Cardno et al. 1999). The situation is, however, much more complicated, as we now appreciate that the transmitted risk is not only for schizophrenia but also for other severe mental illnesses (Gottesman et al. 2010), including autism (Sullivan et al. 2012b). There may be common mechanisms involving genes for proteins important for neurodevelopment and synaptic functioning (Guilmatre et al. 2009). The converse of this lack of specificity is the fact that this genetic risk may remain unexpressed. Even in monozygotic (MZ) twins discordant for schizophrenia, one twin may remain mentally healthy (Gottesman and Bertelsen 1989) even though endophenotypic characteristics may be present (Gottesman and Gould 2003). It is unclear whether the discordance involves an additional resilience factor(s) in the unaffected twin or whether some component of the complete causal constellation is missing.

Although familiar, these findings emphasize the bias in schizophrenia research to search for causes and mechanisms of the disorder rather than resilience and protection. Manfred Bleuler made careful observations on "the offspring of schizophrenics" (Bleuler 1972, 1974) who lived with parents that were resident patients in his asylum, sometimes within nonpatient families accommodated at the asylum. His clinical interests were quickened by the fact that, as Gottesman and Bertelsen showed for MZ twins, many of the offspring of affected parents remained in good mental health throughout much or all of the period of risk of psychosis. His summary has an optimistic air (Garmezy 1977) and presages our current interest in gene–environment interactions. He also alluded to yet a third definition of resilience, namely that of "steeling" or environmental inoculation. This is a particular example of where environment can change resilience not in a moment-by-moment instance, but in the long term and in subsequent, different environments. Bleuler wrote (1974:106):

...despite the miserable childhoods described above, and despite their presumably "tainted" genes, most offspring of schizophrenics manage to lead normal productive lives. Indeed, after studying a number of family histories, one is left with the impression that pain and suffering can have a steeling—a hardening—effect on some children, rendering them capable of mastering life with all its obstacles, just to spite their inherent disadvantages. Perhaps it would be instructive for future investigators to keep as careful watch on the favorable development of the majority of these children as on the progressive deterioration of the sick minority.

This idea of steeling or hardening can be traced back to the mental hygiene movement in the early twentieth century and the roots of child psychiatry (Rutter 1985). It is inherent in developmental views of normal and psychopathological states in childhood and beyond, such as attachment theory and the importance of family and parent–child relationships (Bowlby 1969, 1973, 1980). In the context of adult psychiatry, attachment has gained ground in terms of understanding the emergence of personality disorder, but the long-term steeling effects of early experiences, rather than any deleterious effects, have been more controversial. That said, there have been trials of inoculation or "resilience training" to prevent disorders, such as adjustment reactions and posttraumatic stress, particularly in special groups such as emergency workers (Varker and Devilly 2012).

Thus, the concept of resilience has a long history and has been incorporated into early discussions of schizophrenia. Let us now consider the place of resilience in current thinking about treatments for the disorder.

Incorporating Resilience into Treatment Development in Schizophrenia

Nietzsche was wrong. Well, he would have been wrong had his view that adversity can lead to personal resilience, encapsulated in the opening quote, been applied to schizophrenia. Untreated psychosis leads to a deterioration of outcome (Marshall et al. 2005b) and, following first remission, each relapse leads to an accumulation of residual symptoms and a worsening of functional outcome (Robinson et al. 1999, 2004; Wiersma et al. 1998). Nevertheless, long-term studies indicate that over 20–30 years, a significant proportion of people with the diagnosis can function independently and without medication in normative social roles (Silverstein and Bellack 2008). Short-term studies of psychiatric rehabilitation interventions also indicate that cognitive, interpersonal, and community functioning can be improved through interventions that develop skills (e.g., Silverstein 2000). This is consistent with the focus on recovery of role functioning and identity that is independent of having a mental disorder (e.g., Roe 2001).

Such an approach builds upon the first definition of resilience, where it is viewed as a property that facilitates recovery or prior form or function. Nursing and clinical psychological formulations of schizophrenia routinely incorporate an assessment of an individual's strengths and resources from a psychological and social point of view (Jones and Marder 2008). These resilience factors are woven into a psychosocial treatment package, conventionally used in tandem with antipsychotic drugs. Individuals are encouraged to recognize and develop those factors which, together with knowledge about their condition, can considerably enhance their ability to manage the illness, so as to improve general functioning, maximize benefit from therapies, minimize unwanted effects, identify triggers for relapse, and reduce the risk of such events. This use of strengths and personal resilience in the management of schizophrenia is not specific to the illness. Rather, it is merely an adaptation of modern principles underpinning the effective management of long-term conditions (Goodwin et al. 2010). Resilience factors are naturally incorporated into the multidisciplinary management of schizophrenia as a long-term, complex condition.

There is, as yet, less to say about individual resilience and its interaction with drug treatments or biological approaches. Pharmacogenetic studies, whether of prescribed or illicit drugs, are always couched in terms of risk. No one would encourage a person with schizophrenia to continue taking cannabis on the basis of their genotype; however, most clinicians would doubly stress the importance of abstinence in the face of a putative risk allele, if one could be convincingly identified (Decoster et al. 2012).

Global intellectual ability and performance in individual cognitive domains are known to be positively associated with functional outcome in schizophrenia. Good performance can be considered as resilience. Cognition has become a key aspect of clinical assessment and a target for drug development (Nuechterlein et al. 2008; Kern et al. 2008). However, to date there is no convincing evidence that any particular cognitive profile can be considered a marker of resilience to guide drug therapy, nor that there are useful cognitive enhancers; few trials are adequately designed to even demonstrate useful effects (Kane et al. 2010b; Keefe et al. 2013).

Nondrug approaches to cognition, such as cognitive remediation therapy (CRT), show promise in schizophrenia (Jones and Marder 2008; Wykes et al. 2011; Keefe et al. 2012). There is some evidence that patient ratings of the therapeutic alliance make a difference to outcome of cognitive remediation (Huddy et al. 2012), but this cannot really be considered as, or related to, a resilience factor. Of the few investigations into modification of the effect of psychological therapy by genotype, Greenwood and colleagues (2011) discovered that there was absolutely no association between the catechol-O-methyltransferase val158met polymorphism and cognitive improvement following CRT in schizophrenia.

As markers of resilience, higher IQ or cognitive reserve and age have been examined as modifiers of the effect of CRT. Kontis and colleagues (2012) demonstrated that the effects of CRT were limited in older people with schizophrenia and that cognitive reserve did not influence the relationship of age with CRT efficacy. Higher premorbid IQ was associated with increased practice effects on working memory in younger but not in older individuals. Just as in drug therapy, it is not straightforward to see how cognition can be construed as a marker of resilience to be incorporated into the development of psychological therapies for schizophrenia and to guide their use. However, cognitive ability remains a logical target for such treatments.

Incorporating Resilience into the Prevention of Schizophrenia

Resilience is not a property that will aid recovery or enhance the efficacy of treatments. It is something that will prevent parts of a causal complex, such as stressful events, drug use, or even genetic risk, from adding up to an inevitable pathway to illness. Thus, incorporating resilience into the prevention of schizophrenia seems to be an area that warrants consideration.

Recent epidemiological studies of psychosis have revealed the higher than expected prevalence of psychotic experiences in the general population (Kelleher et al. 2012b), with figures of over ten percent depending on age. Very few of these people go on, however, to develop a clinically relevant psychotic syndrome or illness, even among those who seek help (Morrison et al. 2012). The majority of people who have psychotic experiences could be said to be resilient to their evolution into illness and a diagnosis of schizophrenia. The fact that so many people can be untroubled by hallucinations or delusional beliefs is troubling for psychiatry but less so for clinical psychology. Just as clinical psychology builds resilience into its management, it also incorporates a spectrum of systematization and impact of symptoms, and recognizes that some symptoms can lead to others in a dynamic and even multifaceted way. Anxiety, depression, and psychotic symptoms can form a self-perpetuating cascade of psychopathology in some people, but not in others. Psychiatric epidemiology is only recently beginning to grapple with this level of complexity (Kessler et al. 2012).

Individual differences, including age and developmental stage, may foster resilience (for further discussion, see C. Morgan et al., this volume). General cognitive ability or IQ constitutes, however, a crucial factor. Population-based studies have consistently shown that IQ is lower in children who subsequently go on to develop schizophrenia, and that protection or resilience is also attributable to higher IQ or cognitive reserve (Barnett et al. 2006). A recent systematic review and meta-analysis of the epidemiological studies of this phenomenon (Khandaker et al. 2011) confirms the dose-response relationship between premorbid IQ and schizophrenia, such that there was a 3.7% decrease in risk with every one point increase in childhood IQ. This relationship between early life IQ and psychopathology appears for individual symptoms as well as diagnoses and is relatively marked for psychotic experiences compared with depression and anxiety (Barnett et al. 2012). There is some evidence that this relationship breaks down at very high levels of IQ (Karksson 1970; Isohanni et al. 1999), but the relationship pertains for the vast majority of the population.

The relationships between schizophrenia, IQ in childhood, and early motor development—inefficient in those destined for schizophrenia as adults—are of interest. They may all be manifestations of the effectiveness of neural connectivity, something that allows a parsimonious explanation of their association (Jones et al. 1994; Isohanni et al. 2001; Ridler et al. 2006; Vértes et al. 2012; Alexander-Bloch et al. 2012). Simply put, this suggests that a resilient brain is a well-connected brain, one in which there is an economic balance between the costs of maintaining long-range connections and the efficiency of links between brain regions that share similar inputs (Vértes et al. 2012).

What makes a healthy, effectively connected, well-grown, and resilient brain? This is clearly a question with a complex answer. Genetic and environmental influences are going to be involved, operating from conception through intrauterine life and beyond, to account for the risk for schizophrenia. By environment, I refer both to the physical milieu of nutrients, toxins, and other physiologically important factors as well as electrical activity and experience—things that shape neural structures as well as depend upon them.

There are myriad epidemiological risk factors for schizophrenia. Some are likely to be risk modifiers, ranging from prenatal paternal death (Huttunen and Niskanen 1978) and infection (Khandaker et al. 2012a, b), to excess or lack of vitamin D (McGrath et al. 2010). Like these diverse examples, most risk factors operate in early life, prior to the completion of brain growth in the late twenties. In short, anything which jeopardizes healthy brain growth decreases resilience to schizophrenia. By extension, the promotion of healthy brain growth will increase resilience to schizophrenia, as well as an infinite number of other benefits.

As with so much to do with schizophrenia, Kraepelin addressed resilience and contemplated public health approaches to prevention through its promotion. In his textbook on *Dementia Praecox and Paraphrenia* (1919/1971:253) he addressed prophylaxis when considering children at genetic or behavioral risk for the disorder:

> In children of such characteristics as we so very frequently find in the previous history of dementia praecox, one might think of an attempt at prophylaxis especially if the malady had been already observed in the parents or brother and sisters. Whether it is possible in such circumstances to ward-off the outbreak of the threatening disease, we do not know. But in any case it will be advisable to promote to the utmost of one's power general bodily development and to avoid one-sided training in brain work, as it may well be assumed that a vigorous body grown up under natural conditions will be in a better position to overcome the danger than a child exposed to the influences of effeminacy, of poverty, and of exact routine, and especially of city education. Childhood spent in the country

with plenty of open air, bodily exercise, escalation beginning later without ambitious aims, simple food, would be the principal points to keep in view. Meyer... hopes by all these measures to be able to prevent the development of the malady.

This line of argument regarding connectivity echoes Khan's (this volume) view of schizophrenia as a cognitive disorder. I would argue that the lifelong cognitive aspects of schizophrenia share a common cause with other psychopathology and phenomena seen in the disorder, including developmental and motoric aspects; cognitive aspects occur in parallel with these, but do not underlie them. All are due to a disorder (or variant) of connectivity development, which will have genetic as well as environmental components. It may lead to self-perpetuating and perhaps a catastrophic discrepancy in normal functioning. It will also be sensitive to the developmental stages of the brain throughout the first three decades of life. Most importantly, connectivity development disorders mirror the risk of schizophrenia over the life course of an individual. Kraepelin proposed that healthy development and the growth of a well-connected and resilient brain are well placed to buffer, create, and interact with its environment.

Conclusions

Given our present state of knowledge, which has not progressed much since Kraepelin, resilience to schizophrenia and its prevention can perhaps best be achieved by promoting health in all spheres of life: physical, psychological, and social. Obviously, this would require a massive effort, yet the benefits would be far-reaching, extending well beyond schizophrenia. All of society—whether at political, societal, professional, or individual levels—bear a responsibility for this effort, which can be targeted at the general population (primary prevention) as well as those at risk (secondary prevention). Resilience factors developed in such a way could, by design, be incorporated into the development of psychological treatments at all stages of the disorder: premorbid, prodromal, and thereafter. Even if drug treatments can be tailored to individuals through genetic or other biomarkers, their emphasis is still likely to be on risk, in terms of nonresponse or side effects, rather than on resilience. Biological psychiatry is capable of focusing on individuals at risk, but has yet to add resilience to its therapeutic palette.

7

Insights into New Treatments for Early Psychosis from Genetic, Neurodevelopment, and Cognitive Neuroscience Research

Kristin S. Cadenhead and Camilo de la Fuente-Sandoval

Abstract

Increasingly, schizophrenia research has emphasized the premorbid or prodromal periods of illness with a focus on identifying risk factors for later psychosis and understanding the mechanisms by which the neuropathological changes occur early in the course of illness. Genetic and epidemiological studies have begun to identify specific "vulnerability" genes and environmental risk factors which together may contribute to neurodevelopmental abnormalities and the emergence of psychosis. Neuroimaging and electrophysiological studies demonstrate altered developmental trajectories and evidence of compensatory changes in the early stages of psychotic illness, which perhaps reflects a period of neurotoxicity that coincides with the emergence of psychosis. These unique characteristics of early psychosis coincide with a time of increased brain plasticity, offering a window of opportunity to disrupt the neuropathological processes and remediate the neurocognitive and functional deficits. Insights from genetic, epigenetic, and biomarker studies in early psychosis have identified promising neuroprotective, disease-modifying, and cognitive remediation interventions that have the potential to alter the progressive trajectory of the illness. Adequately powered clinical trials that utilize information gained from biomarker studies are needed in early psychosis patients to determine the most effective individualized interventions. A synergistic treatment approach that offers precision pharmacologic intervention combined with remediation techniques is likely to have the greatest impact during the early course of illness.

Overview and Questions

Schizophrenia is a neurodevelopmental disorder that begins to develop most likely *in utero* and fully emerges during late adolescence or early adulthood. By identifying individuals in the prodromal phase of illness who are at risk for schizophrenia and following these cohorts into the first episode of psychosis, we have started to isolate the brain systems involved across multiple levels of analysis (behavioral, physiological, neurochemical, anatomical, genetic) during this critical pre-psychotic and early psychosis period. The use of multimodal techniques allows the assessment of the association of neurobiological markers to each other over time and provides further insight into the mechanism by which psychosis emerges. With increasing knowledge of the aberrant neurodevelopmental processes at the onset of psychosis, it may become possible to develop better treatment interventions to modify the disease outcome.

In this chapter, we review what is known about neurobiological predictors of psychotic illness, what they reveal about progressive neuropathological changes, and how this informs treatment in early psychosis. Evidence will be presented suggesting that neuropathological changes in the prodrome and first episode of illness, including compensatory changes that emerge at the onset of illness, differ from those found in more chronic forms of the illness. Therefore, specific treatments for prodromal and first-episode patients that target these aberrant processes have the potential to be more effective than the typical treatments used in more chronic forms of the illness, with a greater focus on neuroprotection, disease modification, and cognitive remediation.

The following questions guide our enquiry:

1. How do we best use psychosis risk biomarkers to inform treatment?
2. Is it possible to address the heterogeneity of early psychosis using neurobiological markers to individualize treatment?
3. Which neuroprotective strategies, disease-modifying agents, procognitive interventions, and remediation techniques show promise early in the course of illness?
4. Are there interventions that are unique to early psychosis and the hypothesized changes that occur at the onset of illness?

What Have We Learned about Risk of Psychosis, Mechanism of Disease, and Treatment from Genetic Studies?

With the heritability of schizophrenia estimated at 70–80% (for reviews, see Sullivan et al. 2003), a major proportion of disease risk can be explained by genes. Rapid progress has been made in the identification of genetic variants that confer risk of psychosis (Sebat et al. 2009; Purcell et al. 2009). Recent findings that both rare mutations of large effect and common variants of modest effect contribute to genetic risk for schizophrenia suggest that the disease is characterized by much more genetic heterogeneity than was previously thought. The risk alleles that have been implicated include rare copy number variants (CNVs) and common haplotypes based on single nucleotide polymorphisms (SNPs). Of particular interest in predicting risk for psychosis are mutations with moderate to high penetrance and CNVs that increase risk for schizophrenia by fivefold or more. Genetic information can also be used to inform our understanding of the mechanisms of disease and perhaps subtype individuals to specify treatment.

Genetic Risk Score

One means of leveraging the heterogeneous genetic information is to develop a "genetic risk score" that quantifies the polygenic component of psychosis risk for each individual. Given that less than 40% of individuals who meet the prodromal syndrome criteria (Miller et al. 2003; Yung et al. 2002) are likely to develop schizophrenia or an affective disorder, additional risk factors (such as those provided by genetic information) may improve the positive predictive power of current psychosis prediction algorithms (Cannon et al. 2008), which are primarily based on clinical and family history data, and thus help to determine who would benefit most from preemptive intervention. By including information for the many (>1000) variants, including common putative risk alleles from genome-wide association (GWA) studies, SNPs and CNVs can be weighted proportional to their associated odds ratios to develop a single score that can be used in algorithms of psychosis risk (Mowry and Gratten 2013). First, however, it is essential to determine whether the genetic information provides any added value in identifying disease risk over and above the standardized clinical and demographic criteria currently used to characterize subjects at risk for psychosis.

Pharmacogenomics

The reproducible genetic findings in schizophrenia patients have also provided insight into the mechanism of disease (e.g., genes involved in glutamatergic neurotransmission and neurodevelopment) (Egerton et al. 2012b), supporting and informing translational models and treatment development. The goal is to individualize treatment based on which disease mechanism is present in a particular individual. To date, pharmacogenomic studies have focused primarily on prediction of antipsychotic response and adverse effects using a candidate gene approach based on dopamine and serotonin receptors (for reviews, see Malhotra et al. 2007, 2012; Burdick et al. 2011; Arranz and de Leon 2007). A recent meta-analysis by Zhang et al. (2010) determined that the most robust pharmacogenetic findings are seen in the dopamine receptor D2 (DRD2) promoter region. Carriers of a functional polymorphism (−141C Ins/Del) are half as likely to show a clinical response to antipsychotic medication compared to noncarriers, and this effect was most prominent in first-episode psychosis patients. As noted by Malhotra et al. (2012), it is unlikely that we will attain perfect sensitivity and specificity using genetic data in the near future but it can help to clarify prognosis. We may identify a subgroup of patients who are less likely to respond to standard treatments and thus might benefit from a novel intervention.

Genetic Prediction of Adverse Treatment Effects

Although a number of studies have investigated antipsychotic-associated adverse effects such as clozapine-induced agranulocytosis and tardive dyskinesia, drug-induced weight gain may be the most powerful phenotype in pharmacogenetic studies (for a review, see Correll and Malhotra 2004). In a recent meta-analysis, Sicard et al. (2010) report that carriers of the T allele in a promoter region SNP (−759 T/C) in the 5-hydroxytrytamine 2C receptor (5-HT$_{2C}$) gene had less weight gain compared to those with the C allele. Similar results were recently reported in a study of patients in their first episode of psychosis, in which carriers of a functional promoter region variant (−141C Ins/Del) in DRD2 demonstrated more weight gain than noncarriers after six weeks of treatment, regardless of the antipsychotic. In terms of how this informs clinical practice, individuals at risk for increased weight gain or other adverse events might be given lower doses, adjunctive therapies (psychosocial or pharmacologic), and/or increased monitoring; however, this should be the standard of care for all patients on antipsychotic medication.

What Can Epigenetics Tell Us about Treatment of Early Psychosis?

Environmental risk factors associated with increased risk for psychosis in epidemiological studies include paternal age, hypoxia, urbanicity, migration, maternal infection, obstetric complications, nutritional deficiency, and cannabis use. While many of the environmental risk factors may affect the developing fetus, others appear to be "second hits" that occur in childhood or later adolescence, and may be more informative in determining which interventions are likely to be most effective in late adolescence. Theoretically, the second hits may act by epigenetic modulation of the genome in individuals who already have a genetic vulnerability, or they may potentiate biological pathways implicated in schizophrenia. Two specific environmental factors that illustrate the importance of understanding second hits are stress and drug abuse.

Stress and Psychosis Risk

The higher incidence of schizophrenia in urban areas, being a part of an ethnic minority, and migrant status may all be related to stress or social defeat and lack of social support (reviewed in van Os 2004; Rutten and Mill 2009). Although the link between childhood stress, epigenetic changes, and the onset of psychotic disorders has not been studied in humans, translational studies in animal models have shown that stress can mediate changes in gene expression during key developmental periods via epigenetic mechanisms (reviewed in Rutten and Mill 2009). For example, chronic psychosocial stress (e.g., defeat stress) alters gene expression, particularly of brain-derived neurotrophic factor (BDNF), via a range of epigenetic mechanisms, and this process can be reversed with *tricyclic antidepressant* treatment. Importantly, epigenetic moderation of *BDNF* transcription has been shown to be involved in neuroplasticity, suggesting the potential for preemptive intervention. Other recent advances in the understanding of the biological processes mediating stress have implicated the role of the hypothalamic-pituitary-adrenal (HPA) axis (Walker et al. 2008) as well as neuroinflammation (Meyer 2011).

HPA Axis and Stress Response

Increased HPA activity is associated with psychotic disorders and may increase the activity of dopamine pathways (Van Craenenbroeck et al. 2005; Tsukada et al. 2011; Wand et al. 2007). First-episode patients as well as individuals who meet the prodromal criteria for schizophrenia and later develop psychosis all have elevated cortisol levels relative to normal subjects (Walker et al. 2001, 2010; Guest et al. 2011). In addition, drugs associated with psychosis, including tetrahydrocannabinol (THC), amphetamine, and ketamine, all augment cortisol release in nonclinical and/or clinical populations (Oswald et al. 2005; van Berckel et al. 1998; Munro et al. 2006; D'Souza et al. 2005). Patients with schizophrenia who have the best response to antipsychotic medication show a higher pretreatment cortisol level, raising the possibility that one mechanism of action for antipsychotics may be to suppress the HPA axis.

Treatment implications: Early intervention research in schizophrenia patients has included both psychosocial means of reducing stress and salivary cortisol levels (e.g., yoga, exercise, relaxation) in vulnerable youth (Vancampfort et al. 2012; Cabral et al. 2011; Rocha et al. 2012) as well as pharmacologic interventions that can reduce stress. Antiglucocorticoid agents have been used in the treatment of depression to suppress the glucocorticoid response, but only one pilot study conducted on schizophrenia patients treated with ketoconazole reports improvement in observer-rated depression but no significant alteration of morning serum cortisol levels (Marco et al. 2002). Further studies of therapies such as yoga, exercise, or antiglucocorticoids are needed in prodromal or first-episode subjects to determine whether it would be possible to alter the course of illness.

Neuroinflammation and Stress Response

It has been postulated that early-life exposure to infection and/or inflammation has the potential to induce latent neuroinflammatory abnormalities that can be unmasked by additional exposure to stressful stimuli (Meyer et al. 2011; Bilbo and Schwarz 2009), activating microglia and enhancing the production of proinflammatory cytokines in the central nervous system (Frank et al. 2007; Garcia-Bueno et al. 2008). Brown and Patterson (2011) propose that prophylactic treatments which target maternal infection and associated inflammatory processes could reduce the incidence of schizophrenia and related disorders by one-third. Animal models have demonstrated that the neurodevelopmental effects of prenatal infection/inflammation can be attenuated through interventions which target activated inflammatory response systems or associated physiological processes such as oxidative stress, hypoferremia, and zinc deficiency (Aguilar-Valles et al. 2010; Coyle et al. 2009; Girard et al. 2010; Lante et al. 2007; Pang et al. 2005; Robertson et al. 2007).

Treatment implications: Recent studies in early psychosis patients suggest that anti-inflammatory interventions may attenuate progressive brain changes (Meyer 2011). In an add-on study of *celecoxib* (a preferential cyclooxygenase-2 inhibitor, COX-2) given in conjunction with amisulpride, Muller et al. (2010) found that anti-inflammatory add-on therapy was more effective than antipsychotic treatment alone in treating negative symptoms when initiated in the early phase of schizophrenia. The broad spectrum antibiotic *minocycline*, when administered in conjunction with antipsychotic drugs, also has a significant effect on negative and cognitive symptoms compared with treatment outcomes using antipsychotic drugs alone in early psychosis (Levkovitz et al. 2010). Paralleling its known effects in reducing inflammation and preventing cell death when given after a traumatic brain injury, *aspirin* (COX-1, COX-2 inhibitor) has also been shown to have beneficial effects on symptoms of schizophrenia in patients with less than ten years of illness (Laan et al. 2010). The symptomatic improvement was most marked in patients with the lowest T_H1/T_H2 cytokine balance, suggesting that this treatment is most effective in individuals with relatively high anti-inflammatory cytokine production (Laan et al. 2010). In contrast, anti-inflammatory strategies are not effective in chronic schizophrenia (Rapaport et al. 2005), suggesting that neuroinflammatory processes are active primarily during the early phase of disease and are thus an important target for intervention. *Omega-3 fatty acids* such as eicosapentaenoic acid (EPA) and its derivative docosahexaenoic acid (DHA) have well-documented anti-inflammatory actions (Capper and Marshall 2001). In a randomized, double-blind, placebo-controlled trial conducted in 81 prodromal subjects, Amminger et al. (2010) found that after twelve weeks of treatment, 2 out of 41 individuals (4.9%) in the omega-3 fatty acid group and 11 of 40 (27.5%) in the placebo group had transitioned to a psychotic disorder. Omega-3 fatty acids also significantly reduced positive, negative, and general symptoms and improved functioning compared with placebo. *Antipsychotic medication* has also been shown to affect the proinflammatory cytokine network and immune function in schizophrenia (for reviews, see Pollmacher et al. 2000; Drzyzga et al. 2006), perhaps providing an aspect of disease modification and prevention to the known therapeutic benefits on dopamine regulation.

Cannabis and Psychosis Risk

The epidemiological literature demonstrates an association between the early use of cannabis and later risk for psychotic illness (Andreasson et al. 1987; Arseneault et al. 2002; Weiser and Noy 2005; Moore et al. 2007). In a second-hit model of psychosis, Caspi et al. (2005) demonstrated that carriers of the catechol-O-methyl transferase (COMT) Met versus Val polymorphism (associated with rapid dopamine metabolism, low cortical, and high midbrain dopamine) were more likely to develop psychosis if they used cannabis. Translational studies have revealed the role of cannabinoid (CB) receptors and endocannabinoids in dopamine and glutamatergic regulation, immune function, energy metabolism, and the pathophysiology of schizophrenia (Koethe et al. 2009b; D'Souza 2007; Pacher et al. 2006; Hallak et al. 2011). In clinical studies, anandamide, an endogenous CB1 receptor agonist, has been shown to be elevated in antipsychotic and cannabis-naïve patients with schizophrenia (Leweke et al. 2007b; Giuffrida et al. 2004) and in the prodromal phase of illness (Koethe et al. 2009a). Koethe et al. (2009b) have proposed a model of psychosis in which the endogenous agonists like anandamide may rise in response to increased dopamine transmission and provide neuroprotection. Anandamide reuptake and hydrolysis is inhibited by cannabidiol (CBD), the second most abundant component of *Cannabis sativa* (besides THC), which has weak partial antagonistic properties at the CB1 receptor. Recent studies in animals, healthy humans, and patients with schizophrenia suggest that cannabinoids such as CBD and SR141716 have a pharmacologic profile similar to antipsychotic drugs (Roser et al. 2010).

Treatment implications: Because CBD can reverse many of the biochemical, physiological, and behavioral effects of CB1 receptor agonists, recent studies have explored the possible role of cannabinoids, including CBD, in the treatment of psychosis (Koethe et al. 2009b). While CBD and SR141716 monotherapy has not been found to be efficacious in chronic or treatment-resistant schizophrenia (Meltzer et al. 2004; Zuardi et al. 2006), Leweke et al. (2007a) found clinical benefits of CBD similar to amilsulpride in a preliminary study of 42 acutely ill patients with schizophrenia.

Neurodevelopmental Abnormalities: Can We Intervene?

Accelerated Gray Matter Volume Loss in Early Psychosis

Previous studies in schizophrenia, first-degree relatives, and at-risk subjects have shown reductions in multiple brain regions including prefrontal, superior, and medial temporal lobe gray matter volumes (Borgwardt et al. 2007; Pantelis et al. 2003; Koutsouleris et al. 2009; McCarley et al. 2002; Mechelli et al. 2011). Pantelis et al. (2003) examined gray matter changes over time in prodromal subjects and found that the converted group showed gray matter loss in left inferior frontal, left medial temporal, and cingulate regions at one-year follow-up. Moreover, prodromal subjects who later transition to psychosis have reduced gray matter volume in the left parahippocampal cortex at baseline compared to the nontransition at-risk group (Mechelli et al. 2011).

The neuroanatomical changes in schizophrenia appear to be progressive changes beyond those associated with normal development (Ho et al. 2003; Gur et al. 1998; Jacobsen et al. 1998; Keshavan et al. 1994). Cortical gray matter density declines normally during late adolescent development, resulting in decreased neuropil in the same brain regions implicated in the pathophysiology of schizophrenia (Huttenlocker 1979; Huttenlocker and Dabhokar 1997). As summarized by Pantelis et al. (2005), the available neuroimaging data provides evidence of early (pre- or perinatal) neurodevelopmental changes in schizophrenia which may lead to a vulnerability to postpubertal insults and contribute to the accelerated loss of gray matter and aberrant connectivity in the prefrontal regions. Factors such as substance abuse, stress, and HPA axis dysregulation may lead to neurodevelomental abnormalities which may be neurodegenerative, involving medial temporal and orbital prefrontal regions. Thus, while disturbances of brain structure early in life may be necessary for the future emergence of schizophrenia (Weinberger 1987), neurodevelopmental events during the late adolescent period may participate in psychotic symptom formation via a range of possible mechanisms, including inflammation, glutamatergic or dopaminergic transmission (Weinberger 1987; Feinberg 1982; Keshavan et al. 1994).

Treatment Implications: Pharmacologic and nonpharmacologic interventions have been shown to slow gray matter losses in schizophrenia and related disorders. Eack et al. (2010) used a computer-based *cognitive enhancement therapy* in patients during the early stages of schizophrenia. Compared to those who received supportive psychotherapy over a period of two years, there was greater preservation of gray matter in the left hippocampus, parahippocampal gyrus, fusiform gyrus, and left amygdala in patients who received the active treatment. In a study of treatment-naïve patients with obsessive compulsive disorder (OCD), Hoexter et al. (2012) found that after treatment with either fluoxetine or cognitive behavioral therapy (CBT), gray matter volume loss in the left putamen was no longer detectable relative to controls. Animal models have demonstrated increased neurogenesis, dendritic arborization, and synaptogenesis with *serotonin reuptake inhibitors* (SSRI) (Richtand and McNamara 2008), supporting the notion that these agents may provide an element of neuroprotection. Preclinical and clinical studies also suggest that lithium may exert neurotrophic effects that counteract pathological processes, suggesting protective and potentially regenerative brain effects in the brains of patients with bipolar disorder (Manji et al. 2000; Bearden et al. 2007; Moore et al. 2009; Kempton et al. 2008; Lyoo et al. 2010). Moreover, a preliminary study by Berger et al. (2012) showed a reduction in T2 relaxation time (a nonspecific measure of neuropathological changes) in the hippocampus of putatively prodromal subjects treated with low doses of lithium compared to untreated prodromal subjects. Future early intervention studies using cognitive remediation, CBT, SSRIs, or lithium in the prodrome and first episode of psychosis should incorporate longitudinal analysis of gray matter volume to provide insight into the mechanism of these potential neuroprotective effects and clinical correlates.

Neurochemical Changes in Early Psychosis

While it is a matter of current debate as to whether the accelerated gray matter loss at the onset of psychosis involves (transient) neurodegenerative processes (Archer 2010; McGlashan 2006; McGlashan and Hoffman 2000) or perhaps a progressive excitotoxic process (Bustillo et al. 2010), recent reports using proton magnetic resonance spectroscopy (^{1}H-MRS) have identified neurometabolic changes which may be unique to the onset of psychosis and provide insight into the neuropathological changes (Bustillo et al. 2010; de la Fuente-Sandoval et al. 2011, 2013b; Kegeles et al. 2012; Stone et al. 2009). The dopamine hypothesis has been a useful model in our understanding and study of the psychotic state, but it does not explain the accelerated gray matter loss and deteriorating course in terms of cognition and function seen in the first few years of schizophrenia. Glutamate antagonists are well known to induce positive and negative psychotic symptoms more akin to schizophrenia than the positive symptoms induced by dopamine agonists alone (Javitt and Zukin 1991; Moghaddam and Javitt 2012), and it has been proposed that dopaminergic dysregulation is the final common pathway resulting from an altered glutamatergic neurotransmission early in the course of illness (Carlsson et al. 2001; Olney and Farber 1995a). According to glutamatergic theories, the abnormal developmental trajectory observed in neuroimaging studies could result from reduced elaboration of inhibitory (GABAergic) pathways and excessive pruning of excitatory (glutamatergic) pathways leading to altered excitatory-inhibitory balance in the prefrontal cortex (Lewis and Gonzalez-Burgos 2008). Glutamatergic theories of schizophrenia suggest that an increase in cortical glutamatergic activity, due to genetically or environmentally mediated hypofunction of *N*-methyl-D-aspartate (NMDA) receptors, may lead to a time-limited neurotoxic process and dopaminergic dysregulation at the onset of psychosis (Carlsson and Carlsson 1990a; Javitt and Zukin 1991; Olney and Farber 1995b). The glutamatergic projections are thought to stimulate prefrontal dopamine release directly but inhibit midbrain dopamine neurons projecting to the striatum (via GABAergic interneurons) (Sesack et al. 2003; Sesack and Carr 2002). In support of this hypothesis, de la Fuente-Sandoval et al. (2011) report that antipsychotic-naïve first-episode and at-risk subjects have higher levels of glutamate in the dorsal caudate than normal subjects. In the cerebellum, no group differences were seen, suggesting that high levels of glutamate in the dorsal caudate, a region with prominent projections throughout the cortical mantle, could induce neuronal toxicity leading to a progressive functional and intellectual deterioration. Moreover, antipsychotic-naïve at-risk subjects who later converted to psychosis had higher glutamate levels than those who had not converted at two-year follow-up (de la Fuente-Sandoval et al. 2013b). In the treatment of schizophrenia with antipsychotic drugs, around 60% occupancy of brain DRD2 is required, on average, to produce a therapeutic response (Kapur et al. 2000). However, a substantial proportion of patients still show a poor response even when D2 occupancy is at this level (Pilowsky et al. 1993). This may reflect the importance of nondopaminergic neurochemical dysfunction in the pathophysiology of schizophrenia. A recent study demonstrated that clinically effective antipsychotic treatment normalized glutamate levels in antipsychotic-naïve first-episode psychosis patients (de la Fuente-Sandoval et al. 2013a). These results agree with a recent report (Egerton et al. 2012a) which found that clinically stable first-episode patients had lower glutamate levels compared to patients that were still symptomatic. While studies in medicated patients have shown the same or decreased levels of glutamate compounds compared to controls (Theberge et al. 2003; Tayoshi et al. 2009; Lutkenhoff et al. 2010; Reid et al. 2010; Rowland et al. 2012; Bustillo et al. 2011), patients experiencing psychotic state exacerbations demonstrate elevations of these compounds (Ongur et al. 2008; Ota et al. 2012), suggesting that an improvement in clinical symptoms might relate to decreases in glutamate levels.

Consistent with these findings, a recent ^{1}H-MRS study by Kegeles et al. (2012) found increased γ-aminobutyric acid (GABA) and glutamate+glutamine in the medial prefrontal cortex (mPFC) of primarily antipsychotic-naïve schizophrenia patients, adding support to the theory that dysfunction of fast-spiking parvalbumin-containing GABA interneurons may contribute to high levels of glutamate via pyramidal cell disinhibition (Lewis and Moghaddam 2006). Because there is also evidence that stable-medicated subjects either do not show a difference or demonstrate a decrease in glutamate levels compared with normal subjects (Marsman et al. 2011; Reid et al. 2010; Bustillo et al. 2010), it is tempting to hypothesize that a time-limited neurotoxic process may characterize the early stages of illness (McGlashan and Hoffman 2000; Archer 2010; Lahti and Reid 2011), primarily because excess synaptic glutamate levels are highly neurotoxic (Lau and Tymianski 2010). In addition, activated microglia release substantial levels of glutamate (Barger and Basile 2001), and recent reports suggest that such microglia-mediated toxicity contributes to neuronal damage in the event of neuroinflammation (Block and Hong 2007; Perry 2007; Ransohoff and Perry 2009).

Treatment Implications: Glutamatergic theories of NMDA receptor hypofunction and resulting glutamate-mediated neurotoxicity suggest that glutamate and GABA-modulating agents may prove to be neuroprotective or even capable of modifying the disease early in the course of illness (Moghaddam and Javitt 2012). *LY404039* is a selective agonist for metabotropic glutamate 2/3 (mGlu2/3) receptors which regulate synaptic concentrations of glutamate and other neurotransmitters, including dopamine, GABA, and serotonin (Rorick-Kehn et al. 2007; Seeman and Guan 2009). Recent reports in chronic patients have found that LY404039 improves positive and negative symptoms of schizophrenia (Patil et al. 2007) or has no effect compared to placebo (Kinon et al. 2011). However, clinical trials using glutamate-modulating agents have not been performed in first-episode or at-risk subjects who may most likely benefit from the intervention. Since ^{1}H-MRS permits the *in vivo* study of regional concentrations of various brain metabolites (Di Costanzo et al. 2007), this noninvasive imaging technique may provide important clues into the mechanism of action of many interventions. Berger et al. (2008) assessed the effect of the omega-3 fatty acid, EPA (eicosapentaenoic acid), on brain metabolites with ^{1}H-MRS in the anterior hippocampus of both hemispheres in unmedicated first-episode psychosis patients. This study found that EPA treatment increased levels of glutamine+glutamate and glutathione, and this was associated with negative symptom improvement. Moreover, a study with EPA in depressed bipolar patients, using ^{1}H-MRS in the anterior cingulate cortex (ACC), found a significant increase in N-acetylaspartate levels, presumably induced by a neurotrophic role of EPA (Frangou et al. 2007). Omega-3 fatty acids are essential for normal brain function and development (Bazan 2005; Piomelli et al. 1991) and may also have neuroprotective properties (Lonergan et al. 2002; Lynch et al. 2007). There are also reports on the effects of omega-3 fatty acids on the glutamatergic system, such as modulation of glutamate transporters, glutamate release in hippocampi of aged rats (McGahon et al. 1999), and as a protective agent against neurotoxicity induced by NMDA antagonists (Ozyurt et al. 2007).

Cognitive Neuroscience Insights into Treatment Effects on Neurocognition and Perception

The structural and neurochemical brain changes early in the course of illness reflect changes in cells as well as fibers and extra-parenchymal elements. Abnormalities in the number and distribution of neurotransmitter receptors in these regions are likely secondary to loss of cells, fibers, or neurochemical changes. These early neurochemical changes are hypothesized to lead to gray matter loss, reduced cortical connectivity, sensory, perceptual, cognitive, and global functioning abnormalities in chronic illness (Olney and Farber 1995b; Sharp et al. 2001). Thus, neural circuit dysfunction and associated information-processing abnormalities are likely downstream events that precede the onset of psychotic symptoms, and these may progress with illness onset and provide surrogate endpoints for treatment intervention studies.

Neurocognition in Early Psychosis

Neurocognitive deficits are prominent across the schizophrenia spectrum (Cadenhead et al. 1999b; Cannon et al. 1994; Hawkins et al. 2004; Heinrichs and Zakanis 1998). They are known to predict functional outcomes (Green 1996; Green and Nuechterlein 1999b) and to explain 20–60% of the variance in community functioning, social problem solving, and acquisition of psychosocial skills (Green et al. 2000). Neurocognitive deficits have been shown to be reliable (Faraone et al. 1999; Heaton et al. 2001; Rund 1998), heritable (Ando et al. 2001; Posthuma et al. 2002), and associated with genes linked to schizophrenia (e.g., glutamate signaling, COMT). Substantial cognitive deficits are already apparent in childhood for those individuals who go on to develop schizophrenia, and these tend to exacerbate before the onset of psychotic symptoms and worsen after the initial episode of the illness (Bilder et al. 2006). A number of recent reports (Keefe et al. 2006; Eastvold et al. 2007; Hambrecht et al. 2002; Seidman et al. 2010) have demonstrated that at-risk individuals have neurocognitive deficits across multiple domains that are intermediate to those observed in first-episode patients. In addition, at-risk subjects who later convert to psychosis have greater neurocognitive impairment at baseline compared to those individuals who remain "at risk" at follow-up (Hambrecht et al. 2002; Keefe et al. 2006; Seidman et al. 2010; Eastvold et al. 2007). Longitudinal neurocognitive studies of first-episode subjects show high stability (Addington et al. 2005). The few small longitudinal studies in at-risk subjects found a decline in verbal memory over time; this was most prominent in at-risk subjects who later converted to psychosis (Cosway et al. 2000; Brewer et al. 2005; Whyte et al. 2006; Pukrop et al. 2006; Jahshan et al. 2010).

Treatment Implications

Pharmacogenetic studies: Neurocognition has been used as an outcome measure to assess the cognitive enhancement effects of antipsychotics as well as other promising procognitive agents (reviewed in Burdick et al. 2011). Two studies (Need et al. 2009; McClay et al. 2011) using data from the CATIE (The Clinical Antipsychotic Trials of Intervention Effectiveness) trial have identified SNPs located close to specific genes (e.g., GRM8 [metabotropic glutamate receptor 8], DRD2, and IL1A [interleukin-1-α]) that are associated with greater improvement in neurocognitive paradigms after treatment with antipsychotics. Three small published studies (Bertolino et al. 2004; Weickert et al. 2004; Woodward et al. 2007) have used a candidate gene approach focusing on COMT Val158Met genotype to predict neurocognitive performance after antipsychotic treatment. A Met versus Val homozygote predicted better neurocognitive response, suggesting that it may be possible to define which individuals are likely to show improvement in neurocognitive performance with antipsychotic treatment.

Cognitive enhancement: Given the limited response of neurocognitive deficits to antipsychotic treatments (Mishara and Goldberg 2004), there have been a number of efforts to develop targeted therapies for cognitive deficits in schizophrenia (Barak and Weiner 2011). For example, the *measurement of treatment effects on cognition in schizophrenia* (MATRICS) was developed as a means of identifying cognitive targets and promising molecular targets to enhance cognition (Marder and Fenton 2004). A detailed review by Keefe et al. (2013) found that the majority of cognitive enhancement double-blind add-on studies had been conducted on chronic patients and were underpowered to detect a significant effect. Agents acting at the NMDA receptor have been the most frequently studied compounds, including NMDA receptor modulation, glycine site agonism/partial agonism, and glycine site antagonism. Other trials have included agents that target various mechanisms, including H_3 antagonism, selective activation of hypothalamic regions associated with wakefulness, noradrenergic receptor reuptake inhibition, acetylcholine esterase inhibitors, $α_7$ receptors agonism/partial agonism, $α4β_2$ nicotinic receptors partial agonism, cannabinoid receptor antagonism, D_2 partial agonism + $5-HT_{2A}$ antagonism, and D_1/D_2 agonism. These important studies using promising procognitive agents have yet to provide robust results, perhaps because most studies are underpowered and use chronic rather than first-episode patients who would have greater potential for brain plasticity (Barch 2010). The sole exception noted by Keefe et al. (2013) was a six-month add-on treatment with minocycline versus placebo in young subjects in early phase schizophrenia (Levkovitz et al. 2010). In this study, minocycline (a tetracycline antibiotic with a distinct neuroprotective profile) was found to be superior to placebo in improving cognitive functioning as well as negative symptoms and general outcome. Little is known about the effects of omega-3 fatty acids on neurocognitive performance in schizophrenia, which has been studied more extensively in dementia (Kalmijn et al. 1997; Cole et al. 2009). Accelerated cognitive decline, mild cognitive impairment, and decreased brain volume correlate with lowered tissue levels of DHA/EPA (Tan et al. 2012). Supplementation improves cognitive function early in the course of illness (Mazereeuw et al. 2012). Omega-3 fatty acids as well as other potentially important procognitive agents need to be assessed in well-powered studies of early psychosis patients.

Cognitive remediation, cognitive training: Nonpharmacologic cognitive remediation trials have also shown promise in patients early in the course of illness (Wykes et al. 2007; Eack et al. 2010; Barlati et al. 2012; Breitborde et al. 2011), when intervention is likely to make the greatest impact on the developing brain. Cognitive remediation or training interventions include restorative (e.g., computer-based approaches; Fisher et al. 2009), compensatory (e.g., strategy-based approaches; Twamley et al. 2008, 2011), or environmental adaptation (Velligan et al. 2008). The most recent review and meta-analysis of cognitive remediation techniques (Wykes et al. 2011) found the largest effect sizes (mean effect size of .45 for cognitive improvement, .18 for symptom improvement, and .42 for functional improvement) for compensatory strategy-based approaches in the context of psychiatric rehabilitation. Compensatory strategies or cognitive prosthetics (which teach patients how to "work around" their deficits) can be helpful because they focus on application of appropriate cognitive strategies in the real world. Increasing patients' ability to remember appointments, sustain attention, encode important concepts, and think flexibly may well improve the success of concomitant treatments. Alterations in the environment to decrease cognitive demands and automatize everyday tasks may also be helpful. In essence, compensatory cognitive training provides an intervention which targets healthy neural circuitry to compensate for damaged circuit elements and may even protect this circuitry from future damage (Swerdlow 2011).

It is clear that larger studies are needed in early illness patients to determine whether it is possible to prevent or improve the cognitive deficits early in the course of illness. It is reasonable to expect that younger patients with greater potential neuroplasticity may be optimal candidates for a combined approach using pharmacological and nonpharmacologic intervention, but surprisingly, few data address this question empirically. Although cognitive remediation interventions have been added to augment antipsychotic medication, relatively little is known about the effectiveness of combining pharmacologic interventions (designed to enhance cognition, provide neuroprotection, or facilitate neuroplasticity) with cognitive remediation or CBT. In fact, reviews of the literature on cognitive remediation describe nonpharmacologic and pharmacologic interventions but they are all separate, rather than combined, trials (Goff et al. 2011). It is possible that the procognitive pharmacologic interventions will act synergistically with cognitive therapies to enhance clinical, neurocognitive, and functional outcome early in the course of illness. An analogy comes from anabolic steroids, which increase muscle mass only when used in concert with muscle-engaging activities (Swerdlow 2011). While reducing active psychosis with antipsychotics benefits any cognitive intervention, it is possible that drugs with procognitive effects might more specifically, and perhaps synergistically, enhance the clinical benefits of cognitive therapies.

Electrophysiology and Functional Imaging

Like the MATRICS initiative, Cognitive Neuroscience Treatment Research to Improve Cognition in Schizophrenia (CNTRICS) was developed to identify biomarkers derived from cognitive neuroscience as "surrogate endpoints" (Carter and Barch 2007). The U.S. National Institutes of Health website defines a biomarker as "a characteristic that is objectively measured and evaluated as an indicator of normal biologic or pathogenic processes or pharmacological responses to a therapeutic intervention." A number of potentially important paradigms have been identified which target specific cognitive or affective domains and can be studied in terms of the underlying neural system, animal models, and by using electrophysiological or neuroimaging paradigms. Importantly, for biomarkers to be useful in such models, it is essential that they demonstrate construct validity, reliability, and ease of use in treatment studies. Below, paradigms that hold particular promise, indicated by findings in early psychosis, are discussed as they may reveal neurodevelopmental abnormalities or compensatory processes which may differ from the more chronic forms of schizophrenia.

Prepulse Inhibition

Prepulse inhibition (PPI) is an index of sensorimotor gating in which weak lead stimuli are thought to inhibit the motor response to abrupt startling stimuli (Ison and Hoffman 1983; Graham 1975). PPI is deficient in schizophrenia, first-degree relatives, and schizotypal subjects (Braff et al. 1992; Cadenhead et al. 1993, 2000). In addition, PPI is stable with repeated testing (Cadenhead et al. 1999a; Cadenhead 2011) and is heritable (Greenwood et al. 2007), suggesting its utility as a neurobiological marker for psychosis risk. Genetic studies have identified SNPs that are strongly linked with PPI (Greenwood et al. 2007), including neuregulin-1 (activation of receptors, including glutamate), COMT, serotonin-2A, and DRD3. Translational studies demonstrate the emergence of PPI deficits after developmental manipulations in rodent models (Powell and Geyer 2002), suggesting that it may be useful in understanding a neurodevelopmental disorder such as schizophrenia. Animal studies have identified an extended forebrain/pontine circuit (limbic cortex, ventral striatum, ventral pallidum, pontine tegmentum) that modulates PPI (Swerdlow et al. 1992, 1999). The neurotransmitters active at several levels of this circuitry—dopamine, serotonin, glutamate—cause disruptions in PPI through the stimulation of DRD2s (amphetamine or apomorphine), activation of serotonergic systems, or blockade of NMDA receptors (phencyclidine or ketamine) (Geyer et al. 2001).

Few studies have reported PPI in first-episode psychosis (Aggernaes et al. 2001; Quednow et al. 2008; Kumari et al. 2007; Meincke et al. 2004; Ludewig et al. 2003; Mackeprang et al. 2002). Although the majority found PPI deficits in the first episode, they were not always robust (e.g., only in males, antipsychotic specific). Three studies have assessed PPI in subjects at risk for psychosis: Quednow et al. (2008) and Ziermans et al. (2011) reported that at-risk subjects showed significant PPI deficits. However, Cadenhead (2011) found very different results in 75 first-episode, 89 at-risk, and 85 controls from the Cognitive Assessment and Risk Evaluation (CARE) sample. Unexpected findings included the fact that acutely ill, medication-naïve, first-episode subjects and at-risk subjects who later converted to psychosis had *greater* PPI than medicated first-episode subjects and at-risk subjects who did not convert to psychosis, respectively. This parallels findings from the visual perceptual organization literature (reviewed in Silverstein and Keane 2011; Parnas et al. 2001) that perceptual organization is intact or even superior at first episode. These findings introduce the possibility of early brain changes that diverge from findings in chronic patients early in the course of illness. Although the PPI findings from the CARE study differ from prior studies, it offers an intriguing possibility that there may be compensatory changes in inhibitory processes in response to early neurochemical changes, reflected by greater PPI, early in the course of psychotic illness. This finding may represent an initial change in the neural circuitry regulating PPI prior to the appearance of sensorimotor gating deficits in more chronic forms of the illness. Although preclinical studies show reduction in PPI in response to dopamine agonists and NMDA antagonists (Geyer et al. 2001), compounds such as N-acetylcysteine, which increase extracellular glutamate levels, enhance PPI (Chen et al. 2010). In addition, studies in control subjects have revealed evidence of enhanced PPI under certain conditions (high novelty seeking, specific doses) in response to dopamine agonists (amphetamine, pramipexole) (Talledo et al. 2009; Swerdlow et al. 2009a) and NMDA antagonists (ketamine, memantine, amantadine) (Swerdlow et al. 2002, 2009b; Abel et al. 2003; Duncan et al. 2001). This lends support to the idea of a period of acute glutamatergic dysregulation early in the course of illness leading to increases in PPI. It would then follow that more chronic hypoglutamatergic states would lead to reduced PPI in more chronic patients (Swerdlow et al. 2009b). The neurochemical mechanism by which PPI might be increased in the early stages of psychosis and the location in the modulatory circuitry (Swerdlow et al. 2008) where this occurs is unknown, but this work has implications for treatment development. Clearly, longitudinal studies of early psychosis patients are needed to follow the time course of PPI through the onset of disease to determine whether, for example, there is a window of compensatory changes that would benefit from interventions that reduce glutamate. When combined with other biomarkers (e.g., ^{1}H-MRS, fMRI), it should be possible to tease out the mechanism of disease and identify specific interventions likely to make an impact at this early stage.

Treatment implications: Although more work is needed to define the developmental neuropathology as indexed by PPI in the early stages of psychosis, important translational studies have been performed using the PPI paradigm. Atypical antipsychotics have been shown to reverse PPI deficits in developmental animal models of psychosis and may "normalize" PPI in patients with chronic schizophrenia and clinically normal subjects (Swerdlow et al. 2006, 2008; Vollenweider et al. 2006; Wynn et al. 2007). Feifel and colleagues have demonstrated that *oxytocin* (a neurohypophyseal peptide known to regulate social cognition and affiliation) can modulate PPI deficits induced by NMDA receptor antagonists and dopamine agonists in rodent models (Feifel and Reza 1999; Feifel et al. 2010). Preliminary oxytocin studies in patients with schizophrenia have demonstrated improved positive and negative symptoms, social cognition, and emotional recognition (Feifel et al. 2010, 2012; Pedersen et al. 2011; Averbeck et al. 2011) but the effect of oxytocin on PPI has yet to be reported in schizophrenia patients.

Mismatch Negativity

One emerging view holds that the commonly observed clinical and neurocognitive deficits of schizophrenia patients may arise, at least in part, by dysfunction in the coordination of neural activity at the earliest stages of sensory and cognitive information processing (Green and Nuechterlein 1999a; Phillips and Silverstein 2003). Schizophrenia patients exhibit deficits in basic levels of sensory information processing that are present early in the course of the illness and even precede the emergence of psychotic symptoms. In a passive auditory oddball paradigm, a duration deviant stimulus elicits a mismatch negativity (MMN) response that peaks 100–200 ms after the onset of a stimulus deviance (Naatanen et al. 1978) and is assumed to reflect an automatic, sensory-based deviance detection process (Naatanen et al. 1978; Picton et al. 2000). Deficits in MMN generation using a variety of stimulation parameters (e.g., oddball stimuli that differ in pitch or duration) represent a remarkably robust finding in chronic schizophrenia (Shelley et al. 1991; Light and Braff 2005; Javitt et al. 2000), but the extant literature on MMN in the early stages of the disease is mixed, with some studies identifying abnormalities (Hermens et al. 2010; Umbricht et al. 2006; Devrim-Ucok et al. 2008) while others fail to detect any significant decrements in either duration or pitch of MMN in patients with a psychotic illness duration of less than three years (Valkonen-Korhonen et al. 2003; Salisbury et al. 2002). In a prospective study of first-hospitalized patients with schizophrenia (Salisbury et al. 2007), a strong relationship was found between the progressive reductions of MMN amplitude and left hemisphere Heschl gyrus gray matter volume. In chronic patients, Rasser et al. (2011) report that gray matter reductions are correlated with MMN amplitude. Several studies have now identified deficits in duration of MMN in not only the first episode of psychosis but also the prodromal period of illness (Bodatsch et al. 2011; Atkinson et al. 2012; Brockhaus-Dumke et al. 2005; Jahshan et al. 2012). These findings of MMN deficits in the prodrome contribute to the overall efforts to identify potential markers of vulnerability to schizophrenia as well as to understand the underlying pathological processes leading to the development of the illness.

Treatment implications: Deficits in MMN generation may be associated with impaired NMDA receptor function because phencyclidine (PCP) and other NMDA antagonists inhibit MMN generation in primate models and normal volunteers. Consistent with the links between glutamatergic dysregulation and MMN, *N-acetylcysteine* (NAC, a glutathione precursor) has been shown to enhance MMN in patients with schizophrenia (Lavoie et al. 2008). Lavoie et al. (2008) report improved MMN in response to NAC versus placebo in patients with schizophrenia and suggest that increased levels of brain glutathione improve MMN and, by extension, NMDA function. *Memantine*, a noncompetitive NMDA receptor antagonist, enhanced the amplitude of MMN in normal subjects (Korostenskaja et al. 2007), but the effects have yet to be assessed in schizophrenia. Several studies report the ability to enhance MMN in healthy subjects using a variety of compounds including *serotonin reuptake inhibitors* (Kahkonen et al. 2005; Wienberg et al. 2010), *tryptophan depletion* (Ahveninen et al. 2002; Kahkonen et al. 2005), and *nicotinic receptor stimulation* (Baldeweg et al. 2006). A single case report (Higuchi et al. 2010) on the use of *tandospirone* (a 5-HT_{1A} partial agonist) in schizophrenia demonstrated an increase in MMN that preceded improvement in neurocognition. The effect of 5-HT_{1A} agonism on MMN may be mediated by its influence on glutamatergic and, possibly, GABAergic function (Huot and Brotchie 2011). With respect to antipsychotic agents, large MMN amplitudes predicted good treatment response to clozapine (Schall et al. 1999), although MMN appears to be insensitive to antipsychotic medication in schizophrenia (Umbricht et al. 1998, 1999; Korostenskaja et al. 2005). Event-related potentials (ERPs) could be especially useful in defining subgroups that might benefit from interventions other than dopamine-based treatments, but this requires going beyond group effects and reliably measuring individual differences. Methodological developments to improve the quantification of single-subject data are needed, along with more research which demonstrates that ERP measures can predict treatment response in schizophrenia patients.

Neural Synchrony

Abnormal gamma range (30–80 Hz) synchrony has proved to be an important biomarker for psychosis as it reflects core pathophysiological features of schizophrenia, including cognitive and perceptual abnormalities (reviewed in Gandal et al. 2012). Gamma oscillatory activity is thought to be the mechanism by which neural networks are integrated, facilitating coherent sensory registration. In schizophrenia, gamma abnormalities are evident in first-episode psychosis (Symond et al. 2005), in unmedicated patients (Gallinat et al. 2004), as well as in unaffected relatives (Leicht et al. 2011), suggesting that abnormal gamma synchrony is a heritable feature of schizophrenia. Gamma-band responses have been associated with clinical symptoms, social cognition, neurocognitive performance, and loss of gray matter (Williams et al. 2009a, b), indicating that these measures are likely related to disease pathophysiology.

Treatment implications: Translational models of gamma-band responses have been used in preclinical studies, providing potential targets for treatment development. Several rodent studies have demonstrated that NMDA receptor antagonists (including ketamine, MK-801, and PCP) produce a dose-dependent increase in baseline gamma power (Ma and Leung 2007; Ehrlichman et al. 2009). Behaviorally, this increase in gamma power is associated with locomotor hyperactivity and deficits in PPI in animal models (Ma and Leung 2007; Hakami et al. 2009). Consistent with preclinical findings, ketamine increases baseline gamma power in healthy human subjects (Hong et al. 2010). Mechanistically, it has been proposed that the effect of NMDA receptor antagonists on gamma oscillations (and their psychomimetic properties) is due to reduced excitation of parvalbumin-containing GABA neurons (Lisman et al. 2008). Consistent with this hypothesis, Lewis et al. (2008) assessed MK-0777, a benzodiazepine-like agent with selective activity at $GABA_A$ receptors, versus placebo in 15 chronic patients with schizophrenia. MK-0777 was found to be associated with increased gamma-band power and improved performance on tests of working memory and cognitive control.

Mu Suppression

Mu rhythm suppression in response to biological motion is a relatively new candidate biomarker in schizophrenia research (Singh et al. 2011). Biological motion, as depicted in point light animations, is a well-studied construct in cognitive neuroscience. These displays provide sparse visual input that requires "filling-in" to recover object information to identify the kind of motion (e.g., walking, jumping, dancing) being produced (Keri and Benedek 2009; Blake and Shiffrar 2007). It has been suggested that neural processing of biological motion is an evolutionarily conserved mechanism and that it plays a fundamental role in social adaptation (Blake and Shiffrar 2007). Translational studies have associated biological motion with neural activity in the mu (8–13 Hz) range over the right sensorimotor cortex and are thought to index the activity of "mirror" neurons based on studies in primates (Bonini and Ferrari 2011; Keuken et al. 2011). Mu rhythms measured from this brain region show reliable, dose-dependent suppression when the subject perceives biological motion (but not nonbiological motion). Thus, mu wave suppression is an easily quantifiable operational measure of the neural processing of biological motion. In a recently published study, Singh et al. (2011) showed that neural mu wave suppression induced by biological motion is impaired in first-episode patients, and that the neural impairment is inversely correlated with negative symptoms and social adjustment, providing construct validity for the mu suppression paradigm as an operational measure of social cognition in patients with psychosis. In a related study, McCormick et al. (2012) recorded and analyzed mu rhythm suppression over the sensorimotor cortex during observed and actual hand movement in actively psychotic patients and found evidence of increased suppression during observed movement that was correlated with positive symptoms.

Treatment implications: Keri and Benedek (2009) and Perry et al. (2010) have recently demonstrated that intranasal *oxytocin* significantly enhances detection of biological motion compared to intranasal placebo in normal subjects. Although oxytocin has been shown to improve social cognition and emotional recognition in patients with schizophrenia (Averbeck et al. 2012; Feifel et al. 2010; Pedersen et al. 2011), the effects of oxytocin on mu suppression in schizophrenia patients have yet to be reported. In an innovative study of *neurofeedback training* in high functioning autism, Pineda et al. (2008) report that individuals with autistic spectrum disorders who have mu suppression abnormalities can renormalize mu suppression and improve sustained attention after training to the mu frequency band. Like other cognitive remediation interventions, neurofeedback training offers a potential nonpharmacologic intervention which can target core information-processing abnormalities that contribute to social functioning deficits in patients with schizophrenia.

Functional Neuroimaging

The most robust fMRI findings associated with schizophrenia are altered PFC, ACC, and temporal lobe activation, particularly during the performance of tasks which engage executive functions, such as verbal fluency paradigms. Fusar-Poli et al. (2007) examined studies of first-episode psychosis and individuals at high risk (schizotypal, genetic high risk, at risk) for psychosis. First-episode patients showed significant PFC abnormalities with most studies reporting reduced activation in the dorsolateral prefrontal cortex (DLPFC) during cognitive tasks. Some authors have suggested that hypofrontality in the DLPFC may be a specific feature of schizophrenia at the time of the first psychotic episode. Only one study reports greater prefrontal activation in first-episode patients: Mendrek et al. (2005) found that first-episode patients had greater DLPFC activation during the easy level of a working memory task, but less activation when task demands were high.

In general, high-risk subjects display neurophysiological abnormalities in cortical regions that have also been observed to be dysfunctional in first-episode psychosis (Broome et al. 2010; Allen et al. 2012). In contrast to most first-episode studies, however, a number of genetic high-risk studies reported relatively greater prefrontal activation than in controls (Seidman et al. 2006; Callicott et al. 2003; Thermenos et al. 2004). Studies in prodromal subjects (Sabb et al. 2010; Allen et al. 2011, 2012) also found greater activation in specific brain regions: Sabb and colleagues found increased neural activity in the bilateral mPFC, left inferior frontal (LIFG) and middle temporal gyri, and

ACC in at-risk subjects compared to controls (Sabb et al. 2010; Allen et al. 2011). Further, increased activity in the superior temporal gyrus, caudate, and LIFG distinguished those at-risk subjects who subsequently developed psychosis from those who did not. Using a combined fMRI and PET study, Allen et al. (2011), reported that at-risk subjects who later develop a psychotic episode show increased activation in bilateral PFC, brainstem (midbrain/basilar pons), the left hippocampus, and greater midbrain-PFC connectivity during a verbal fluency task. Furthermore, exploratory analysis of [18F]-DOPA PET data showed that transition to psychosis was associated with elevated dopaminergic function in the brainstem region. These interesting findings of increased activation in genetic high risk and putatively prodromal subjects have been hypothesized to reflect a compensatory response to volumetric reductions in gray or white matter to maintain adequate performance (MacDonald et al. 2005a) or "cortical inefficiency"(Callicott et al. 2000).

Multimodal neuroimaging during the prodrome and first episode of psychosis offers the potential to delineate the causal relationship between key pathophysiological processes in the evolution of psychosis to determine if the unique findings of hyperactivation reflect compensatory changes related to the hypothesized window of neurotoxicity. For example, the combination of structural MRI, fMRI, PET, SPECT, or ^{1}H-MRS can address the relationship between glutamate or dopamine and changes in gray matter or cortical activation. In an elegant series of studies, Fusar-Poli and colleagues found that alteration in prefrontal activation in at-risk subjects in a verbal fluency task was related to elevated striatal dopamine using PET (Allen et al. 2011, 2012; Fusar-Poli et al. 2010, 2011a, b). In an fMRI study using a working memory paradigm, the same group found a positive correlation between frontal activation and fluorodopa uptake in the associative striatum in controls but a negative correlation in the at-risk group (Fusar-Poli et al. 2011a, b). The key finding from these studies is that, for individuals at very high risk of schizophrenia, altered prefrontal activation during a task of executive/working memory function was directly related to striatal hyperdopaminergia. This provides evidence of a link between dopamine dysfunction and the perturbed prefrontal function, which may underlie the deficits in cognitive processing evident in people with prodromal symptoms of psychosis and predate the first episode of frank psychosis.

Treatment Implications

Although a number of functional neuroimaging biomarkers are being developed as part of initiatives, such as the CNTRICS study, there is little consensus in the literature on treatment effects, such as antipsychotic effects, on the BOLD signal (Carter and Barch 2007). In a literature review on antipsychotic effects, Roder et al. (2010) report that there does not appear to be any common underlying mechanism of action of antipsychotic drugs that influence the BOLD signal in a systematic way in all areas of the brain in the same direction. Some studies find differences in BOLD signal with treatment but others do not, and there is no clear difference between first- and second-generation antipsychotics on BOLD signal. The most consistent finding in the literature is that haloperidol decreases BOLD signal in cortical and subcortical structures.

In terms of nonpharmacologic interventions, both CBT and cognitive training have been found to improve working memory performance as well as brain connectivity (Kumari et al. 2009, 2011; Vinogradov et al. 2012). Kumari et al. (2009) examined changes in working memory performance in response to CBT using the N-Back and found stronger DLPFC activity. In addition, DLPFC-cerebellum connectivity during the highest memory load condition (2-back > 0-back) predicted post-CBT clinical improvement. Using a facial expression task, Kumari et al. (2011) found that the CBT group showed attenuation of fMRI BOLD response to fearful and angry expressions at follow-up relative to baseline. Preliminary cognitive training studies (Vinogradov et al. 2012) in schizophrenia patients have shown that computerized auditory training significantly improves verbal memory performance as well as early magnetoencephalographic responses in auditory and prefrontal cortices that are positively associated with quality of life six months later. Vinogradov et al. (pers. comm.) examined the association between computerized auditory training-induced behavioral improvements and changes in brain activation in an fMRI paradigm during a 2-back, verbal working memory task in patients with schizophrenia. At baseline, during the 2-back working memory task, patients showed impaired performance, reduced activation in bilateral DLPFC, and no significant associations between brain activation and 2-back performance. After cognitive training, patients significantly improved their performance on the task and showed increased DLPFC activation. These preliminary CBT and cognitive remediation studies demonstrate the importance of nonpharmacologic interventions in treating cognition deficits of schizophrenia. The functional brain measures provide an important tool to evaluate brain connectivity in response to psychosocial as well as cognitive enhancing drugs. Many important studies are needed to compare treatments as well as the potentially synergistic effect of combined treatment in early psychosis.

Given What We Know Now, Can We Alter the Pathological Processes in Early Psychosis?

As eloquently reviewed by Swerdlow (2011), an increasingly detailed image of neural- and molecular-level dysfunction in schizophrenia has emerged to reveal failures of early brain maturation, dysfunctional neural circuitry, and failure to develop appropriate connectivity across widely dispersed brain regions. These circuit abnormalities are complex, vary across individuals, and are hard wired, illustrating the challenge of developing treatment that can effectively alter the course of the illness once it has reached the chronic or even acute phase. As knowledge of the mechanisms underlying the emergence of psychosis increases, the picture becomes even more complex with the identification of each new gene or epigenetic contribution and the resulting compensatory changes in hard-wired neural circuitry. Based on our current models of treatment for schizophrenia, it does not seem possible to reverse a process that has been developing for two decades; however, it may be possible to prevent or modify the identified neurobiological processes that occur at disease onset and improve outcome.

Clinical research has shown that the longer the duration of untreated psychosis, the poorer the treatment response (Addington et al. 2004; Melle et al. 2004), thus suggesting that earlier intervention may improve the outcome of the illness. In a comprehensive review, Berger et al. (2003) outline how altered regulatory mechanisms of progenitor cell generation and death could be targeted for neuroprotection or disease modification in early psychosis. Although researchers previously believed that stem and progenitor cell generation in mammals was only possible in early life, recent research suggests that the hippocampi (Kornack and Rakic 1999), periventricular zone, (Steindler and Pincus 2002) and olfactory bulbs (Byrd and Brunjes 2001) retain the capacity to generate progenitor cells which differentiate into neurons. A number of compounds reviewed in this chapter—which show potential in altering neurobiologically defined surrogate endpoints and also modulate apoptosis pathways (lithium, sodium valproate, BDNF, clozapine, quetiapine, lamotrogine, omega-3 fatty acids), block necrosis pathways (vitamin E)—increase synaptogenesis (SSRIs) or block the inflammatory response (COX-2 inhibitors, aspirin) (Jacobs et al. 2000; Malberg et al. 2000; Vaidya et al. 1997), and provide evidence of neuroprotective properties in preclinical and clinical studies.

Predicting Treatment Response

Ultimately, the goal of treatment in early psychosis patients is to modify active neuropathological changes and associated functional disability. With a greater understanding of aberrant neural systems in early psychosis, new treatments will be introduced. A number of psychosocial and pharmacologic interventions have great potential as neuroprotective, disease-modifying, or procognitive interventions in early psychosis (Tandon et al. 2011). Given the heterogeneity of the prodromal period and the first episode of psychosis, the importance of treatment "precision" is evident. If we can identify which type of treatment can best target the abnormal neural system of a particular individual, treatment is likely to be more effective (Vesell 1978; Foster et al. 2010; Wilke and Dolan 2011). Ideally, with the use of various risk factor and biomarker assessments it will be possible to develop a neurobiological profile to predict treatment response. In line with the Research Domain Criteria (Insel et al. 2010) proposed by the National Institutes of Mental Health, it may be optimal to focus on neural systems to target treatment as opposed to diagnostic and statistical diagnosis (cf. Carpenter, this volume).

Where We Are and Where We Need to Go: Future Directions in Treating the Early Phase of Schizophrenia

It is astounding that the "dopamine hypothesis of schizophrenia" has been in the mainstream for over sixty years and that pharmacologic management of schizophrenia is still based on antagonists or partial agonists of the dopamine D2 receptors (for a review, see Howes et al. 2009). Despite extensive genetic, epigenetic, developmental, and cognitive neuroscience literature on schizophrenia, which reveals that many innovative ideas and directions are being pursued, it has been difficult to identify new pharmacologic interventions that take the treatment of schizophrenia much beyond first- or second-generation antipsychotic medication. The majority of treatment studies have been performed in chronic patients, and many are underpowered or plagued by methodological differences which complicate the reliable merging of data across studies.

It is clear that adequately powered studies of early psychosis patients are needed to assess the extensive armamentarium of neuroprotective, disease-modifying, and procognitive compounds already identified. The anti-inflammatory agents, including COX-2 inhibitors, omega-3 fatty acid, and minocycline, already offer promise in first-episode patients and may affect functional outcome by more effectively targeting negative symptoms and neurocognition. SSRIs and lithium may decelerate the loss of gray matter in the early stages of illness by promoting neurogenesis. Glutamate-modulating agents may be particularly important if the hypothesized window of neurotoxicity (revealed by possible compensatory changes in brain function, sensorimotor gating, and brain metabolism) proves to be present in early illness.

Psychosocial and cognitive remediation techniques have emerged as some of the most effective interventions to target neurocognition, functional capacity, and functional outcome. Empirically supported treatments for psychotic disorders now include a variety of psychosocial interventions, such as CBT, social skills training, vocational rehabilitation, and cognitive remediation. Substantial research indicates that CBT changes brain function in brain disorders such as OCD (Baxter et al. 1992; Schwartz et al. 1996; Saxena et al. 2009), and we now have preliminary evidence in schizophrenia (Vinogradov et al., pers. comm.; Kumari et al. 2011). CBT has been shown to reduce symptoms and improve long-term functioning in patients with chronic (Granholm et al. 2007) and first-episode schizophrenia (Power et al. 2003; Petersen et al. 2005) as well as those in the prodromal phase of illness (Morrison et al. 2004). Cognitive remediation offers the potential to reinforce healthy circuits to compensate for areas of cognitive deficits and perhaps reduce gray matter loss and improve brain connectivity at the same time.

Although psychosocial treatment interventions have been added to augment antipsychotic medication, relatively little is known about the effectiveness of combining pharmacologic interventions designed to enhance cognition, provide neuroprotection, or facilitate neuroplasticity with CBT, cognitive remediation, or exercise. Future clinical trials in early psychosis patients should ideally compare nonpharmacologic and pharmacologic interventions as well as assess whether a combination of therapies is more effective than any one alone.

Clinical and functional outcomes are important in clinical trials but ongoing work to develop biomarkers linked to functional outcome, treatment response, and pathological circuitry as surrogate endpoints represents an innovative approach which should be pursued. Most importantly, if reliable neurobiological measures can be developed for the clinical setting to assist in the specification of treatments for a particular patient, it should be possible to truly individualize care based on brain function, risk factors, and prediction of response.

Acknowledgments

K. S. Cadenhead is supported by National Institutes of Health (NIH) grants R01 MH60720, U01 MH082022, and K24 MH76191

8

From Epidemiology to Mechanisms of Illness

John McGrath and Andreas Meyer-Lindenberg

Abstract

Schizophrenia research encompasses many different categories of observation: (a) genetic research, which examines variants in single base pairs, (b) cellular and applied neuroscience, including animal models, (c) clinical research representing a broad spectrum of patient-centered research, and (d) population-based epidemiology and health services research. Each field of research has a natural tendency to become more specialized and, as a consequence, more inward looking. Meta-research, the study of the process of research per se, shows that creativity tends to occur at the boundaries of disciplines and research areas. This chapter examines ways to facilitate this type of cross-disciplinary translational research. Examples are provided of collaborative scientific programs that have used clues from fields such as epidemiology and genetics, and these clues are explored via the prism of various neuroscience platforms (e.g., molecular, cellular, behavioral, animal models, brain imaging). Cross-disciplinary projects have the potential to catalyze new discoveries in neuroscience. Our field needs to build efficient shared discovery platforms to encourage greater cross-fertilization between schizophrenia research and the general neuroscience research community.

How Can We Optimize Discovery? Research on Research

There is a natural order in the way scientific disciplines evolve: complex areas of enquiry require highly specialized and focused research skills. Within each disciplinary niche, different cultures emerge in a healthy and appropriate fashion. Local dialects develop within a group and world views are shared within the tribe. These cultures are handed down to the next generation of scientists. Although this process introduces efficiencies within a field, it can also lead to inward thinking and creative stagnation. There is a general awareness that the sociology of science can hinder as well as advance scientific process. For example, important discoveries in one arcane field may not be immediately appreciated by the general research field. As a consequence, new discoveries may not be efficiently translated into distant fields.

In recent years the sociology of science has itself been the focus of research, addressing the central question of how scientists can optimize discovery (Lehrer 2009). Leaving aside rate-limiting steps, such as adequate research funding and quarantined research time (versus administrative, teaching, and clinical duties), there are interesting lessons to be learned. For example, researchers need to understand the dangers of "failure-blindness" (i.e., we need to appreciate findings that contradict our assumptions). A key feature of productive research groups relates to intellectual biodiversity. The meta-research evidence shows that research creativity is optimized when we actively "seek out the ignorant." For example, when we are required to talk to those who are unfamiliar with our experiments (other disciplines, students, the general public), we sometimes reframe research findings in a fresh perspective. Similarly, talking to colleagues from other disciplines can spark the creative exchange of fresh metaphors or provide missing pieces of the intellectual jigsaw puzzle. These notions have been influential in the formation of new research clusters, such as the Howard Hughes Medical Institute Janaelia Farm site (Cech and Rubin 2004).

Optimizing Discovery in Schizophrenia Research

People who enter schizophrenia research tend to be incurable optimists. In spite of the bewildering heterogeneity of our target phenotype, and in the face of limited knowledge of the neurobiological correlates of schizophrenia, we remain confident that the questions we ask are tractable and that progress is being made. Self-proclaimed "decades of the brain" come and go and still clinical outcomes for people with schizophrenia are suboptimal. There is, however, good cause for optimism in light of examples of excellent clinical research that is fuelling discoveries in basic neuroscience and vice versa.

We present examples where ideas from within one field of schizophrenia research have been efficiently translated into other fields. We acknowledge that there are many examples of such types of research. Our selection is intended to prompt further debate on this topic.

From Scottish Pedigrees to Neuronal Hub Proteins: DISC1

In the early 1970s, observant clinicians linked a chromosomal translocation involving chromosome 1 with a range of neuropsychiatric outcomes in an extended Scottish pedigree (Blackwood et al. 2001). The translocation disrupted a protein coding gene, which was subsequently labeled "disrupted in schizophrenia 1" (DISC1). Mindful that this structural variant was associated with other clinical outcomes, the ability to explore the function of the protein in transgenic models quickly revealed that the large protein coded by this gene was involved in an unexpectedly wide range of functions in the developing and adult brain (Porteous et al. 2011). Based on research conducted over the last few years, this protein has now been linked to a very wide range of molecular and cellular functions (Hayashi-Takagi et al. 2010; Seshadri et al. 2010). The protein acts as a hub for a large number of protein interactions.

Regardless of how prevalent this particular structural variant is in the general population, and regardless of what proportion of all schizophrenia is linked to mutations in this particular gene (probably very little), there is no doubt that this discovery has triggered important advances in basic neuroscience.

From Mental Health Registers to *De Novo* Mutations: Advanced Paternal Age

Epidemiologists use population-based studies (e.g., cohorts, mental health registers) to search for gradients within and between groups as well as across time. This category of research is good for generating clues (e.g., links between a particular disease and different candidate risk factors), but it is limited with respect to (a) exploring the underlying biological mechanisms and (b) proving causality. Indeed, in the absence of randomized controlled trials, clues from observational epidemiology (e.g., a cross-sectional study that links a candidate risk factor with a disease outcome) are notoriously prone to the influence of unmeasured confounding (Davey Smith and Ebrahim 2001). A good example of how epidemiology can drive neuroscience discovery relates to the work by Malaspina et al. (2001), who reported an association between advanced paternal age and an increased risk of schizophrenia in offspring. Importantly, this paper suggested that *de novo* mutations in the male germ cell may contribute to this finding. The epidemiology research community quickly replicated and extended the finding to a range of other health outcomes, including childhood and adolescent behavior, intelligence, bipolar disorder, and autism.

Resultant clues from epidemiology were then examined in rodent models (Garcia-Palomares et al. 2009; Smith et al. 2009; Foldi et al. 2010); these studies reported altered behavioral and brain structural outcomes in the offspring of older sires. The use of inbred rodent models allowed for prompt testing of the underlying hypothesis regarding *de novo* male germline mutations. Experimental studies based on the mouse confirmed that the offspring of older sires had significantly more *de novo* copy number variants (Flatscher-Bader et al. 2011). Remarkably, the study found that the mutations involved genes previously linked to autism and schizophrenia. There is now convergent evidence linking copy number variant (CNV) load with schizophrenia (O'Donovan et al. 2008). Thus, within schizophrenia research, there has been an unexpected convergence between risk factor epidemiology and genetic studies.

With the advent of affordable high throughput genetic sequencing as well as access to mother–father–offspring schizophrenia trios, the relationship between paternal age, *de novo* mutations, and risk of schizophrenia can now be explored. Recent deep sequencing studies have confirmed the association between paternal age and *de novo* mutations (Kong et al. 2012). This type of research may help define subgroups within the heterogeneity of schizophrenia.

From Place of Birth to Functional Magnetic Resonance Imaging

Prior epidemiology data suggests a two- to threefold increase in schizophrenia risk in individuals brought up in urban environments (Krabbendam and van Os 2005). The relationship follows a dose-risk response function: the longer individuals are exposed to highly urban environments during childhood and adolescence, the greater the risk of developing schizophrenia in adulthood (Pedersen and Mortensen 2001). Not surprisingly, adverse effects of urban upbringing are moderated by risk genes (Krabbendam and van Os 2005; van Os et al. 2008) with excessive rates of incidence in genetically vulnerable individuals brought up in the city. Similarly, first- and second-generation immigrants have a twofold increase in risk for schizophrenia independent of the specific characteristics of a given ethnicity or host country (Bourque et al. 2011). Both urbanization and migration processes challenge the capacity of an individual to cope with complex social stressors, such as disintegration of family networks, tightened competition, and discrimination. Epidemiological data suggest that the incongruence of subject-specific and environment-specific features is particularly crucial: the more an individual stands out from the social milieu in terms of minority status, social fragmentation, and socioeconomic status, the higher the risk is to develop schizophrenia (Zammit et al. 2010b). It has been proposed that social stress plays a key role in mediating these effects, possibly via dysregulation of the hypothalamic-pituitary-adrenal (HPA) axis and sensitization of the mesolimbic dopamine system (Pruessner et al. 2004; van Os et al. 2008). On the epigenetic level, an important mechanism for the effects of adverse environmental exposure during development involves hypermethylation of the promoter region of the glucocorticoid receptor gene (*NR3C1*), which reduces the expression of *NR3C1* in brain and promotes the manifestation of increased sensitivity to stress and HPA dysregulation in adulthood (McGowan et al. 2009). The neural system correlates in humans, however, are largely unexplored.

Lately, a new line of neuropsychiatric research aims to delineate these social-environmental risk effects in brain. Recent functional neuroimaging work, for example, examined the effects of urban upbringing on social evaluative stress processing in human social-emotional circuits (Lederbogen et al. 2011). In this study, the functional integrity of the neural stress response system was challenged using cognitive tasks presented in the context of disapproving video feedback from investigators. This work provided evidence for a link between early-life urbanization and anterior cingulate cortex (ACC) function during social stress processing, a key region involved in the regulation of limbic activity and negative emotion. Robustness and specificity of the effects of urbanization were confirmed in supplementary studies examining ACC function in the context of a different social stress paradigm and during cognitive processing without stress, respectively (Lederbogen et al. 2011).

From Genetic Clues to Brain Functioning: "Genetic Imaging"

Twin, family, and adoption studies clearly indicate that genetic factors contribute substantially to the risk for psychiatric disorders. Heritability estimates, such as 81% in schizophrenia (Sullivan et al. 2003) and 37% in major depression disorder (MDD) (Sullivan et al. 2000), reflect the varying ratio of genetic and environmental factors which jointly determine risk or resilience (Caspi and Moffitt 2006). There is an obvious interest in identifying the gene variants underlying this hereditary component, since they promise valuable insights into pathophysiology, diagnosis, and treatment of the associated disorders (Hyman 2007). However, attempts to reveal the "culprit" genes by linkage analysis turned out to be of little value, although they had been applied to Mendelian disorders very successfully (Gottesman and Gould 2003). Apparently, the effects of psychiatric risk variants were too modest to be detectable by linkage. As a promising solution to study such subtle effects, association studies were introduced (Risch and Merikangas 1996). These studies apply a candidate gene approach requiring a priori defined genes. Accordingly, these studies are intrinsically prone to a bias in the selection of candidates, which usually focus on genes that are known to code for key player proteins involved in neurotransmission and suspected to be linked to mental illness, such as the monoaminergic system in mood and anxiety disorders (Levinson 2006) and the dopaminergic or glutamatergic system in schizophrenia (Owen et al. 2004).

This approach resulted in a plethora of studies reporting associations between single candidate genes and clinical or treatment-related phenotypes of mental illness, such as MDD (Levinson 2006; Kato and Serretti 2010) or schizophrenia (Owen et al. 2004; Arranz and de Leon 2007). However, the initial gold rush in the search for candidate genes has been followed by a disillusioning decade of failed replications and considerable disagreements among scientists (Abbott 2008). In fact, many genetic associations were likely overestimated by initial studies (Trikalinos et al. 2004) or may have even been chance findings given the common practice of selective reporting (Sullivan 2007).

With the advent of genome-wide association (GWA) approaches, it is now possible to test associations of more than one million DNA variants simultaneously. This technique holds the promise of hypothesis-free gene discovery for common disorders by mapping the whole genome with common markers. However, there are several limitations to this technique, such as statistical compromises which have to be made, given the incredibly high number of investigated genes (Psychiatric GWAS Consortium Steering Committee 2009; Cichon et al. 2009). This notion is reflected in the weak support of GWA studies with regard to traditional candidate genes and the limited agreement among the increasing number of GWA studies (Pezawas and Meyer-Lindenberg 2010; Bosker et al. 2011).

After the first report of an association between genetic variation and a neuroimaging measure in 2000 (Heinz et al. 2000), imaging genetics has developed into a leading research strategy in neuroscience. Countless studies have demonstrated the influence of risk alleles on neural intermediate phenotypes which, in turn, relate to different psychopathological manifestations and diagnostic entities (Bigos and Weinberger 2010; Domschke and Dannlowski 2010; Meyer-Lindenberg 2010b; Scharinger et al. 2010). In contrast to several candidate endophenotypes, which turned out to be equally complex as behavioral phenotypes, recent meta-analyses indicate that neural intermediate phenotypes satisfy the premise of increased penetrance (Gottesman and Gould 2003; Munafo et al. 2008; Mier et al. 2010). For instance, a polymorphism in the promotor region (5-HTTLPR) of the serotonin transporter gene (*SLC6A4*) has been shown to account for up to 10% variance of amygdala activation, whereas its role in predicting behavioral phenotypes such as neuroticism, MDD, or antidepressant treatment response is at least one order of magnitude lower (Serretti et al. 2007; Munafo et al. 2008, 2009; Clarke et al. 2010; Taylor et al. 2010). Accordingly, imaging genetics may eventually provide one of the tools needed to decipher the polygenic heritability of psychiatric disorders as anticipated by Gottesman and Shields more than four decades ago (Gottesman and Shields 1967).

These genome-wide significant variants are opening up new avenues to risk pathways of executive function in imaging genetics. In the first such study, a sample of healthy individuals was used to verify a variant (rs1344706) in the zinc finger protein 804A gene (*ZNF804A*) that has been implicated in schizophrenia by GWA (Esslinger et al. 2009). Remarkably, healthy carriers of the risk variant exhibited unfavorable prefrontal-hippocampal functional connectivity in a pattern characteristic for schizophrenia (Meyer-Lindenberg et al. 2005; Esslinger et al. 2009). Following this approach, several new results emerging from GWA studies using clinical or neurocognitive phenotypes have been confirmed by imaging genetics methods, such as variants in *HOMER1*, *CACNA1C*, or *SCN1A* (Bigos et al. 2010; Rietschel et al. 2010; Papassotiropoulos et al. 2011).

From Influenza Epidemics to the Impact of Maternal Immune Activation on Brain Development

Soon after the renaissance of the neurodevelopmental hypothesis of schizophrenia (Murray and Lewis 1987; Weinberger 1987), various researchers proposed that the offspring of mothers exposed to influenza may have an increased risk of schizophrenia (McGrath and Castle 1995). While the evidence linking exposure to this particular infectious agent has been mixed, there is now a large body of research which suggests that prenatal infection to a wide range of early life infections is associated with increased risk for schizophrenia in the offspring (Brown and Derkits 2010).

Using rodent models and noninfectious agents designed to trigger immune responses (e.g., agents that mimic the bacterial cell wall or RNA polymers that resemble viral RNA), experimental studies have uncovered previously unexpected reciprocal interactions between immune pathways and brain development (Meyer et al. 2009; Patterson 2009). This research converges with evidence from (a) genetics to link regions of the genome critical for immune response to schizophrenia (Ripke et al. 2011), and (b) developmental neurobiology to implicate mechanisms initially thought to be restricted to immune pathways with brain development and function (Boulanger 2009). These discoveries are now able to feed back into more focused and hypothesis-driven analytical epidemiology.

General Reflections and Recommendations

The need to support translational research that facilitates discoveries in basic science into clinical settings is now widely recognized by funding agencies. With respect to the care of people with schizophrenia, there is a need for this type of research, just as there is a need to ensure that known effective treatments are delivered to those in need. However, in poorly understood fields of research, we first need to do the basic science in order to fuel the subsequent translational pipeline. We argue that complex brain disorders such as schizophrenia require continued investment in research which takes clues from various fields of schizophrenia research and feeds them back into high-quality neuroscience. Put bluntly, if we want to fix broken brains, we first need to understand how healthy brains are built and how they work.

The need to facilitate the fertile intersection between schizophrenia epidemiology and developmental neurobiology has been detailed elsewhere (McGrath and Richards 2009). Because neuroscience is such an intensely productive and fast-moving field of research, trying to engage with the field as an outsider is akin to "sipping from a fire hose." Despite this, we argue that it is critical for schizophrenia research to be firmly anchored to a neurobiologically informed framework. Schizophrenia researchers have the skills to generate candidate exposures and to identify neuroanatomical, neurochemical, or behavioral phenotypes of interest to clinical research. Rodent models (Arguello and Gogos 2006), zebrafish, or invertebrates such as *Drosophila* and *Caenorhabditis elegans* (Burne et al. 2011) can provide powerful and efficient research platforms to explore key research questions for both genetic and nongenetic risk factors and to help identify the function of genetic candidates.

From our current perspective, one of the more exciting developments in this field has been a renewed focus on the social world and its evolutionarily honed counterpart, the social brain. In a recent review (Meyer-Lindenberg and Tost 2012), we concluded that the existing evidence, while preliminary in nature, supports a causal role for the social environment in risk, resilience, and manifest illness, suggesting that everyday social interactions are both actor and stage for mental illness. Novel translational research strategies are needed to delineate the neural outcomes of the complex underlying gene–environment interactions. An in-depth understanding of these mechanisms holds the prospects of novel strategies for pharmacology, psychotherapy, and social policy that target and converge on the identified neural circuits. In the "decade of psychiatric disorders" (Bassett et al. 2010), a renewed focus on social neuroscience has therefore much to offer for scientists, patients, and therapists alike.

The challenge is to optimize links between researchers from (a) the diverse fields of schizophrenia research and (b) the even more diverse fields of neuroscience. How can we engineer future research between these groups to "set traps for discovery"? Building shared research platforms between groups with different skills is clearly an important step. As Cech and Rubin (2004:1167) note:

> The spark of transdisciplinary approaches and insights requires "productive collisions" between people in different disciplines, just as atoms and molecules must undergo productive collisions to react. If engineers, biologists, and computer scientists live apart, they need to make an appointment in order "to collide."

Shared research platforms need to be engineered to encourage "collisions" between diverse scientists. We argue that schizophrenia research needs to take a more assertive stance in driving neuroscience research. Too often we have been passive recipients of "leftover" neuroscience. Neuroscience needs us, just as much as we need neuroscience (McGrath and Richards 2009).

First column (top to bottom): Craig Morgan, Heike Tost, Aristotle Voineskos, Robert Bittner, Steve Silverstein, Heike Tost, and Robert Bittner
Second column: Michael O'Donovan, Kristin Cadenhead, Steve Silverstein, John McGrath, Michael O'Donovan, Kristin Cadenhead, and Steve Silverstein
Third column: John McGrath, Peter Uhlhaas, Heike Tost, Kristin Cadenhead, Peter Uhlhaas, Craig Morgan, and Aristotle Voineskos

9

How Can Risk and Resilience Factors Be Leveraged to Optimize Discovery Pathways?

Craig Morgan, Michael O'Donovan, Robert A. Bittner,
Kristin S. Cadenhead, Peter B. Jones, John McGrath,
Steven M. Silverstein, Heike Tost,
Peter Uhlhaas, and Aristotle Voineskos

Abstract

Based on wide-ranging discussions and specific examples drawn from the interests and expertise of the group, this chapter addresses the question of how knowledge of risk and resilience in relation to the etiology of schizophrenia can be leveraged to optimize discovery of preventive and therapeutic approaches. It explores the challenges and gaps in knowledge that have emerged as a result of recent, significant progress in understanding the factors that confer risk for schizophrenia. The fuzzy boundaries of schizophrenia and overlap in risk factors between schizophrenia and other mental disorders are highlighted, as is the predominant focus on risk rather than on resilience. Examples of research in genetics (including epigenetics) and neuroimaging are provided which examine putative mechanisms and pathways that could be leveraged to develop novel interventions.

Implications for prevention and intervention are considered from the point of view that heterogeneity and nonspecificity in schizophrenia present opportunities both to disentangle shared pathways that underpin a wide range of disorders and to develop novel approaches to prevention and intervention. The chapter concludes with recommendations that highlight key areas for future research.

Introduction

If we want to move closer to the prevention of complex disorders like schizophrenia and implement effective treatment, we must first understand the matrix of risk factors that underlies the etiology and pathogenesis of such syndromes.

Considerable progress has been made in understanding the factors that confer risk for schizophrenia and, more broadly, psychosis since the disorders were first described, beginning with family, adoption, and twin studies and now including studies of molecular genetic and environmental factors. Nonetheless, considerable gaps remain in our knowledge of individual risk factors, of how these combine and interact across levels to increase risk, and of the developmental pathways and mechanisms through which they impact on neural systems and circuits to produce the clinical phenomena of psychosis. These gaps limit our ability to utilize existing data to inform the development of strategies to promote resilience and reduce risk, at both population and individual levels.

How can current knowledge of risk and resilience factors be leveraged to optimize discovery pathways, and thereby better inform prevention and intervention? From the start, consideration of this question forced us to ponder what is currently known about schizophrenia and to set this in the context of past assumptions and future challenges.

Past assumptions:

- Schizophrenia was a relatively homogenous disease construct, albeit with subtypes.
- Schizophrenia affected men and women equally and had a flat epidemiological profile across time, place, and persons.
- Schizophrenia had a small, manageable set of risk factors.
- Neuroscience would reveal a readily interpretable mechanism of action, which would lead to effective treatments.

Current understanding:

- Schizophrenia is a poorly understood group of disorders that defies ready simplification based on symptoms, putative neurobiology, or etiopathogenesis.
- Schizophrenia affects men more than women, and the incidence of the disorder varies significantly by place and social group (e.g., within nations and between nations, between ethnic subgroups).
- Risk factors for schizophrenia (e.g., genes, prenatal exposures) are associated with a wide range of other brain-related adverse health outcomes (especially neurodevelopmental disorders).
- Common mental disorders like anxiety and depression often precede and coexist with schizophrenia.
- Isolated and transient psychotic experiences are prevalent in the community.

Future challenges:

- Acknowledge that genetic and nongenetic risk factors linked to schizophrenia will probably be shared with many other mental health outcomes (i.e., lack of specificity for exposures).

- Acknowledge that psychotic experiences are shared with a subgroup of the general population and a range of common mental disorders.
- Acknowledge that individuals can pass through a pluripotential phase of illness evolution, at which stage a number of outcomes are possible, and that a clinical staging model may offer new options for treatment and prevention.
- Consider heterogeneity and nonspecificity not as problems but rather as opportunities to unravel shared pathways that underpin a surprisingly wide range of brain-related outcomes.
- Acknowledge that interventions, which target these nonspecific outcomes, may deliver attractive and cost-efficient benefits with respect to overall disease burden.

Simply put, the agenda for schizophrenia research needs to be recontextualized. We must widen the category of observation and generate new metaphors and semantic labels to help leverage this perspective.

Constructs of Schizophrenia and Psychosis

Before we can understand risk and resilience, it is necessary to define the target disorder(s) or syndrome(s) of interest. Should a narrow (e.g., DSM-IV schizophrenia) or broad (e.g., nonaffective psychosis, all psychotic disorders including bipolar disorder) focus be taken? This is a key issue for future research and is discussed by Corvin et al. (this volume). Our discussions focused on the *broader spectrum of psychosis* in light of (a) the robust evidence from genetics and risk factor epidemiology that indicates a shared risk architecture across the psychosis spectrum, (b) clinical overlap and uncertain boundaries between different types of psychotic disorders, especially in the early phases of disorder (Murray et al. 2004), and (c) evidence from general population samples that isolated and transient psychotic experiences are common and are associated with similar risk factors to those identified for clinically defined disorder (van Os et al. 2009).

For some research questions, it may be appropriate to take an even broader perspective. For example, why do some copy number variants (CNVs) increase risk for a broad range of neurodevelopmental disorders, such as learning disability, epilepsy, autism, and schizophrenia (Van Den Bossche et al. 2012)? When referring to existing research, we are necessarily bound to use the groups and categories that were the focus of study (i.e., in some specifically schizophrenia, in others all psychotic disorders, whereas in still others psychotic experiences in nonclinical samples).

Risk and Resilience

Within the fields of medicine, human genetics, and epidemiology, there is an understandable tendency to focus on the identification of factors that increase the probability of adverse health outcomes (i.e., risk factors). Often, the identified variables are risk indicators or proxy markers (i.e., variables that index exposure to risk-increasing exposures) rather than factors that directly impact on risk. Some of the identified exposures in schizophrenia and psychosis research are at this broad level; for example, migrant or minority ethnic status (Fearon and Morgan 2006) and urban birth (Krabbendam and van Os 2005). The task for the research community is to use these broad markers or clues to help identify the direct risk-modifying factors. Insofar as the primary focus has been on risk, only limited attention has been paid to resilience and protective factors. As discussed by Jones (this volume), resilience may be most usefully defined as *the degree of adaptability when faced with adversity*. As such, both protective factors and resilience can be conceptualized as factors that reduce risk following exposure to a candidate risk factor (i.e., statistically as effect modifiers). This noted, risk dominates the existing literature to such an extent that most of the data are framed in terms of risk rather than protection or resilience.

Incidence and Risk

It is now established that the incidence of schizophrenia and other psychoses varies markedly across and within populations (McGrath 2007). Most notably, the incidence is higher in men, in densely populated urban areas, and in some migrant and minority ethnic populations; that is, in groups of people who happen to comprise an ethnic minority in a given geographical region (McGrath et al. 2004). Further, observational epidemiological studies have identified a large number of putative risk factors and risk indicators at multiple levels (from the societal to the molecular), with risk and odds ratios (OR) commonly ranging from around two (e.g., obstetric complications; Clarke et al. 2006) to around ten (e.g., family history; Mortensen et al. 1999). Specifically in relation to genetic risk, molecular genetic studies currently provide strong support for associations with at least some single nucleotide polymorphisms that confer weak increments on risk (OR < 1.2) and at least some CNV deletions and duplications which are rare but confer much stronger effects on risk in the small proportion of cases who are carriers (OR > 3) (e.g., deletions at 22q11 are associated with a 30-fold increased risk; Sullivan et al. 2012a). The list of candidate factors is extensive and, in addition to genes, a nonexhaustive list of the most robust includes older paternal and maternal age, obstetric complications (especially hypoxia), developmental delays, childhood adversity (especially abuse and bullying), and cannabis use (especially at a young age and with variants high in THC). The range of nongenetic candidate factors has been

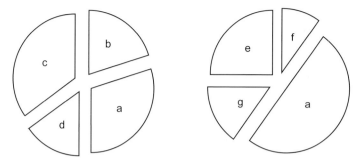

Figure 9.1 Hypothetical examples of clusters of risk factors that together may constitute a sufficient cause of psychosis: (a) genes, (b) trauma, (c) adversity, (d) substance use, (e) paternal age, (f) obstetric complications, and (g) viral infection.

regularly summarized in the literature (Murray 2003; van Os et al. 2010) and tends to converge on early development (i.e., childhood preadolescence) as a key phase for exposure.

In parallel, a large number of studies have identified cognitive and biological markers of risk: cognitive deficits in a number of domains that pre-date onset by many years (Kates 2010; Reichenberg et al. 2010; Welham et al. 2009); brain structural abnormalities, notably reduced gray matter volume prior to onset and ventricular enlargement (Steen et al. 2006); sensitization of the mesolimbic dopaminergic system (Collip et al. (2008); and HPA axis (Mondelli et al. 2010b; Pariante et al. 2004 axis. Despite these gains in our understanding of the risk architecture of psychosis, known risk factors explain only a small fraction of the liability. Psychosis is evidently multifactorial, with roots in early neuro- and sociodevelopment. No single factor, as far as we know, is either sufficient or necessary to cause onset. Instead, clusters of (overlapping) causes (see Figure 9.1) most likely work together (in varying combinations) to bring about the disorder (Schwartz and Susser 2006). Moreover, it may be that heterogeneity and overlap of clinical presentation mirrors hetereogeneity and overlap in clusters of causes that lead to onset.

Specificity

A key issue related to incidence and risk is specificity. Many of the risk factors for psychosis are nonspecific and overlap with other disorders and syndromes (e.g., bipolar disorder, depression, anxiety, and posttraumatic stress disorder). As illustrated in Figure 9.2, continuities (e.g., shared genes) and discontinuities (e.g., premorbid IQ) in risk exist between schizophrenia and bipolar disorder (Demjaha et al. 2012). Many of the social risk factors recently implicated in psychosis are associated with a wide range of disorders and other adverse outcomes. For example, childhood adversity, broadly defined, is associated with nearly every mental disorder as well as with a wide range of negative outcomes,

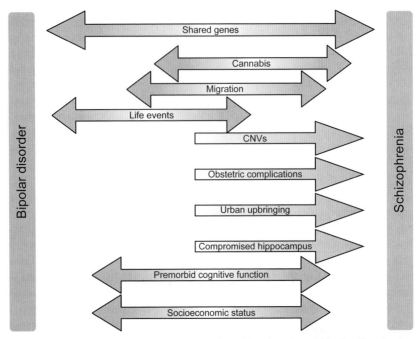

Figure 9.2 Shared and distinct risk factors for schizophrenia and bipolar disorder (reprinted from Demjaha et al. 2012, with permission of Oxford University Press).

including school exclusion, poor educational attainment, subsequent unemployment, re-victimization, substance use and abuse, and offending behavior (Kessler et al. 2010; McLaughlin et al. 2010). Complicating the picture further, much of the evidence that implicates social risk factors (e.g., trauma) has been based on studies of psychotic experiences in general populations (Varese et al. 2012). These studies show that psychotic experiences in these samples are strongly associated with common mental disorders, primarily depression and anxiety (Varghese et al. 2011). Similar overlaps have been observed in relation to cognition and biomarkers (Kelleher et al. 2012a).

There has been a tendency to view this heterogeneity and overlap in risk as a problem and a challenge for efforts to understand distinct disorders. However, nonspecificity is common across medicine; risk factors often operate across multiple diseases (e.g., cardiovascular, diabetes, cancer). Nonspecificity is generally observable in nature and as such should be viewed and embraced as an opportunity to understand shared pathways and interventions. For example, adverse environmental experiences and genes which lead to impaired brain function may combine to create a generalized vulnerability platform, or pluripotent risk state. Over time, early nonspecific symptoms and signs may develop from which (depending on the presence of other risk or protective factors) more specific clinical disorders may emerge, either alone or as multiple comorbidities. This implies that if interventions could be targeted at this early state (i.e., before the signs of specific disorders appear), substantial public health benefits spanning a range of health outcomes could result (McGorry 2007).

Interactions and Causal Pathways

Science strives to find simple, parsimonious hypotheses that are better suited to the scientific method (e.g., falsification). However, those which examine one risk factor at a time do not reflect the reality of biology. Interactions can exist between (a) genes (epistasis), (b) genes and the environment, as well as (c) two or more environmental risk factors. Risk factors can be linked with additional contingencies that result in unexpectedly complicated pathways. This necessitates moving beyond efforts that isolate independent causal factors to consider interactions and causal pathways; that is, to elucidate webs of causation along pathways to psychosis.

Currently, there is intense interest in exploring interactions and causal pathways, most notably in relation to gene–environment interactions (van Os et al. 2008). Not surprisingly, a growing body of data points to complex interrelationships between many of the candidate factors noted above, including putative though not robustly supported interactions between genes and environmental exposures, e.g., AKT1 and cannabis use (van Winkel 2011); environment–environment interactions, especially across levels of analysis, e.g., cannabis use and urbanicity (Kuepper et al. 2011); cumulative impacts of multiple exposures, e.g., trauma and social adversity (Morgan et al. 2008; Varese et al. 2012); and mediation along causal pathways, e.g., sexual abuse via revictimization and affective dysregulation (Bebbington et al. 2011).

Recognizing these complex interrelationships constitutes an important initial step toward the dissection of the risk architecture for psychosis. They have the potential to inform us on the contexts (e.g., fragmented neighborhoods) within which individual-level exposures (e.g., social isolation) impact on risk, on how specific risk factors cluster and add up to increase risk, and on the developmental trajectories which, at each point, increase the probability of disorder. The potential implications for prevention are clear: efforts could be targeted at key stages of development and at specific groups or areas. However, modeling these interrelationships is statistically complex and controversial (with notable potential for Type I error), requiring large samples with data on a range of exposures.

In relation to public health, we need to be particularly alert for interactions between two or more risk factors that result in "qualitative" or crossover interaction. Zammit et al. (2010a) cite the example of paternal antisocial personality traits and childhood conduct problems. When paternal antisocial personality traits are present, the more time a father spends with a child, the higher the risk of conduct problems. In the absence of paternal antisocial personality traits, however, the more time a father spends with a child equates to a lower risk of conduct problems (Jaffee et al. 2003). Such an example seems intuitive, but one can also envisage that, at a molecular level, if there are optimal levels of a certain bioactive molecule (i.e., best function is achieved by not too little and not too much), then for individuals with low constitutive levels of that molecule, additional exposures which tend to elevate that molecule's abundance would be protective. In others with optimal or high constitutive levels of that molecule, such exposures would, however, be damaging. There are no unequivocally demonstrated examples of this in psychosis, although the apparent existence of an optimal dopamine level for some aspects of prefrontal cortical function suggests such a possibility (Mattay et al. 2003; Vijayraghavan et al. 2007; Williams and Goldman-Rakic 1995).

Another example is given by the methionine (or Met) allele from the brain-derived neurotrophic factor (BDNF) val66met polymorphism, which leads to reduced secretion of the BDNF propeptide (Egan et al. 2003). While decreased secretion of the neurotrophin and the resulting impediment of neuroplasticity may imply a resulting predisposition to mental disorder, this is not necessarily the case. In the face of a stressor, the Met allele may be protective, as the amount of BDNF available to exert potentially negative plastic effects on the brain may be reduced (i.e., the variant acts as a buffer against the stressor). On the other hand, the same stressor in an individual with the valine (or Val) allele at the same locus may elicit a response that results in a neuronal or cellular response that is qualitatively different. The Val variant of the BDNF val66met is more likely to engage a plasticity pathway when compared with the Met variant. This could result in maladaptions of the brain in the context of stress. To complicate this further, the greater plasticity associated with the Val allele may also make individuals more responsive to protective factors, such as social support and psychotherapy. According to this differential susceptibility concept (Belsky et al. 2009), the genetic profile of an individual thus shapes the plasticity or responsiveness of the brain to environmental influences in general, thereby challenging the traditional view that susceptibility variants are inherently bad.

Risk Prediction in Populations

The overlapping and distinct risk factors for schizophrenia and a range of other disorders pose a number of challenges, especially in terms of utilizing this knowledge to predict onset, to identify high-risk groups, and to guide prevention and intervention. For example, the identification of individual risk factors (or indicators), each of which may have contributed only a minimal amount to overall risk, is of limited value in developing interventions or for the purposes of prevention. This is especially true for environmental factors and is illustrated using the example of cannabis.

Robust evidence from prospective cohort studies indicates that early use of cannabis is associated with a significantly increased risk of psychosis and related outcomes (Arseneault et al. 2004; Moore et al. 2007). This makes cannabis an attractive candidate for public health interventions (Degenhardt et al. 2009). However, because the effect size is relatively modest (e.g., Moore et al. 2007 report a twofold increase in risk) and because schizophrenia has a relatively low incidence (about 15 per 100,000 per year according to McGrath 2007), as do other psychotic disorders (Kirkbride et al. 2006), the population attributable fraction associated with this exposure is disappointing. Based on the best available epidemiological data, Hickman et al. (2009) estimated the number of individuals who would need to stop using cannabis to prevent one incident case of schizophrenia (technically referred to as a "number needed to prevent" or NNP). In people aged 20 to 24 with heavy cannabis use, they found that the NNP for men was 2800 whereas for women it was 5470. Estimates for people who use less cannabis is about four to five times higher. Considering that the best available public health interventions related to cannabis cessation have weak outcomes (i.e., these interventions themselves have high NNP), leveraging cannabis use as a means to reduce the incidence of schizophrenia becomes much less attractive. Alternative strategies related to cannabis use and the risk of psychosis may relate to (a) identifying individuals who are at increased risk due to other factors (e.g., genetic susceptibility, exposure to other risk factors, onset of academic decline, transfer to special education class due to behavioral problems), (b) reducing access to potent forms of cannabis, and (c) public health campaigns targeted at young teenagers to encourage delayed onset of first cannabis use.

In other areas of medicine (e.g., diabetes, cardiovascular), multiple risk factors have been combined into risk scores (e.g., Framingham Risk Score for cardiovascular disease), with varying degrees of complexity, aimed at predicting disease outcome (D'Agostino et al. 2001). The assumption in such models is that risk factors combine to increase the likelihood of disease or of poor outcomes. If brief and readily applicable in clinical settings, such tools may be of particular value in efforts to identify individuals at high risk of disorder and/or poor outcome.

Our discussions considered whether the development of such tools for risk factors in relation to psychosis was feasible. The various environmental factors implicated thus far are a mixture of risk indicators and risk factors measured at different levels (e.g., trauma, ethnicity, social fragmentation). As noted, further work is needed to establish how these various factors relate to each other. The limitations are illustrated in a study using data from the British 1946 Birth Cohort, in which Jones and Van Os (1998) found that combining a number of neurodevelopmental risk markers yielded a positive predictive value of only 1.2% for schizophrenia.

Based on data available presently, it appears that we need to determine whether (a) risk factors (or their combination) with stronger predictive validity exist and/or whether (b) additive models of risk assessment may not apply for psychosis, and so new concepts, analytic techniques, and algorithms may be necessary. A more restricted approach, focusing on a narrower domain of risk, may be more productive at this stage.

Take, for example, polygenic risk scores: Given the effect sizes of typical common variants in schizophrenia, if such alleles are to contribute to risk prediction it will be through examination en masse of large groups of markers rather than individual associations. One way of applying these large sets of alleles is through a method known as polygenic score analysis. As initially applied in schizophrenia by the International Schizophrenia Consortium (ISC), this approach was used to test the hypothesis that schizophrenia risk with respect to common alleles is distributed across very large numbers of genetic variants with small effect size (Purcell et al. 2009). The process involves designating, as putative schizophrenia risk alleles, those alleles that are "associated" with schizophrenia at extremely relaxed thresholds (e.g., $p < 0.5$) in discovery or risk score "training" genome-wide association (GWA) data sets. In subsequent independent "test" data sets, individuals are then assigned "polygenic scores" based on the average number of "risk" alleles weighted by their effect sizes in the training data set, and the scores for cases and controls are compared. Data from the ISC revealed that such scores were highly significant predictors of affected status in the independent schizophrenia data sets, and indeed also predicted bipolar disorder (Purcell et al. 2009). Based on the ISC training data set, the effect size for predicting case status was extremely small, but the ISC study estimated that larger training GWA studies might achieve more robust predictive values at a level that, while not of diagnostic value, might identify individuals at substantially elevated risk of the disorder at a level equal to or better than family history. If this prediction turns out to be correct, and enough data become available over the next few years to test it, such polygenic scores might be deployed to identify individuals at relatively high risk.

Ultimately, it would be optimal to combine risk factors from many domains. The high-risk paradigm offers another potential avenue for prediction, with emerging evidence for specific clinical and demographic factors that predict transition to psychosis (Yung and McGorry 1996; Demjaha et al. 2012). For example, a family history of psychosis, a recent decline in functioning, and a new onset of sub-syndromal psychotic experiences are associated with an increased risk of transition in up to 40% of individuals in the peak age (15–25 years) for the development of psychotic disorder (Cannon et al. 2008; Murray et al. 2004). In one study of individuals at high risk, when factors such as family history, a decline in social functioning, drug abuse, and delusion-like symptoms were combined, around 80% of cases who made the transition to psychosis were predicted (Cannon et al. 2008; Murray et al. 2004). Development of neurobiological measures that index vulnerability to psychosis (e.g., gray matter volume, cortisol, neurocognitive profile, event-related potentials) is also a promising area that may lead to the development of other assessment tools for predicting risk.

From Risk Factors to Risk Pathways and Mechanisms

How can knowledge of risk and resilience factors be leveraged to optimize discovery pathways? That is, how can we elucidate the complex causal matrices and pathways through which identified risk factors impact on neural circuits in the pathogenesis of disorder? Achieving this necessarily requires research across multiple levels, ultimately from the societal to the individual to the molecular. Important examples exist to illustrate how this can be achieved (see McGrath and Meyer-Lindenberg, this volume) as well as the significant obstacles that can limit progress. Our discussions inevitably focused on examples drawn from our areas of interest and expertise. We do not suggest that these are the only or most important ones. Other examples include links between stress and the HPA axis (Mondelli et al. 2010a); links between stress and brain chemistry, notably dopamine (Howes and Kapur 2009; Howes et al. 2012b); and links between exposure to threat and cognitive pathways (Garety et al. 2001). One of the key themes for us was how the basic neurosciences could be more effectively engaged in the study of psychosis. Related to this, Andre Fenton (pers. comm.) provided the following "view from a basic neuroscientist":

It will be generally valuable to recruit basic neuroscientists to study problems that are directly relevant to schizophrenia. In particular, this recruitment will be necessary to characterize the neurobiological consequences of genetic alterations that have been identified in schizophrenia. However, most basic neuroscientists will continue to be reluctant to study animal models based on genetic risk factors for the simple reason that the relevance of a particular animal model to schizophrenia is questionable. This will be particularly true for genetic models derived from genetic screens if the penetrance of the mutation is low. In this case, studying the particular gene or genetic alteration will also be of uncertain value and thus of low interest.

There are three basic ways to encourage the desired recruitment:

1. Demonstrate that the target mutation is clearly and importantly involved in schizophrenia. To run a basic neuroscience research program the animal model needs to be both well defined and relevant. Meeting this condition would allow the researcher to explore pathophysiological and behavioral consequences with confidence that the findings have relevance to schizophrenia.
2. Demonstrate that a particular behavioral or pathophysiological endpoint is crucial to a core aspect of schizophrenia—understanding that the endpoint will likely represent one or a small number of features of

the syndrome, as opposed to an animal model of the full syndrome. In this way, the research program can proceed without needing to know whether the model being studied is relevant to schizophrenia per se.
3. Provide a clear set of theories. Given theory, the researcher can proceed by making predictions and evaluating experimental outcomes against the theory. This will allow the go/no go decisions to be made as the research program evolves.

Examples in Genetics

One major challenge in the genetics of schizophrenia lies in translating the identification of common variation into a deeper understanding of the biological pathways involved. This challenge exists because odds ratios for identified common genetic variants are low, and thus they may not induce robust changes that can easily be modeled in cellular or animal systems. This limits the use of these findings and poses challenges for engaging the wider neuroscience community. From a neuroscience perspective, high-penetrance variants provide a much more promising basis than low penetrance variants for investigating biological mechanisms and pathways through cellular and animal studies. At present, known high-penetrance variants for psychosis are restricted to CNVs, where the typical molecular lesions span multiple (often very many) genes, any one or more of which might be relevant (Sullivan et al. 2012a). Therefore, the reliable identification of high-penetrance single-base mutations may offer more precision in modeling. Strategies are now underway to identify these smaller molecular lesions; for example, whole exome and genome sequencing of case-control samples, and sequencing of mother–father–offspring families for variants that arise in affected persons as new mutations. While many variants of potential interest have been identified through sequencing (Xu et al. 2012), thus far none has been demonstrated to have an etiological role in schizophrenia. This work is in its early stages and should these variants exist, there is good reason to be optimistic that some will be identified.

Although it is difficult to model effects at the individual level, common genetic variants still offer opportunities for engaging neuroscientists with different requirements. Existing approaches seeking multiple weak variants in biological systems have pointed to targets for investigation and even possible therapy in Alzheimer's disease (Jones et al. 2010b). While it is possible, or even likely, that the complexity of psychosis means these associations are distributed across more biological processes or functions (which reduces power), there is already evidence for (a) enrichment of common genetic risk factors in bipolar disorder in genes encoding types of calcium channels (Sklar et al. 2011), (b) schizophrenia risk factors in a set of genes whose expression is regulated by microRNA-137 (GWAS Consortium 2011), and (c) rare variants in a set of genes affiliated to the glutamate NMDA complex (Kirov et al. 2012). The identification of the broad processes involved has the potential to generate specific hypotheses, which can then be more readily exploited by basic neuroscientists. Moreover, although it remains to be established, it is at least plausible that common weak and rare strong variants will often converge on the same genes, or the same biological process. This means that modeling weak genetic effects through more robust genetic lesions (e.g., gene knockout) may be a tenable approach to deriving insights into the relevant mechanisms.

Other avenues for exploiting genetic findings have not yet been fully explored. Available sample sizes are probably inadequate for gene–gene interaction studies. Samples with both rich data on environmental exposures and extensive genetic data are even smaller, limiting the possibilities to explore gene–environment interplay. With respect to genes and the environment, some large studies are underway (e.g., EU-GEI 2008).

Genetics offers further avenues for investigating pathways to disorder, including the incorporation of nongenetic data. The identification of high-risk individuals through molecular methods (see above discussion on polygenic risk scores), for example, can be expected to facilitate any number of study designs that look at trajectories to disorder and, in particular, risk and resilience factors which distinguish those at high risk who go on to develop the disorder versus those who do not or whose outcome is more or less severe. One concrete example for which there are many analogous approaches is to follow a large sample of people with a defined molecular lesion that confers high risk (e.g., a specific CNV) with detailed longitudinal phenotyping. Such approaches may not just identify risk and resilience factors per se, but the detailed trajectories (e.g., EEG changes or time courses of cognitive and social interaction changes) can inform the work of basic neuroscientists in generating and exploiting model systems (cf. analogous changes during animal brain development), a process which might be iterative with the model systems informing designs of human high-risk studies.

Another area that is attracting considerable interest is epigenetics, which promises to produce novel insights into the dynamic interplay of genes and environments. While researchers often use the shorthand of labeling risk as "genetic" or "environmental" (i.e., nongenetic), it has long been accepted that this simplistic dichotomy does not reflect the transactional nature of biology. In particular, it does not capture the contingencies that occur between information derived from the DNA sequence (which we inherit from our parents) and instructions from the environment (which can range from basic chemical requirements for life, to mother–infant bonding and the family unit, to broad, system-level components at the level of society). The science of epigenetics aims to capture some of the mechanisms that mediate the interaction between these two broad domains. While the boundaries of this field are still being refined, it is clear that environmentally mediated factors (e.g., altered nutrition, stress) can change the tissue-specific and developmentally specific modification of DNA (e.g., via mechanisms related to methylation, histone coding, chromatin packaging). These mechanisms allow environmental exposures to lead to persistent changes in the patterns of gene transcription, analogous to those that result from genetic variation, which may have profound implications for cellular properties and resultant emergent properties of this tissue. Within the field of schizophrenia research, there is considerable hope that this category of observation may provide mechanisms that link an exposure (e.g., prenatal famine, early life stress exposure) and biologically relevant phenotypes (e.g., neuroendocrine responsiveness, neurotransmitter properties) (Labrie et al. 2012; Oh and Petronis 2008; Petronis 2010; Pidsley and Mill 2011; Rutten and Mill 2009; Toyokawa et al. 2012).

Examples in Neurobiology

Measures of brain structure and brain function, including structural and functional MRI as well as electrophysiological paradigms, offer particular promise for studies of pathways and mechanisms linking environmental risk factors to psychosis. In particular, when measured using these tools in longitudinal fashion in animal models and humans, the effects of environmental risk factors, genetic risk factors, or both can provide an index of risk factor effects (i.e., mechanisms of effect) on brain structure and brain function. Furthermore, these types of studies can provide a platform where preclinical studies of novel therapeutics can be tested on brain structure and brain function, as well as behavioral deficits. As outlined by Cadenhead and de la Fuente (this volume), all of these measures demonstrate evidence of change during the prodrome and first episode of psychosis, perhaps revealing early brain changes at the emergence of psychosis.

In the preceding decade, a particular focus of neuroimaging research in psychiatry has been the identification of neural correlates of genetic risk variants for schizophrenia in the brain using imaging genetics—a research strategy that combines molecular genetics and neuroimaging techniques (Meyer-Lindenberg and Weinberger 2006). One of the main tenets of this approach is the idea that genetic susceptibility effects are not directly expressed at the behavioral level; instead, they are mediated by molecular and cellular mechanisms that shape the structural and functional properties of neural circuits. Compared with behavior, risk-related genetic effects likely have a higher penetrance for more direct indices of these structural and functional changes, and may be studied in healthy volunteers in the absence of illness-related confounds such as medication. At the beginning, studies focused on candidate genes. Recently, attention has shifted to the examination of genome-wide significant schizophrenia risk variants, where the link to the syndrome itself has been established with sufficient confidence.

For example, a promising systems-level risk phenotype is altered functional connectivity of the dorsolateral prefrontal cortex (DLPFC) and hippocampus during working memory. Influential pathophysiological models of schizophrenia propose that genetic and environmental risk factors disturb the normal developmental maturation of pathways that interconnect these structures (Murray and Lewis 1987; Weinberger 1987), which is thought to promote deficits in experience-dependent plasticity, abnormal functional and structural connectivity, as well as psychosis in adulthood (Harrison and Weinberger 2005; Meyer-Lindenberg 2011). Consistent with this, disturbed prefrontal-temporal functional connectivity is evident in chronic, first episode, and prodromal samples (Crossley et al. 2009; Meyer-Lindenberg 2011; Meyer-Lindenberg et al. 2005; Rasetti et al. 2011; Wolf et al. 2009). In addition, anomalies in functional connectivity of DLPFC and hippocampus have been detected in unaffected relatives of patients with schizophrenia (Rasetti et al. 2011), healthy carriers of a genome-wide supported schizophrenia risk variant (Esslinger et al. 2009; Paulus et al. 2013; Rasetti et al. 2011), and genetic animal models of schizophrenia (Sigurdsson et al. 2010). A genome-wide supported risk variant for schizophrenia and bipolar disorder in *ZNF804A* is particularly interesting in this context, as the genetic association to altered DLPFC–hippocampus functional connectivity per se has been replicated in an independent sample (Meyer-Lindenberg 2010a).

Recently, efforts have recently been extended to investigate the effects of established (but complex) social-environmental risk factors in the brain (Meyer-Lindenberg and Tost 2012). One example of the potential for interrogating epidemiology and neuroscience research is the characterization of the neural effects of urban upbringing, an established environmental risk factor for schizophrenia. Using functional MRI to examine brain response during social evaluative stress processing in healthy volunteers, a recent study (Lederbogen et al. 2011) detected an association of urban upbringing and functional alterations in the perigenual cingulate cortex (pACC), a key brain region for the regulation of negative emotion and stress (see Figure 9.3). Prior data from epidemiology suggests that the adverse effect of urban upbringing is modulated by genetic risk factors, with an excess rate of psychosis in genetically vulnerable individuals brought up in urban environments (van Os et al. 2004). From a conceptual point of view, it appears most plausible that certain genetic and environmental risk factor constellations gain their clinical momentum through converging adverse impacts on the functionality of shared neural systems. Direct proof for adverse gene–environment interactions in the brain can be provided by probing the identified functional systems in individuals stratified by genetic and social background.

In addition to the identification of neural correlates of genetic risk, efforts have been made to use neuroimaging and electrophysiological measurements (ERPs, neural synchrony, prepulse inhibition) for predicting risk status for the development of psychosis (see Cadenhead and de la Fuente, this volume; Atkinson et al. 2012; Bodatsch et al. 2011; Jahshan et al. 2012; Tost and Meyer-Lindenberg 2012). Studies with electro- and magnetoencephalography, for example, have the advantage over functional imaging of capturing neuronal dynamics with a millisecond temporal resolution. However, such approaches

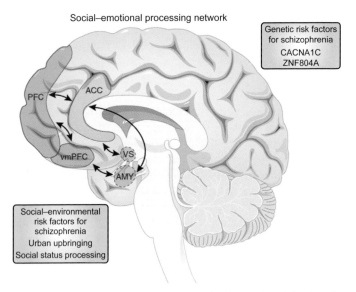

Figure 9.3 Effects of social and environmental risk factors for schizophrenia on regulatory circuits of human social-emotional processing (Tost and Meyer-Lindenberg 2012).

face several important conceptual and methodological challenges: intra-site reliability for longitudinal studies and inter-site reliability and variability of functional measures in the general population.

Analogous to the molecular genetics field, the identification of the effects of risk factors of small effect size as well as complex gene–gene or gene–environment interactions in the brain require the availability of large data archives. To test the potential of systems-level neuroscience measurements for risk prediction and biomarker discovery fully, future research needs to be conducted in multicenter studies with standardized quality assurance measures, data acquisition and processing schemes (including task paradigms), and analysis algorithms.

Research Challenges

Consideration of the previous examples (which, again, reflected our group's experience) and our review of what is known about the risk architecture of psychosis from epidemiology led us to consider the challenges associated with an attempt to develop ambitious research programs aimed at integrating findings across multiple levels. Two implications were clear: (a) very large samples are necessary and (b) detailed information on these samples is needed across the full range of putative risk and resilience factors, from the environmental (including both individual level and area level exposures) to neurobiological to molecular. This means that research efforts need to be significantly scaled up and broadened. Sampling and measurement (broadly defined to include the full range of assessment tools, from self-report questionnaires to neuroimaging to biological samples) constitute crucial areas for future research to address.

Sampling and Samples

What is the optimal strategy for the most efficient generation of new, large samples that will collect data across multiple levels of analysis?

A useful starting point is population-based sampling, which can set the principles for the selection of large representative cohorts and case-control samples. Population-based sampling broadly refers to the generation of a random sample (using a suitable sampling frame) from a known population (such that each person within that population has an equal chance of being selected to participate in the study) and, if not subject to selection bias, provides (relatively) precise estimates of the prevalence of exposures. This, then, sets the optimal standard for sampling to estimate risk and exposure prevalence. In some northern European countries, most notably Denmark and Sweden, the availability of register data on the whole population, and the facility to link these data to, for example, information on health service contacts and hospital admissions, sidesteps the issue of sampling altogether, as the sample is the entire population. Such systems provide considerable opportunities for studies that utilize data on whole populations across a wide range of domains, and studies from these countries have already produced a series of seminal findings that have advanced our understanding in a number of areas (e.g., Pedersen and Mortensen 2001).

Other sampling strategies are also important. Use of the Internet, for example, permits rapid generation of large samples with specific characteristics. At the Cognitive Assessment and Risk Evaluation (CARE) program at the University of California San Diego, the Internet has become an increasingly important means of recruiting early psychosis participants. Domingues et al. (2011) report that 16% of 223 subjects enrolled over a ten-year period were identified via the Internet. The number of subjects recruited per year via the Internet increased each year during the course of the study. The primary Internet site that refers to the CARE program is schizophrenia.com, a site dedicated to providing high-quality information on schizophrenia to the general public. On this site there is a link to a "Schizophrenia Screening Test and Early Treatment Resources," which includes a screening instrument developed by Yale University and a list of prodromal psychosis programs worldwide. This method of recruitment is similar to that of many early psychosis programs worldwide.

In general, sampling for studies of biomarkers and neurobiological risk pathways have tended to be more ad hoc and purposive than described above, in part because the required sample sizes tend to be smaller and in part because of the relative practical difficulty of completing assessments (e.g., neuroimaging studies always require individuals to attend research facilities within which scanners are based) and consequent burden to participants. The aims are not usually to estimate main effects, but to explore pathogenic processes. Consequently studies may seek, for example, to identify individuals at the extremes of distributions (e.g., top 10% genetic liability and bottom 10%) or with and without exposures of interest (e.g., sexual abuse). This way more detailed assessments and samples relating to hypothesized psychological and biological pathways can be conducted (e.g., MRI, fMRI, cortisol, PET). Still, systems-level neuroscience research faces specific issues related to sampling bias. This arises from various sources such as technique-specific contraindications (e.g., nonremovable metal implants or electrified devices in the case of MRI), inconveniences (e.g., claustrophobia-provoking space restrictions), and the preferred location of high-end research equipment in urban areas. This can promote an underrepresentation of older, lower-educated participants with general somatic comorbidities from rural areas.

It may seem obvious, then, that the optimal strategy is to construct large population-based samples (either cohorts or case-control) in which sub-studies of psychological, biological, and genetic mechanisms based on selected subsamples can be nested within the larger study of risk and resilience factors (and their interactions). Data collected on all participants recruited to the epidemiological study can be used purposefully to identify individuals with particular characteristics; for example, individuals at high or low risk of disorder (defined according to prespecified criteria, such as cognitive decline and family history of disorder) as well as at the extremes of genetic liability. These individuals can then be assessed in greater detail on a wider range of cognitive tests (neural and social) and (potential) biomarkers. Recruitment to these nested studies will be no less challenging than recruitment to such studies generally; still, the fact that participants are drawn from a known sample means that nonrecruitment bias can be quantified in relation to all variables collected on original participants. In short, the findings from nested sub-studies of mechanisms will be more readily generalizable to a wider population (or at least the limits on generalizability will be more apparent). It is not just studies of neural pathways that can benefit from such an approach. In relation to environmental assessments, the necessarily more cursory and crude measures of environmental exposures used in large samples can be validated using more detailed assessments in subsamples (which may include independent corroboration, e.g., obstetric events, child abuse). This model is increasingly being adopted in existing cohorts (e.g., Avon Longitudinal Study of Parents and Children or ALSPAC, Dunedin, Environmental Risk Longitudinal Twin Study).

Two other design considerations came to the fore in our discussions. First, cohort studies to investigate psychosis are difficult and expensive; psychotic disorders are relatively uncommon (compared with anxiety and mood disorders) and relevant exposures or biomarkers often occur or are evident long before onset. This has, in part, fuelled interest in extended psychosis phenotypes (i.e., transient and mild psychotic experiences, schizotypy) and endophenotypes (e.g., cognitive performance), which are more common. A number of cohort studies are ongoing that have relevant data (e.g., Avon Longitudinal Study of Parents and Children, Dunedin, Environmental Risk Longitudinal Twin Study, Christchurch Study). Individually, each study may not be large enough to provide sufficient numbers of individuals who develop a psychotic disorder. As far as we are aware, what has not been considered is combining samples, such that as participants pass through the age of risk for psychosis, the numbers meeting criteria for clinical disorder may allow for meaningful analyses. Genetics has led the way in showing how large-scale collaborations can yield samples that would have been otherwise impossible but which are essential to achieve sufficient power. Epidemiology needs to follow suit!

Second, the case-control studies noted above should not be discarded. Given what we now know about the genetic architecture of the disorders, they are likely to be the mainstay of primary identification of genetic risk factors for the disorder for which there is still a pressing need but now a clear pathway. Cheaper and more efficient than cohort studies, they offer additional advantages:

- They allow for studies of clinical disorder.
- It is possible to collect more detailed information on a wider range of exposures.
- It is possible (as above) to nest within them studies of mechanisms.
- They allow for the simultaneous study of area-level factors and their impact on incidence rates and, using multi-level modeling, for analyses of the relative impact of area and individual factors on risk.

Of course, there are many pitfalls with case-control studies; most notably, they rely on retrospective assessment of exposures, a particular problem when the recall of exposures may vary by case-control status. Various strategies can, however, be adopted to minimize recall bias (including use of corroborative evidence and, where available, contemporary records). In addition, for some exposures (e.g., child abuse) which cannot reliably be assessed at the time of their occurrence, case-control studies may be the only feasible design. Finally, case-control designs can be readily extended to include siblings and other relatives to reduce markedly unmeasured residual confounding.

In short, this points to the need for both (a) a scaling up and a mixed economy of research, in which studies of neurobiological pathways are nested within population-based studies (i.e., cohort or case control), and (b) sharing epidemiological data sets in order to have sufficient power to explore complex causal pathways (e.g., gene–environment interactions, environment–environment interactions). Regarding the latter point, while study-level meta-analysis is now well established in schizophrenia research, individual-level meta-analysis is often hindered by complex, time-consuming ethical and legal constraints related to data sharing and protecting confidentiality (van Os et al. 2009; Walport and Brest 2011). Robust network and "cloud-based" systems, however, can now perform pooled analyses of individual-level data without sharing data; that is, individual-level data is returned to a secure central hub for "virtual" pooling but is never committed to disk nor stored on the server at any point (Wolfson et al. 2010). After careful harmonization of variables and data structure, this methodology has recently been implemented by the International Collaboration for Autism Registry Epidemiology.

Concepts and Measurement

> Measure what is measurable, and make measurable what is not so.—attributed to Galileo (1564–1642)

The proposition that social risk factors are important in the etiology of psychosis is now widely accepted, largely as a result of recent studies which show that incidence is socially patterned and that various contextual (e.g., social fragmentation, ethnic density) and individual-level (e.g., trauma) exposures are strongly associated with psychosis. The conceptualization and measurement of social risk factors in psychosis research remains crude, with only limited attention paid to, for example, the nature, timing, duration, and severity of exposure. It is not uncommon, for instance, for studies of child abuse to be based on single questions with no information about age of abuse, severity, frequency, or perpetrator(s). Similarly, our understanding of the processes that are indexed by proxy variables, such as population density (the usual way in which urbanicity is operationalized) and ethnicity, remains limited. It is unfortunate that as genetics and neuroscience develop ever more sophisticated technologies for interrogating molecular and neural processes, measurements of environmental exposures remain crude and outdated. This inevitably limits efforts to delineate the precise social processes that increase risk for psychosis and, without improvement, will thwart efforts to move from (social) risk factors to neurobiological mechanisms and pathways.

A useful starting point from which to move beyond the current situation may be a taxonomy to characterize the various types of socioenvironmental factors implicated in psychosis, as a basis for better understanding interrelationships between them and for developing more (or identifying already existing) sophisticated assessments. One possible schema distinguishes:

- *social position* (status) or variables that relate to an individual's or household's place within a social hierarchy (e.g., social class, ethnicity, gender);
- *social experience* or variables that relate to events or difficulties such as abuse, trauma, life events, and daily hassles;
- *social interactions* or variables that capture social connections and breakdowns (e.g., social networks, support); and

- *wider social contexts* ranging from schools and neighborhoods to regions and societies, which may exert independent effects on health and which may modify the impact of variables measured at other levels.

Delineated this way, it is evident that we have lumped together very different exposures under umbrella terms with limited meaning: social defeat (Selten and Cantor-Graae 2005), social disadvantage (Morgan et al. 2008), etc. This points to the need for tools that more fully capture the nature of exposures. Where these are not available (and we should look first; see below) there is significant work to be done developing them. Of particular note here, and by way of an exception to the above, is the use of *in vivo* experience-sampling techniques, which capture in real time daily hassles and emotional responses, thus illustrating precisely the type of innovations required (Myin-Germeys and van Os 2008; Myin-Germeys et al. 2001).

While the above discussion primarily relates to social factors, a more general point should be made regarding measurement, which may once again seem obvious but merits highlighting: standardized paradigms are needed for neurobiological assessments and procedures (e.g., structural and functional MRI, electrophysiological measurements, neurocognitive batteries). Standardization, however, is not enough. Measurement of specific processes (e.g., perception, memory) must be accomplished without confounds from other processes (e.g., poor attention, low motivation) or from generalized performance impairments or other factors such as smoking or poor nutrition (Knight and Silverstein 2001; Silverstein 2008).

Rapprochement with the Social Sciences

The above leads into consideration of the relevance of the social sciences to efforts to understand the impact of social factors on risk of psychosis. There is undoubtedly considerable skepticism within psychiatry about the value of the social sciences to understanding the etiology of psychosis. This stems from a now untenable view that the onset of schizophrenia and other psychoses is unaffected by environmental factors, as well as from the legacy of mistrust that developed when many social scientists sided with, and provided ammunition for, the amorphous antipsychiatry movement during the 1960s and 1970s, which challenged the very existence of mental illness (Morgan and Kleinman 2010). As a result, despite select examples where collaborations have been enormously fruitful (e.g., for depression, Brown and Harris 1978; for social class and schizophrenia, Hollingshead and Redlich 1958), researchers have been slow to draw from the social sciences in seeking to further investigate the crude social factors recently implicated in psychosis. In relation to concepts and measurement, there is much in fact that could be gained from a rapprochement with the social sciences, most notably sociology, which is primarily concerned with precisely the factors and processes being considered in relation to psychosis: urbanicity, social class, ethnicity, and all forms of social adversity. In short, to optimize discovery pathways, there is as much a need to engage with the social sciences as with the neurosciences.

Implications for Prevention and Early Intervention

In considering possibilities for utilizing knowledge on etiological pathways to psychosis for prevention, our discussion was shaped by Geoffrey Rose (1985:33):

> I find it increasingly helpful to distinguish two kinds of aetiological questions. The first seeks the causes of cases, and the second seeks the causes of incidence. "Why do some individuals have hypertension?" is a quite different question from "Why do some populations have much hypertension, whilst in others it is rare?" The questions require different kinds of study, and they have different answers.

This seminal paper (Rose 1985) suggests that strategies for prevention can be separated into those that seek to reduce (prevent) incidence rates of disorder in populations and those that seek to identify individuals at high risk of disorder and prevent individual cases of disorder.

General Populations

In relation to psychosis, we are not close to being able to predict risk (incidence) at a population level (or at least determinants of incidence at a population level), as already noted above in the discussion of risk prediction tools. Based on a systematic review and meta-analysis of studies of the incidence of psychosis in England, Kirkbride et al. (2012) have done some initial work on predicting incidence rates given knowledge of the sex, age, and ethnicity of populations and population density. Such systematic reviews are particularly useful for service planning and resource allocation. However, they do not provide any information about how incidence rates may be modified.

This noted, some of what we now know about pathways to the development of psychosis (via premorbid cognitive and functional decline), and the common occurrence of symptoms of depression and anxiety and isolated psychotic experiences prior to onset (i.e., a pluripotent risk state in which a number of adverse outcomes are possible), point toward generalized strategies to intervene to prevent more severe outcomes during childhood and adolescence. This was the basis for the staging model of mental disorder developed in Melbourne by McGorry et al. (2006). From this perspective, and further considering the nonspecific nature of many of the environmental factors implicated in psychosis, broader public health interventions that aim to reduce exposure to risk factors, and perhaps promote resilience or protective factors, may impact on incidence of psychosis (along with other disorders and adverse outcomes).

Clearly, better prenatal care, improved nutrition, reduced pollution, more green space for exercise, and public health campaigns to reduce tobacco and drug use are positive interventions as a whole. This extends to interventions such as school-based educational programs and interventions that target children who exhibit difficulties or a decline in performance at school. A further rationale for broad-based environmental efforts to reduce serious mental disorder later in life comes from animal studies, where exposure to enriched environments in adolescence has been shown to prevent psychotomimetic drug-induced behavioral, social, and cognitive changes thought to model aspects of schizophrenia (Koseki et al. 2012). By educating school officials and providing early interventions, it might be possible to identify earlier those at greater risk for not only psychosis but also mood, anxiety, behavioral, or learning disorders and alter the potential course.

School-Based Interventions

One strategy for prevention involves targeting risk indicators (e.g., poor interpersonal skills, poor social problem-solving skills, poor stress tolerance, poor self-regulation, decline in academic function) in children who show signs of risk for future mental health problems. Such interventions can be delivered in the school classroom as well as in other settings. Importantly, children receiving these services are not specifically identified as being at-risk for psychosis; intervention may have positive effects on a range of outcomes (some of which have already been demonstrated), consistent with the idea that intervening during the pluripotent risk state may be more effective than intervening at the later stage of high risk for a specific disorder.

For example, there is a developing evidence base for social-emotional learning interventions (delivered by trained teachers as part of regular classroom curricula) that shows significant effects on positive behavior, improved social emotional competencies and academic performance as well as decreases in conduct problems and emotional distress, and these are increasingly being implemented in school settings (Durlak et al. 2011). To date, however, the effects of such programs on preventing the development of serious mental disorders that continue into (or emerge in) adulthood are not known. Recent evidence from the animal literature, however, suggests that prophylactic training of specific cognitive functions might reduce the negative effects of a later-onset schizophrenia-related brain abnormality (Lee et al. 2012). Thus, the effects of very early intervention services will be important to explore, especially in light of the relative failure of current "ultra high-risk" identification and treatment efforts to delay psychosis by more than one year (Yung and Nelson 2011). In implementing school-based programs at a relatively early age (e.g., 8–13) for children identified as being in a pluripotent risk state, two critical issues and potential barriers involve (a) avoidance of labeling and stigmatization and (b) funding for adding interventions to the school curriculum.

At-Risk and Early Psychosis Populations

When we narrow our focus from the general population to those individuals who are at greater risk of psychosis, or who are already showing early signs of psychosis, a number of novel and potentially important pharmacologic and nonpharmacologic interventions have been identified that address risk domains (e.g., stress response, inflammatory processes, nutrition) but have yet to be studied as preventative or disease-modifying strategies in the early course of psychotic disorder (for a review, see Cadenhead and de la Fuente-Sandoval, this volume). Similarly, informative biomarkers that provide insight into mechanisms of illness or serve as putative predictors of psychotic illness can serve as surrogate endpoints in clinical trial designs and provide new directions for biomedical research. To illustrate the use of neuroimaging markers in treatment development and prevention, consider the following examples:

1. Once identified, and sufficiently established through independent replication, neuroimaging phenotypes related to genetic and/or environmental risk for schizophrenia may serve as neural systems markers that may be targeted for treatment development, in a "top-down" approach to treatment. For example, knowledge from neuroimaging and neurophysiological studies that demonstrate impaired structure and function in the DLPFC (see above) provides a rationale to target this particular brain region, and its associated functional neural circuitry, while studying the effects of novel pharmacological compounds or other therapies. One example for a novel therapeutic approach is repetitive transcranial magnetic stimulation (rTMS). Prior evidence suggests favorable effects of rTMS in DLPFC on negative symptoms, cognitive function, and intermediate neural markers linked to both this particular brain region and functional alterations in schizophrenia (Barr et al. 2013; Boroojerdi et al. 2001; Gromann et al. 2012; Prikryl et al. 2012; Rounis et al. 2006). Here, established neuroimaging markers of genetic or environmental risk, such as DLPFC-hippocampus coupling, may be used as functional readouts to examine whether novel therapies are efficient in modifying this particular neural system. If so, these markers may further be used to optimize these treatment approaches (e.g., by finding the optimal range of stimulation intensity in the case of rTMS, or the optimal dose range in the case of novel compounds). Notably, the fact that the risk marker itself is biological in nature does not mean that these features can only inform primarily biological interventions. In all these efforts, the brain is best conceptualized as an intermediate observation level where genetic and environmental risk factors converge and increase illness risk by their complex combined effects on shared neural subsystems. The same principle applies to the validation and optimization of psychotherapeutic approaches; for example, in the

context of the effects of behavioral therapy on established neural risk markers, such as amygdala hyperactivity in depression and anxiety disorders (Bryant et al. 2008; Siegle et al. 2006).

2. To give an example of a bottom-up approach, the effects of an environmental risk factor on brain structure or function might be useful for informing, or supporting, disease prevention strategies. Following the discovery of such a risk factor in a population-based sample, neuroimaging can be used to identify the effects of this in the brain (e.g., the effects of the complex phenomenon "urbanicity" on pericingulate function; Lederbogen et al. 2011). Further decomposition of these complex environmental risk factors into causal subcomponents (vs. epiphenomena) is certainly necessary to inform true preventative approaches. Here, neuroimaging risk markers may add an additional level of observation that can be exploited to guide these efforts (e.g., if significant associations of pericingulate function with social support measures, but not socioeconomic status, were to be detected). Naturally, not all of the identified neural functional (and presumably causal) risk subcomponents will be immediately susceptible to manipulation in real-world environments to modify disorder risk. However, the combination of evidence for a risk factor from a population-based sample, coupled with follow-up validation demonstrating effects of this risk factor on the brain, can provide a powerful platform to inform public health strategies for prevention.

There are, however, a number of impediments to the effective implementation of clinical trials for novel interventions in early psychosis, such as lack of interest by pharmaceutical companies in older drugs that would not provide profits (e.g., aspirin, minocycline). In addition, there are safety concerns regarding the use of children or teenagers who are represented in at-risk populations. The latter point raises a significant issue; namely, by separating services for children and adolescents from services for adults, barriers are created that impact research, intervention, and prevention.

Interface with Child Psychiatry

There is a clear disconnect between disciplines that address child and adult psychiatry across all countries represented at the Forum: Australia, Canada, Germany, Great Britain, Ireland, The Netherlands, Switzerland, and the United States. Breaking down the boundaries between the clinical disciplines of child psychiatry, developmental disabilities, and adult psychiatry carries with it a number of potential benefits.

As noted, psychosis is a developmental disorder, the early signs of which are often evident long before the emergence of positive symptoms and behavioral disturbance. Those who go on to develop psychosis may be known to child services. Partitioning of services into separate child and adult systems discourages combined approaches that might enhance power by marshalling all relevant forces (including datasets), inhibits the use of expertise to develop the best developmentally appropriate tools of clinical (and other) measurement for research, and is a barrier to information flow about the outcome of research. Given the evidence that certain disorders in childhood are either risk factors for, or early manifestations of, adult disorders, it can be expected that the research agenda can be fostered by longer-term clinical perspectives, even at the level of case reports that seed experimental or more detailed observational studies. The discontinuity in service provision inhibits a longitudinal perspective, both among clinicians and in more formally designed studies that aim, for example, to identify which disorders have childhood precedents, who is most at risk, and even optimal treatment of the subset of people with childhood disorders who go on to develop psychotic (as well as other) disorders. As noted, there is evidence that preventative or ameliorative interventions for adult disorders may require delivery by those who conventionally work in the childhood arenas.

In addition to a more seamless integration of child and adolescent and adult psychiatry, prevention efforts could be enhanced by closer ties between mental health experts and the following groups: special education teachers, juvenile justice program staff, social workers, developmental psychologists, and family therapists. For example, special education teachers, by definition, work with children with serious emotional disturbances and/or cognitive/academic difficulties, and a substantial proportion of whom can even be considered to be at high risk for developing a serious mental disorder. Although these teachers have many skills for improving the social and academic performance of these children and adolescents, these skills are essentially unknown to child psychiatrists and psychologists. In addition, these teachers typically lack training in the identification of risk factors for serious mental disorder and in interventions developed within psychiatry. Similarly, a stressful upbringing (e.g., dysfunctional family environment, economic disadvantage) has been shown to increase risk for schizophrenia in both genetic high-risk (e.g., Tienari et al. 2004) and non high-risk (Wicks et al. 2005, 2010) populations, and it has been suggested that insights from the fields of neuroscience, genetics, psychology, and studies of the social world could be integrated into formulations focusing on interlevel interfaces, with profound implications for training, practice, and research in the field of family processes and therapy (Sluzki 2007). We do not wish to imply that there is a problem with different professionals possessing different skills. However, it is a problem if these professionals work separately and in relative isolation, without informing each other's work or treatment/education plans.

Recommendations

Expanding on our initial set of questions, we propose the following directions for future research:

1. Research needs to target the identification of risk factors—both genetic and nongenetic—scaled up by (a) combining existing cohorts and (b) constructing large population-based samples with nested studies of neural pathways and mechanisms.
2. Related to point 1, there is a need for stronger efforts, incentivized by funders, to make available large relevant epidemiological data sets, with a high priority for data sets that have genetic data from which to draw individuals at higher risk, to the research community much earlier (following the lead taken in genetics).
3. There is a need for much stronger interdisciplinary ties and fully integrated research programs between the neurosciences, the social sciences, developmental psychology, and immunology (neuroimmunology).
4. To move beyond crude markers of environmental risk (e.g., urbanicity, ethnicity), we need to develop (or identify from the social sciences and use) more sophisticated concepts and social assessment tools to capture the complexities of social contexts and experiences over time.
5. To allow better developmental modeling by neuroscientists and the genesis of theories which then become testable in those models, comprehensive longitudinal characterizations need to be developed at the earliest possible stage of cognitive, psychological, neurophysiological, social, and environmental profiles of individuals at high risk (identified through the population and other high-risk study designs).

With respect to prevention and treatment, we identified the following needs:

1. Emphasis needs to be given to the importance of implementing public health strategies that impact on important risk factors for psychosis.
2. Promising interventions that may have failed in patients with long-standing disorders need to be tested in at-risk and early psychosis populations.
3. The effectiveness of school-based interventions that target young people in a pluripotent risk state needs to be examined in terms of improving cognitive, academic, and social functioning, and for reducing behavioral disturbance and later incidence of serious mental disorder.
4. Child and adolescent services need to be merged with young adult services to better target the population at greatest risk for psychosis and follow them through the full developmental course.

Conclusions

The scientific map of schizophrenia is not a blank sheet. Over the last few decades, major advances in understanding genetic and nongenetic risk factors have been achieved, although some of the discoveries have been a source of frustration for those looking for quick and simple solutions. For example, the genetic architecture has not delivered common polymorphisms with large effect. Early cannabis use appears to be associated with an increased risk of schizophrenia, but population-based cannabis reduction will probably not prevent many new cases. Nevertheless, genetics has finally started to shed light on possible disease mechanisms, whereas research related to trauma exposure, migrant and minority ethnic status, and city birth have put the somewhat neglected area of stress and socially mediated risk factors firmly back on the table for our field. Although much more work remains, we do not face a *terra incognita*, upon which we are doomed to stumble. Rather, we have a map, albeit a very incomplete one. Given what we now know, and with the exponential growth in neuroscience and steady access to new technology, there are good reasons to believe that the challenges we face in schizophrenia research are tractable.

To the junior researchers who are contemplating entering or remaining in this field, we wish to reassure you that you should not feel intimidated by the uncertainties surrounding psychosis, a condition that exposes some of the farthest reaches of what it means to be human. Tenacity and creativity are required to add momentum to our field, and we encourage you to participate in this important and exciting world of research.

Models

10

Human Cell Models for Schizophrenia

Ashley M. Wilson and Akira Sawa

Abstract

Research of mental disorders that affect mainly unique human traits or higher brain function will benefit greatly from the introduction of live human tissues relevant to account for the phenotypes. Human neuronal cell models allow for precise molecular and functional characterization of patient phenotypes and genetic backgrounds. Sources of human cell types discussed here include cellular reprogramming of patient somatic cell lines (either first to pluripotency or directly to neuronal cells) and biopsy of olfactory tissue. Induced pluripotent stem (iPS) cells are particularly useful to study developmental trajectories and functional activity in many disease-relevant cell types. In fact, several attempts have been made to use iPS cell-derived neurons to study schizophrenia and other psychiatric disease. iPS cell technology consists of very high-cost and laborious experiments that may be ameliorated by a recent, more short-term cell conversion technique to obtain directly induced neuronal (iN) cells from somatic cell lines. Moreover, neuronal cells from olfactory epithelium (OE) biopsy have yielded promising research in that they serve as a reasonable surrogate for the brain without adding any genetic manipulation. These human cell models should be integrated with current clinical psychiatric and functional characterizations as well as animal models to progress the translational and clinical applications of basic research.

Why Do We Need Human Cell Models for Schizophrenia Research?

Human cell models offer a promising strategy to study the biology that underlies schizophrenia and can serve to complement animal and computational models. We define human cell models as central nervous system (CNS)-relevant cells that are enriched or reprogrammed or directly converted from biopsied tissues of patients and normal controls.

The Gap between Animal Models and Human Pathology/Biology

Although rodent models are very useful in addressing some key biological mechanisms that are potentially related to human brain disorders, it is not clear whether mouse or rat neurons can faithfully replicate the pathologies of human brain disorders due to the substantial species differences in neurons, including the following:

1. The unique features of enlarged cerebral cortex in humans are formed through distinct developmental mechanisms to generate cortical neurons, compared with those of rats and mice (Hansen et al. 2010).
2. There is evidence that even the same molecule has a differential spatiotemporal expression pattern in neurons in humans compared to rodents. For example, MeCP2, a molecule responsible for Rett syndrome, is known to have a differential expression pattern in human and mouse brains (Shahbazian et al. 2002).
3. A very recent study reports that development and structures of synapses, basic physical compartments in neuron–neuron communications, are different due to the evolutionary changes regarding the *SRGAP2* gene (Charrier et al. 2012).

Thus, human neurons (if they are available) would be very important to elucidate human-specific characteristics of neurons, which may not be fully covered by rodent models alone. Consequently, such cells may be crucial to clarify molecular mechanisms of brain disorders, especially neuropsychiatric disorders in which human-specific traits may be impaired, and to build assay systems for translational use.

Downfalls of Autopsied/Postmortem Brain Studies

Analysis of autopsied human brains has made important contributions to the field. Transcriptome-profiling experiments show widespread, yet specific, gene expression disturbance across the brain, within multiple cell types and biological processes. Data from these gene expression profiles are utilized in rodent models for the study of disease-relevant molecular cascades (Lin et al. 2012). Information from these vast studies can be gateways to animal model research and extremely informative to tissue culture studies and drug development (Horvath et al. 2011).

Postmortem brain studies are useful, but they have many confounding factors: lifestyle differences, age at time of death, cause of death, varying length of disease/age of onset, and various environmental influences including the effects of medication and substance abuse. All of these factors are especially prevalent among schizophrenia samples. Furthermore, we cannot gain any understanding of functionality from the postmortem brain, and there is no tight link to developmental processes, which likely play a key role in the pathophysiology of schizophrenia (Cascella et al. 2007).

Thus, human cell models may provide complementary approaches to obtain disease-associated molecular and cellular changes.

Utility of Human Cell Models

Recent advances in reprogramming and cell culture technologies have allowed us to obtain human-derived neuronal cells (Yang et al. 2011; Dolmetsch and Geschwind 2011). The advantages of this resource should have great implications for psychiatric disorders, such as schizophrenia, that are uniquely human. Findings from research such as the aforementioned rodent studies and human postmortem brain analyses can provide a foundation for studying human cellular phenotypes *in vitro*. Alterations in neuronal properties, such as arborization, synaptic density, neuronal migration, neuronal connectivity, and signaling, have been found in postmortem and rodent model studies for psychiatric disorders. Thus, functional changes in these biological paradigms may be tested by using human-derived neuronal cells. Furthermore, by employing unbiased assays, especially those for molecular profiling, to human neuronal models, we may be able to build novel hypotheses to unravel the pathophysiology of complex psychiatric diseases (Figure 10.1).

The direct utility of neuronal cells from living patients is vast. Functional characterization and cellular properties can be integrated with unique patient

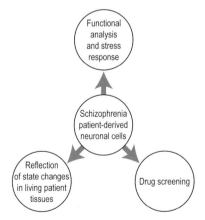

Figure 10.1 Cell models for schizophrenia research. Live sampling of patient tissues provides researchers a unique opportunity to observe functional phenotypes of human neurons. Mechanisms of existing patient abnormalities can be effectively characterized and evaluated at baseline in response to external stimuli and stress. Possible reflection of state changes at the time of biopsy would be useful for longitudinal design, cellular response to treatment as well as identification of biomarkers for diagnosis and prognosis. In addition, live patient tissues would be useful for drug screening and personalized medicine.

attributes such as genes and symptoms. Using electrophysiological recordings, we can address cell autonomous changes as well as functional connectivity among neurons. Easy availability allows for precise molecular and mechanistic characterization of cell lines. Currently, we can link specific genetic abnormalities found naturally in patient samples or through genetic manipulations to live neuronal phenotypes. The advantages of human neuronal models are fully utilized when we wish to address the mechanisms of gene–environment interaction, which is well understood to play a significant role in the etiology of mental illness. Cellular response to stressors or detrimental environmental effects can be implicitly monitored in human cell models. In addition, the translational utility of animal models for bringing new drugs for mental illness to market has fallen short. Testing human cellular response to novel compounds may help to synergize efforts for effective drugs.

Sources of Live Human CNS-Relevant Cells

In this section we describe representative methodologies for human cell models, including their advantages and limitations (see also Table 10.1).

Table 10.1 Comparison among available human neuronal tissues: induced pluripotent stem (iPS) cell, induced neuronal (iN) cell, and olfactory-derived neurons have unique advantages and can be compared to traditional postmortem brain analysis for a better understanding of schizophrenia and other psychiatric diseases. Each cell system also has disadvantages which must be overcome.

	Advantages	Disadvantages
Postmortem brain	• Whole brain • All cell types present	• Many confounding factors • Functional assay not available • Less link to developmental trajectories
iPS cell and embryonic stem cell-derived neurons	• Live neurons • *In vitro* functional study • Examination of developmental trajectories • Directed differentiation of cell type	• Long-term, laborious, and high-cost • Heterogeniety • Epigenetic memory of lineage
iN cells	• Live neuronal cells • *In vitro* functional study • Short-term experiments • Potentially high-throughput	• Limited directed differentiation of cell type • Low conversion rate and maturity of neurons
Olfactory tissues	• Live immature neuronal cells to homogeneity • Homogenous cell population • *In vitro* functional study • Easy for preparation	• May not completely represent brain neurons • May not be able to chase several developmental phases

Induced Pluripotent Stem Cells

Easily accessible patient cells, such as skin fibroblasts, can be reprogrammed to a pluripotent state similar to embryonic stem cells. Theoretically, these iPS cells allow for the production of any cell type from somatic cells. Specifically, iPS cell-derived neurons can give us resources with which to examine developmental trajectories and neuronal functions, in addition to traditional molecular and histochemical tissue analysis.

Advancement of the Technology

First derived from mouse fibroblasts, iPS cells were induced to a pluripotent state by the application of four embryonic stem cell maintenance factors: Oct3/4, Sox2, c-Myc, and Klf4 (Takahashi and Yamanaka 2006). This was quickly followed by iPS cell production from human somatic cells with the four factors: Oct4, Sox2, NANOG, and Lin28 (Yu et al. 2007). Since these first studies, iPS cells can be derived from multiple resources of somatic cells and differentiated into multiple cell types found in brain tissue (Dolmetsch and Geschwind 2011).

Transduction methods have progressed since the early stages of this technology. The most common method of transduction—integration of transcription factors by viral infection—is adequate for disease modeling purposes, though not for potential transplantation applications (Han et al. 2011). Alternative methodologies, such as those using nonintegrating vectors, excisable lentiviral vectors, proteins that are taken up into the cells to facilitate cell reprogramming, or synthetic mRNA, have since been successfully tested in iPS methodology (Warren et al. 2010; Kim et al. 2010; Kaji et al. 2009; Soldner et al. 2009; Yu et al. 2009). Also, small molecules can be used in addition to or in replacement of factors, but small molecule-only transduction is not yet available (Li et al. 2009). If iPS cells are to be used to study reprogramming or developmental mechanisms, it is best to use the most robust reprogramming method to ensure the most efficient transduction. This is usually accomplished by retrovirus or lentivirus transductions. It is desirable to use robust reprogramming methods when using iPS cells for disease modeling and drug screening. However, nonintegrative methods would reduce the amount of heterogeneity and tumorigenicity among resulting cells. Thus, when considering the use of iPS cells for cell therapy, nonintegration methods are much safer.

iPS cells are most frequently generated from fibroblasts, but have also been established from patient blood. Cells collected from fresh peripheral blood can be used to produce iPS cells efficiently using similar methods to fibroblast reprogramming technology, allowing for an additional patient resource to be easily accessed (Loh et al. 2010; Seki et al. 2010; Staerk et al. 2010). iPS cells have also been produced from immortalized blood cell lines (Choi et al. 2011;

Rajesh et al. 2011). Although this is not yet well established, it is an important advancement because it is the resource most commonly stored in most genetic repositories. Therefore, it is very important for the field of disease research to advance this specific reprogramming technique.

Differentiation into disease-relevant cell types (mainly neurons and glia) and subtypes depends on careful protocols that guide cells throughout developmental stages to specific lineages relevant to disease. Neurons are typically produced from iPS cells by first going through a neural progenitor stage, followed by directed differentiation to neuronal subtypes using neuronal transcription factors and inhibitors of other developmental pathways. Resultant cell types are dependent on factors used and precise timing of culture conditions (Han et al. 2011). iPS cells can be differentiated into glutamatergic, dopaminergic, GABAergic, and motor neurons, as well as astrocytes and oligodendrocytes. Careful cell type specification from iPS cells to neuronal tissues of interest is important for the application of this technology to psychiatric disease. Researchers should aim to produce cell types that are directly relevant to the disease of interest to best elucidate disease-relevant mechanisms (Hansen et al. 2011).

Disease Application

Live neuronal cells derived from humans will be particularly useful for studying developmental trajectories and neuronal characteristics of cells from patients with psychiatric disease. Characterization of iPS cell-derived neurons from patients with psychiatric disorders in the autism spectrum and with schizophrenia has been accomplished (Marchetto et al. 2010; Brennand et al. 2011; Pasca et al. 2011). Although the sample size is extremely small, a pioneering study suggests that neurons from patients with schizophrenia show reduced neurite number, overall connectivity, and levels of glutamatergic receptors and postsynaptic density proteins (Brennand et al. 2011).

The study of patient samples with rare genetic mutations may also be a useful avenue for using patient-derived neuronal cells in schizophrenia research. For example, neuronal cells derived from patients with Rett syndrome, an autism spectrum genetic model, showed morphological, electrophysiological, and early developmental deficits when compared to controls (Marchetto et al. 2010). Similar methods could benefit schizophrenia research through the application of established genetic susceptibilities.

iPS cells also allow for analysis of completely different cell types from the same patient. This is especially useful in the study of systemic disorders. In patients with Timothy syndrome, a rare genetic disorder caused by a mutation in the calcium channel $Ca_v1.2$, iPS cells have been used to show abnormalities in both cardiac and neuronal cells (Pasca et al. 2011; Yazawa et al. 2011). In relationship to this study design, increasing evidence supports the notion that schizophrenia might be a systemic disorder, instead of a mere brain disease (Kirkpatrick 2009). Understanding how genetic abnormalities affect different systems and cell types may help to unify some hypotheses and models of schizophrenia.

Technical Limitations

iPS cell technology may have a profound effect on the field of schizophrenia research, but it also has substantial limitations. Production of iPS cells and subsequent cell types requires long-term cell culture, which means that it is a very laborious and high-cost technology. Furthermore, the lifetime of iPS cell cultures is limited, and thus the maturity of neurons produced is also limited. There is a balance between the capabilities of long-term cell culture and maturity of those cells. Improvements to the robustness of conversion through upcoming transduction methods will reduce the strain on research in the future. Commercially available iPS cell lines may help accelerate research in laboratories as well.

There are also limitations within iPS methodologies. Arguably, the most prominent is the heterogeneity between iPS cell clones and the developmental differences which arise from them. Conversion from somatic cell to iPS cell is low within a cell line, and many iPS cells will be cloned from one converted cell. Therefore, any differences that exist between originally converted cells are amplified. Careful selection of clones can limit the impact of this.

Epigenetic memory of iPS cells and cells derived from them constitutes another limiting factor. Although neurons derived from iPS cells show implicitly neuronal phenotypes, it has been shown that, when compared to embryonic stem cells, epigenetic structure of DNA is still related to the somatic cell type of origin and not fully matched to natural stem cells. These DNA methylation signatures can be altered to resemble more closely the signatures of true stem cells through the use of chromatin-modifying drugs or serial reprogramming and differentiation (Kim et al. 2010).

Future Perspectives for Schizophrenia Research

The association of environmental factors in the etiology and manifestation of psychiatric diseases, including schizophrenia, is well established. However, direct links between diseases and stressors have not yet been determined. Therefore, at least for the immediate future, it is best to use cell lines from patients with a defined genetic background, to be certain that the observed cell phenotypes are directly associated with disease phenotypes. Such research may be expanded to genetic models of mental illness, such as 22q11 and 16p11 mutations.

Direct Cell Conversion

Influenced by the idea that somatic cells can be modified to pluripotency, stem cell biologists attempted to directly induce fibroblasts to other differentiated cells, including neurons. In the area of neuroscience, the first successful example was to convert mouse fibroblasts to functional neuronal cells (Vierbuchen et al. 2010).

Advancement of the Technology

Wernig and colleagues (Vierbuchen et al. 2010) produced iN cells from embryonic and postnatal mouse fibroblasts through lentiviral induction of neuronal transcription factors. Screening of combinations of 19 neuronal and epigenetic reprogramming transcription factors led to the discovery of a three-factor system for reprogramming: the combination of Ascl1, Brn2, and Myt11. These iN cells express multiple neuronal markers such as βIII-Tub and MAP2, generate action potentials, and form functional synapses.

Conversion of human fibroblasts to iN cells was soon accomplished by using the same three factors used in mouse experiments (Pang et al. 2011). Ascl1, Brn2, and Myt11 will also convert human iPS cells directly to neuronal cells; however, the addition of transcription factor NeuroD1 was necessary to induce conversion of fetal and postnatal human fibroblasts directly to neurons. In addition to immunohistochemical staining for neuronal markers, iN cells generate action potentials, and matured cells make synaptic contacts. The increased complexity of reprogramming from mouse to human cell lines is an example of the vast evolutionary changes to human biology. This further highlights the importance and utility of obtaining human neuronal cells.

The direct conversion of human somatic cells to neuronal cells is a great advancement for psychiatric disease research, but methods need to be made more efficient and robust for further studies. Higher conversion rates, especially of adult cell lines, and improved maturity of iN cells will be important for schizophrenia studies consisting primarily of an adult population with complex molecular and functional phenotypes to be studied. The field has begun to make efforts in the right direction. For example, miRNA-mediated conversion of fibroblasts to iN cells improves efficiency of human iN cell conversion (Yoo et al. 2011). When combined with the transcription factors NeurD1, Ascl1, and Myt11, miR-9 and miR-124, which have both been shown to be important for neuronal differentiation and development, yield efficient conversion of fibroblasts to iN cells. The addition of small molecules that regulate important pathways in neuronal development has also led to a robust improvement in iN cell conversion (Ladewig et al. 2012).

In regard to neuronal cell type, most investigators used a nondirect approach, including those described above, to produce cultures that coincidently contain a majority of glutamatergic neurons, while obtaining other neuronal types, such as GABAergic and dopaminergic neurons, more rarely. Few groups have directed their conversion to a neuronal fate of interest. However, the addition of the transcription factors FoxA2 and Lmx1a to the original three-factor system directly yields dopaminergic (tyrosine hydroxylase-positive) iN cells (Pfisterer et al. 2011). Furthermore, a minimal set of transcription factors—Ascl1, Nurr1, and Lmx1a—is sufficient to produce induced dopaminergic neuronal cells (Caiazzo et al. 2011). Motor neuronal cells can be produced with the addition of eight motor neuron specification factors (Son et al. 2011).

Disease Application

Widespread application of iN cell technology has not yet occurred in psychiatric disease research. However, Qiang et al. (2011) have observed the conversion of cell lines to iN cells from patients with familial Alzheimer's disease. This is an important proof of concept that adult human iN cells can demonstrate a neuronal cell-specific pathology that can be characterized by histological and electrophysiological methods.

Technical Limitations

iN cell technology is similar to iPS cell technology in its limitations: it still consists of laborious and high-cost experiments, despite having the advantage of being a short-term cell culture. Efficiency of conversion is often low, and maturity of neuronal cells is often limited. In addition, there is a large gap in conversion rate of samples from fetal or newborn patients and adult patients. Thus, the field will need to greatly advance the technology of adult patient cell lines before schizophrenia research can be done well in iN cells. More efficient experiments will eventually relieve this burden from laboratories and make analysis of cultures more informative. Furthermore, epigenetic memory and partial conversion of cells is even more apparent than in iPS cells. In fact, partially converted iN cells can seem to take on an uncharacteristic morphology, yet be positive for neuronal markers like MAP2 (Yang et al. 2011). Researchers may be able to look to improvements in iPS cell technology for help with this. In one iN study, Ladewig et al. (2012) used small molecule inhibitors common to iPS cell neuronal induction for more efficient direct conversion to neuronal cells. iN cell technology is also limited in its ability to reach cell type-specific conversion at high efficiency. For example, in addressing the pathophysiology of schizophrenia, a subclass of GABAergic neurons is needed in parallel to glutamatergic neurons. This may be overcome by referring to novel methodologies that are developed in cell type-specific differentiation of iPS cells.

Future Perspectives for Schizophrenia Research

Cell type-specific conversion will help potential application of iN cells to schizophrenia research. In addition, possible conversion to glial cells, such as astrocytes, microglia, and oligodendrocytes, may also be important for psychiatric disease research and may help to recapitulate schizophrenia as a whole brain disorder. Generation of induced microglia, for example, could benefit a neuroimmune hypothesis of schizophrenia. Furthermore, recent generation of induced neural stem cells from mouse and human fibroblasts could allow for the straightforward study of human neuronal development *in vitro*, omitting the need to begin with a pluripotent state (Ring et al. 2012).

Nasal Biopsy and Olfactory Cells

For over two decades, olfactory neurons via nasal biopsy have been expected as a possible surrogate tissue to study the brain (Trojanowski et al. 1991; Talamo et al. 1989). Due to technical barriers, as described below, this technique has not been widely utilized. Paradoxically, after the limitations of iPS and iN cell technologies became known, the significance of nasal biopsy and olfactory cells has been revisited and underscored.

Advancement of the Technology

The olfactory epithelium (OE) is an easily accessible, direct resource of patient-derived neuronal cells that can be obtained through a simple and relatively noninvasive procedure (Cascella et al. 2007). Early characterizations of OE showed that it is composed of several cell types: structural/supportive cells, neuronal cells which express distinct neuronal markers (e.g., neural cell adhesion molecules and microtubule associated proteins), and basal stem cells that are supposed to give rise to new olfactory neurons (Trojanowski et al. 1991). Thus, immunohistochemical study of OE tissue sections has been used to examine neurodevelopmental processes and disease-relevant molecular changes occurring within the OE tissue of patients and controls (Arnold et al. 2001, 2010). Investigators have also tried to develop methodologies of culturing as well as of differentiating and distinguishing neurons from OE-biopsied tissues.

Following initial studies, improved biopsy methods have increased the amount of neural tissue that can be obtained from one OE biopsy. Introduction of endoscopic sampling from patients have improved the efficiency and quality of samples that had originally been obtained "blindly." Furthermore, biopsy from the dorsoposterior regions of the nasal septum has been shown to increase the probability of obtaining neuronal cells from the tissue (Feron et al. 1998).

Improvements to culture conditions of the biopsy have set the stage for new experiments with human olfactory neurons (Feron et al. 1998). Methodology to culture OE slices has been used to examine *in vitro* cell death, mitosis, neuronal density, and response to a neurotransmitter in both healthy controls and individuals with mental illness (Feron et al. 1999). In addition, OE slice cultures have been used to investigate cell cycle alterations in culture and expression profiles by microarray (McCurdy et al. 2006).

In parallel, efforts to use dissociated OE cell culture for further characterization of the neuronal cells have been made by several groups. Functional activity of olfactory receptor neurons obtained by dissociation of OE tissue has been addressed by measuring intracellular calcium in response to odorants (Rawson et al. 1997; Restrepo et al. 1993). This method was applied to evaluate possible differences in cells from patients with bipolar disorder and controls (Hahn et al. 2005). A more recent study reported that cultures of dissociated OE cells can include neuronal cells that are mature enough to express odorant and neurotransmitter receptors and active signaling mechanisms (Borgmann-Winter et al. 2009).

Cultures from OE tissues can produce neurospheres (i.e., clusters of cells consisting of multipotent progenitors), which in turn generate cells expressing neuronal markers (such as MAP2) and some glial markers (Roisen et al. 2001). Cyclic-AMP, retinoic acid, forskolin, sonic hedgehog, and other media nutrients have improved the neural differentiation and maturation of neuronal cells from neurospheres (Zhang et al. 2004, 2006; Roisen et al. 2001). Interestingly, retinoic acid, forskolin, and sonic hedgehog can elicit motor and dopaminergic characteristics of the neuronal cells, suggesting that olfactory neuronal cells *in vitro* are sensitive to cell fate directions without direct genetic manipulation (Zhang et al. 2006). Isolated cultures of neurospheres produce more numbers of new neurospheres and continue to proliferate over time, and the progenitors from them are restricted to neuronal and glial cell fates (Othman et al. 2005). Progenitor cells obtained from OE neurospheres have characteristics consistent with other stem cells, such as retained telomerase activity and stability of apoptotic activity in culture over time (Marshall et al. 2005a). Neurosphere-derived cells have been used to study gene and protein expression as well as neuronal cell functional activity (Matigian et al. 2010; Fan et al. 2012).

Methods to purify a unique cell population to near homogeniety from biopsied tissues have recently been explored. By using laser capture microdissection, it is possible to purify neuronal layers in which an olfactory neuron receptor marker OMP can be enriched up to thirtyfold more than whole OE tissue (Tajinda et al. 2010). In addition, a protocol that can enrich immature neuronal cells to near homogeneity has also been established (Kano et al. 2012).

Disease Application

Olfactory neuronal cells provide a good surrogate system to study brain disorders, including mental illness. Initially, OE-derived resources, such as cells, were used in research in Parkinson's disease and Alzheimer's disease. Alzheimer's disease is known to accompany odor detection deficits, which provided justification to study OE. In fact, one of the earliest studies with OE found unique pathological changes in the tissue from patients with Alzheimer's disease (Talamo et al. 1989). More recent studies have found that the altered phenotypes found in the OE tissue from patients with Alzheimer's disease correlate with the whole brain pathology that is characteristic of the disease, including amyloid-β accumulation (Arnold et al. 2010). Olfactory dysfunction is also a robust symptom of Parkinson's disease (Doty 2012). Moreover, cells from OE neurospheres from patients with Parkinson's disease show dysregulated gene expression of mitochondrial function, oxidative stress, and xenobiotic metabolism pathways (Matigian et al. 2010), which have previously been linked to Parkinson's disease pathology (Henchcliffe and Beal 2008).

OE tissues could be particularly useful for psychiatric disease research. The psychological processes of motivation, emotion, and fear are closely associated with olfaction as well as the negative symptoms of schizophrenia (Zald and Pardo 1997; Andreasen 1982). Indeed, olfactory deficits have been reproducibly associated with schizophrenia, especially negative symptoms of the disease (Turetsky et al. 2009). A question that arises is whether the olfactory phenotype is due to molecular alteration in OE cells/neurons, or due to a more complicated mechanism, including upstream olfactory circuitry, or both (Sawa and Cascella 2009). A recent study reported that individuals with schizophrenia and their first-degree relatives have different odor detection thresholds for two odorants that differentially activate intracellular cAMP-mediated signaling, indicating that molecular deficits in OE cells/neurons are likely to be, at least in part, associated with the disease pathophysiology (Turetsky and Moberg 2009).

OE biopsies have been used by several groups to study *in vitro* alterations in cells from patients with schizophrenia and other psychiatric disorders. Neuronal cells from the OE tissue show differences in cell adhesion, proliferation, and death in schizophrenia compared to controls (Feron et al. 1999). OE tissues from patients with schizophrenia show reduced density of p75NGFR positive basal cells and increased density in GAP43 positive immature olfactory neurons, as well as increased ratios of immature olfactory neurons or olfactory marker protein positive mature neurons to basal cells, indicating altered development and differentiation within patient tissues (Arnold et al. 2001). In dissociated OE tissue cultures, decreased calcium signaling was observed in cells from patients with bipolar disorder when compared to controls (Hahn et al. 2005). In addition, cell cycle alterations have been found in OE neurosphere-derived cells from patients with schizophrenia and bipolar disorder (McCurdy et al. 2006). Cells showed increased mitosis and expression of cell cycle proteins in patients with schizophrenia, and increased cell death and phosphatidylinositol signaling pathway proteins in those with bipolar disorder. Proliferation rate was also shown to be increased in neurosphere-derived cells from patients with schizophrenia (Fan et al. 2012). Finally, gene and protein expression profiling of OE neurosphere-derived cells from patients with schizophrenia show dysregulated neurodevelopmental pathways (Matigian et al. 2010). In addition to the cellular phenotypes described above, epigenetic profiles of immature olfactory neuronal cells have revealed alterations in oxidative stress response pathways in schizophrenia compared to controls (Kano et al. 2012).

In conjunction with the practicality of OE-derived cells for disease research, OE tissues and cells may also be beneficial for development of therapeutics. The OE itself is a useful area for drug delivery to the brain as it is one of the few areas of the CNS that is readily accessible (Kandel et al. 2000). Therefore, drug development using OE cell culture would provide a direct way to assess the effect of a drug on neural tissue. Consequently, OE-biopsied tissue has previously been used as a tool to evaluate the pharmacological effects of a CNS-acting therapeutics and has revealed biological activity of the astrocyte-targeted drug, thiamphenicol (Sattler et al. 2011).

Technical Limitations

OE tissue can be used to study human neuronal mechanisms and disease characteristics, but it has some drawbacks. First, nobody has fully validated OE-derived cells as CNS neurons that are physiologically relevant at the cell autonomous level or in the context of synaptic connectivity. Second, although OE-derived cells represent a diverse array of molecular signaling pathways relevant to studying brain diseases like schizophrenia, unbiased and extensive studies of whether and how these cells resemble CNS neurons have not yet been conducted.

Future Perspectives for Schizophrenia Research

Given that OE-derived cells are higher throughput, less laborious, less time-consuming, and much less expensive resources when compared to iPS cell-based models, the utility of OE needs be enhanced. To achieve this goal, the resolution of the technical limitations described as above becomes very important. Establishment of further protocols to prepare/enrich homogeneous cell populations, hopefully to fully mature neurons with relevant synaptic formation, is expected. As olfactory deficits are a key phenotype of schizophrenia, especially its negative symptoms, it is very important to study how cellular and molecular changes in OE-derived cells can represent these higher functions.

Link to Animal Models

Gene expression profiling studies of human cells and tissues from patients with schizophrenia and other mental illnesses can provide clues of disease-relevant molecular changes (Lin et al. 2012). However, a major limitation in such human studies is that the information cannot encompass neural circuitry-mediated disease pathology. To compensate for this limitation, human cell study should be linked to research with animal models for the following reasons: First, molecular information obtained from human cell research can be utilized to generate new genetically engineered models, which may be useful in studying the biology that underlies the disease pathology. Second, it will be informative to examine currently available animal models for molecular changes observed in patient cells.

The use of rodent models for schizophrenia research is discussed by O'Donnell (this volume). In addition to rodent models, nonhuman animal models remain an important tool for neuroscience research. With their extremely well-established nervous systems, small animals (e.g., fly, nematode, zebrafish) provide a well-defined substrate for correlates between molecular and cellular processes and behavior (Burne et al. 2011). For example, *Drosophila* (fruit fly) is very commonly used in genetic manipulation studies and can provide relatively high-throughput gene–behavior relationship data for neuroscience studies. For DISC1, transgenic flies have been linked to effects on behavior and pathways for gene transcription (Sawamura et al. 2008). The nervous system of *Caenorhabditis elegans* (nematode) is completely defined down to the cellular level, including its nervous system (White et al. 1986). Although very simple, its nervous system contains neurons that act very similarly to mammalian cells and interact via common neurotransmitters such as glutamate, GABA, and others (Burne et al. 2011). Mechanistic understanding of molecular pathways that are important for psychiatric disease research, such as for DISC1, can be easily observed in the animal (Brandon and Sawa 2011). Furthermore, investigators have paid attention to zebrafish: due to their transparent bodies, brains in these small animals can be observed in intact, behaving animals. In addition, genetic and molecular manipulations of the zebrafish nervous system can be manipulated in the same way as invertebrates, but their nervous system structure and function is much closer to the mammal. Deletion or duplication of the 16p11.2 chromosomal region in humans has neurocognitive effects, which can produce effects of macro- and microencephaly when the human transcript is inserted into the zebrafish genome (Golzio et al. 2012).

Beyond Human Cell Biology: How Can Human Cell Technology Be Used in a More Translational Sense?

As human cell engineering technologies, such as iPS cells, iN cells, and olfactory cells, continue to advance, the molecular signatures associated with schizophrenia should be able to be clarified. Nonetheless, human brain imaging is crucial to address the important question of how such molecular changes at the cellular level affect the brain function and molecular disposition of the same individual from whom those cells were obtained.

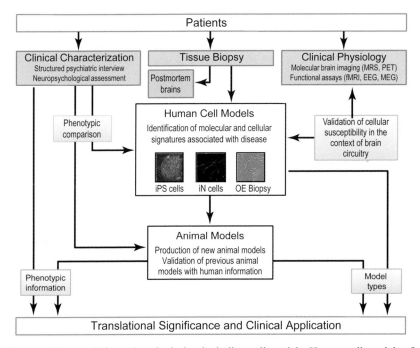

Figure 10.2 Multifaceted study design including cell models. Human cell models of psychiatric disease can be easily integrated with other techniques for better translational research and clinical applications. Aspects of clinical physiology and psychological assessment can be intricately examined at the molecular and cellular levels to clearly identify biological signatures of disease. These cell models can in turn influence and be assimilated with current and future animal models for further understanding of the neurocircuitry and behaviors which those biomarkers represent.

To address brain region-specific molecular changes, positron emission tomography and magnetic resonance spectroscopy are useful modalities. Correlation of molecular and cellular changes in iPS, iN, and olfactory cells with clinical, neuropsychological, and electrophysiological measures will provide us with important information for translational use. A multifaceted study design (see Figure 10.2) involving human cell models represents a promising major approach for schizophrenia research and should be actively pursued. Furthermore, human cell models should be utilized for mechanism-oriented compound screening.

Acknowledgments

Akira Sawa is supported by NIH, NARSAD, Stanley, RUSK, and MSCRF.

11

How Can Animal Models Be Better Utilized?

Patricio O'Donnell

Abstract

Although animal models of schizophrenia have been around for some time and new ones are proposed regularly, their usefulness is still questioned. Many current concepts on schizophrenia pathophysiology have been driven by animal research, yet when these concepts were translated into novel therapeutics, the results have been less than promising. This chapter reviews many of these models and new concepts, and argues that the problem has been that animal models were not used enough in preparation to clinical trials. Furthermore, a great deal of animal work has been directed to establishing their validity—a misguided and far from useful effort. Validity concepts are outdated and not adequate for research relevant to a disorder for which its etiology and pathophysiology are unknown. Models need to be appreciated based on their usefulness: for a disease without a clear pathophysiology, animal models are essential tools to test specific hypotheses about neurobiological and behavioral outcomes of manipulations that produce pathophysiological conditions. Novel targets should only be translated into clinical efforts after comprehensive work in animal models has been conducted to allow establishing mechanisms of action, biomarkers to identify optimal populations to be targeted, and even whether those targets are better thought of as adjuvants or sole treatments. Recognizing what animal models can and cannot achieve will go a long way in benefiting schizophrenia research.

Introduction

Modeling complex psychiatric disorders such as schizophrenia in animals is certainly a challenge. In fact, it could be argued that reproducing this uniquely human disease in animals, and particularly in rodents, is an almost impossible task. However, many different animal models have been proposed and studied over the past few decades. With the advent of genetic models, this field has grown further and new models are proposed almost every month. Schizophrenia research has gained important insight from animal work, and many pathophysiological scenarios previously proposed for this disorder have either been reinforced or dismissed based on animal model studies. For example, the early emphasis on the dopamine hypothesis gave way to the current focus on excitation-inhibition balance, glutamate receptors, and GABA interneurons in cortical circuits. Novel pharmacological approaches have been sought based on animal model work. Unfortunately, these new treatments have failed to provide conclusive results, and some of them are now on the verge of being dismissed. Did we miss the mark? Are animal models misinforming the field, and are we on a wild goose chase? Or is it still too early to jump from animal work to novel therapeutics? Here I will argue that animal models will be extremely important in driving the field forward, but we need to drastically change the manner used to conceptualize them.

Can We Truly Model Schizophrenia in a Rodent?

Perhaps the primary problem with the current use of animal models in schizophrenia research is that we took the concept of modeling disease from the neurology realm. In that field, models are used for their ability to reproduce the disease in animals. For example, Parkinson's disease has several powerful animal models (e.g., 6-hydroxy dopamine in rats, MPTP in mice and monkeys) that reproduce the critical pathophysiology: loss of dopamine cells. These models have contributed to a better understanding of the timing of dopamine loss and its consequences, the role of oxidative stress, and other cellular damaging processes, etc. In Huntington's disease, several different mouse and rat transgenic models with poly CAG repeats in the *huntingtin* gene reproduce a genetic change strongly associated with disease etiology. In both cases, animal models were designed with the goal of very closely reproducing the disease. In these and other areas of medicine, three litmus tests of validity for animal models were developed:

1. face validity, or the ability to reproduce manifestations of the disorder;
2. construct validity, or the fidelity in reproducing disease etiology or pathophysiology; and
3. predictive validity, or the ability to show beneficial effects of drugs that work in the human condition.

While these validity criteria have been the boon of neurology research, they are the bane of psychiatric research. If we as a field believe we can reproduce schizophrenia in a rodent, we are deluded (pun absolutely intended). How can we talk about construct validity for a disease for which we do not know the etiology and have little clues about pathophysiology? Most confusing, why do we emphasize predictive validity when we try to assess aspects of the disease that are not treated well by current medications? We have obtained droves of information with the existing proposed models, and novel models are constantly being added to address genetic, environmental, and developmental factors. It is due time to leave behind the neurology legacy and think about animal models in the frame of psychiatric disorders that need work to elucidate their neurobiological mechanisms. Specifically, we can use animal models as reagents to test defined hypotheses about risk factors or possible pathophysiological scenarios. We can use them to assess what kind of biological processes can be related to the deleterious impact of certain gene variations, environmental insults, or developmental anomalies with the goal of gaining a better understanding of clinically relevant biomarkers. We can test the neurobiological underpinnings of endophenotypes observed in patients (imaging, neurophysiology, and even postmortem) with manipulations that generate specific cellular, synaptic, or circuit alterations in animals and assess whether they yield similar imaging, physiological, or behavioral alterations. All these efforts will be most productive, however, if we do not kid ourselves into thinking that the models reproduce a disease as complex as schizophrenia. Thus, if we move beyond the limiting concept of validity, we can use animals to test hypotheses efficiently in a manner that can help us accept or reject ideas about schizophrenia etiology and pathophysiology, which can then be advanced to human studies.

Despite heated arguments about their validity, many different animal models have indeed provided important insight on possible mechanisms that may contribute to the disease. Several models have been proposed to address environmental, developmental, and genetic factors, as well as the role of specific transmitter systems and brain regions. Experimentalists have been conducting research that provided useful information all along while conceptualizing their research in a house of cards framework of validity. There is no perfect model, and if we are able to escape the validity trap, we can learn something from practically every model proposed. Below I will review some of these models, addressing their usefulness and ability to test schizophrenia-related hypotheses.

Pharmacological Models

Noncompeting NMDA receptor antagonists have been extensively used, and they have provided critical information that led to the formulation of a currently popular hypothesis on schizophrenia pathophysiology: cortical disinhibition. Agents such as phencyclidine (PCP), ketamine, or MK-801 have been used in several species, including humans, to study mechanisms associated with the psychotomimetic effect of PCP that was initially reported in the 1950s (Luby et al. 1959). Although there has been an argument regarding whether acute or chronic NMDA blockade is the more "valid" model, studies with either single dose or repeated treatment have provided data indicating that NMDA blockade results in enhanced glutamate levels in the cortex (Moghaddam et al. 1997), increased pyramidal cell firing, and decreased interneuron firing (Homayoun and Moghaddam 2007). As imaging data have been reinterpreted in the 2000s to qualify the old "hypofrontality" functional concept as the result of a higher level of baseline activity and reduced capacity in prefrontal networks (Callicott et al. 2000), the notion that a psychotomimetic agent such as NMDA antagonist would cause disinhibition seemed to fit well. Therefore, the view that cortical disinhibition may be responsible for cognitive deficits in schizophrenia was driven by animal model work. This is an example of fruitful use of an animal model; of course, now we need to move beyond the initial observations and pose specific hypotheses addressing mechanisms that could result in such a disinhibited state. For example, open questions include whether NMDA receptors in cortical inhibitory interneurons are primarily targeted by NMDA antagonist, causing increase pyramidal cell firing, whether the excitation-inhibition imbalance is the result of a larger network effect instead of selective effects on inhibitory interneurons, and whether cortical disinhibition can be causal to cognitive or other behavioral deficits. All these are testable hypotheses. Only with a better understanding of cellular and synaptic mechanisms yielding a disinhibited cortex will we be able to design better therapeutic tools. In addition, the NMDA antagonist findings have been frequently interpreted as indicating there is something wrong with NMDA receptors in schizophrenia. However, obtaining schizophrenia-related outcomes with a pharmacological blockade of NMDA receptors does not necessarily mean that NMDA receptors are impaired in the disease; reducing function in the receptor population targeted by these antagonists may have a downstream effect that could reproduce schizophrenia pathophysiology without requiring abnormal NMDA receptors in the disease. NMDA antagonist models have been extremely useful, regardless of their validity, and the data obtained with them have driven the field to establish new hypotheses. An example of the leads that NMDA antagonists have opened is the role of immune activation and oxidative stress in vulnerable neuronal populations, as parvalbumin (PV) interneurons are altered by NMDA antagonists in a manner that requires interleukin-6 and oxidative stress (Behrens et al. 2008). The field is now ripe to challenge those hypotheses with further experiments and, in doing so, we may gain insight about neurobiological processes that could play a role in schizophrenia.

Developmental Models

Although NMDA antagonists have provided support for several current concepts regarding the pathophysiology of schizophrenia, these models lack a developmental component. It is now commonly accepted that schizophrenia is a developmental disorder in which a combination of predisposing gene variations and environmental factors may alter neural circuits with a protracted developmental trajectory (Waddington 1993; Pantelis et al. 2005). Although there are cognitive deficits prior to diagnosis, full-fledge symptoms do not appear until late adolescence. This could be due to either delayed deleterious effect of a persistent condition that eventually produces enough changes to alter behavior, or alterations put into evidence late in development by the protracted maturation of cortical circuits. Animal models, again irrespective of how well they fit validity criteria, can be used to test these possibilities. Several models are being used in which a perinatal manipulation is introduced so that behavioral, neurochemical, anatomical, and electrophysiological anomalies emerge during adolescence. The two most extensively used models are the antimitotic methylazoxymethanol acetate (MAM) during gestational day 17 in rats and the neonatal ventral hippocampal lesion (NVHL).

The NVHL model and its variations (intrahippocampal injection of tetrodotoxin, TTX, or lipopolysaccharide) is widely used and, with near 150 publications over the past several years, it is probably the most extensively explored (Tseng et al. 2009; O'Donnell 2012). This model was developed in the early 1990s to test the hypothesis that an altered early postnatal developmental trajectory in a brain region linked to schizophrenia (the hippocampus) results in behavioral anomalies with a delayed onset (Lipska et al. 1992). This is another example of a useful model that provided data beyond simple validation, yielded important information about prefrontal cortical synaptic processes, and added a developmental perspective to the disinhibition hypothesis. At the time the model was generated, the notion that schizophrenia is a developmental disorder had been proposed, but the only evidence available was from postmortem studies which showed altered cytoarchitecture (Kovelman and Scheibel 1984). As those human findings could not be replicated, the neurodevelopmental hypothesis of schizophrenia required testing to affirm its plausibility. Lipska and Weinberger decided to explore the impact of neonatal lesions of the ventral hippocampus and other brain regions on adult behavior as a way to assess whether early alterations could result in deficit with an adult or adolescent onset. The ventral hippocampus in rats was chosen because this region corresponds to the anterior hippocampus in primates, and the early postnatal period was selected for the lesion because it corresponds to the third trimester of pregnancy in terms of brain development. A narrow window was identified in which a lesion would yield adult rats with several behavioral anomalies: postnatal day (PD) 6–8. Adult rats with a NVHL show hyperlocomotion, exaggerated response to stress and stimulants, prepulse inhibition deficits, loss of social interactions, and a variety of cognitive deficits including poor working memory, set-shifting deficits, and reversal-learning deficits, and most of these deficits are only fully observed in adult, not preadolescent, animals (Swerdlow et al. 2001; Brady et al. 2010; McDannald et al. 2011). Furthermore, there have been reports of altered prefrontal cortical circuit physiology, also with adolescent onset. In particular, prefrontal cortical fast-spiking PV-positive interneurons fail to acquire the periadolescent changes in modulation by dopamine (Tseng et al. 2008), rendering adult prefrontal circuits in a state of disinhibition. Indeed, cortical disinhibition can be evidenced in excessive firing of pyramidal neurons during epochs in a choice task that correspond to decision making and high levels of dopamine cell firing, as well as in the loss of beta oscillations during those epochs (Gruber et al. 2010). This is a remarkable convergence with what was previously identified with NMDA antagonists, but is now a consequence of an early developmental manipulation. Again, this is another example of a good use of a model, with a manipulation designed not to produce a disease state but to test a specific hypothesis about pathophysiological processes. In recent years, a great deal of effort was placed on assessing cognitive phenomena in the NVHL model, with the goal of determining whether cognitive constructs altered in schizophrenia show deficits in the model as well and, if so, whether novel therapeutic ideas could be beneficial in these animals as a way to test these new approaches in a diseased brain. Thus, a model that has been frequently sidelined (despite being extensively studied) because of the perceived lack of validity due to the "lesion" aspect has been extremely useful in demonstrating that early developmental perturbations can indeed yield late onset behavioral deficits; it has also reproduced cognitive deficits that can be linked to phenomena observed in schizophrenia. This model has serious shortcomings in terms of validity (a lesion is not normally part of schizophrenia), but it has nonetheless been extremely useful. Indeed, the model should not be interpreted as reproducing hippocampal pathology in the disease; its consequences are most likely due to the impact of altering hippocampal function influence on the development of downstream structures such as the prefrontal cortex (PFC). The NVHL model has provided important information on cellular and systems elements that contribute to adult cognitive deficits and is providing interesting data on the potential role of immune activation and oxidative stress in interneuron deficits (O'Donnell et al. 2011). This model may be useful in addressing such open questions as whether cortical fast-spiking interneurons are an early factor that, when affected, drives altered excitation-inhibition balance; whether cellular processes (including, but not limited to immune activation and/or oxidative stress) are responsible for the behavioral deficits; and whether other interneuron types may be affected and give rise to the deficits. Currently the NVHL model is also used to screen for efficacy of novel compounds targeted to improve cognition in schizophrenia. Thus, in spite of validity shortcomings, this and other models have been useful for testing hypotheses and gaining insight.

Another valuable developmental model is the administration of the antimitotic MAM at gestational day 17 in rats. For decades, the administration of MAM at early gestational dates was used to study cortical development; in the 2000s, a slightly later date of administration proved to cause delayed onset of behavioral deficits similar to those observed with the NVHL model (Flagstad et al. 2004). The impetus for the MAM model was to test whether a developmental manipulation that did not entail an explicit lesion could produce the emergence of schizophrenia-related anomalies in adolescence and early adulthood. Although it could be argued that by avoiding a lesion, a "shotgun" approach of impaired microtubule function in the entire brain was introduced, data indicate that the deficits seem prominent in the hippocampus and PFC regions, suggesting a degree of selectivity on the impact of the MAM treatment (Moore et al. 2006). As in the NVHL model, here we have a developmental manipulation designed to test the impact of early deficits on adult behaviors. Adult offspring of MAM-treated dams exhibit hyperlocomotion, enhanced reactivity to stress, prepulse inhibition deficits, loss of PV immunostaining, loss of high-frequency oscillations, and cognitive deficits (Flagstad et al. 2004; Gourevitch et al. 2004; Moore et al. 2006; Penschuck et al. 2006; Lodge et al. 2009). This model is also extremely useful in providing the opportunity to link early developmental deficits with adult dysfunction in dopamine systems. The ventral hippocampus is critical in driving the activity of subcortical dopamine projections, and the altered VH function induced in adult rats by the gestational MAM treatment results in excessive activity in subcortical dopamine systems (Gill et al. 2011). This model is also used for drug screening, and it is another example of clever experimental design to address the possible contribution of biological processes to altered functions that may be relevant to schizophrenia. Although the MAM model has validity issues, it has proven extremely useful in testing specific hypotheses and has provided insight regarding possible pathophysiological processes and their behavioral consequences.

Environmental Models

Several models have been developed to test the possible impact of environmental factors hypothesized to play a role in schizophrenia. Epidemiological data indicate a strong association between schizophrenia and maternal or perinatal infection or parasitic disease. It has been hypothesized then that immune activation during early development may yield altered brain circuitry that could be relevant to schizophrenia (Brown 2006). Several animal models were designed to test this hypothesis, including gestational administration of the viral particle poly I:C or the bacterial endotoxin lipopolysaccharide (LPS). Adult offspring of treated dams express a variety of behavioral deficits such as reduced prepulse inhibition, altered latent inhibition, and several other indicators of cognitive function (Zuckerman et al. 2003; Meyer et al. 2006). Furthermore, recent work with these models reveals loss of PV immunostaining in prefrontal cortical regions (Meyer et al., unpublished data), providing a remarkable convergence in key pathophysiological observations with several other models. Immune activation has strong epidemiological support, and testing its impact in animals may reproduce a causal or predisposing factor (Meyer and Feldon 2012). However, beyond the real or perceived validity of these models, their usefulness resides in their ability to test specific hypothesis about the neurobiological impact of a factor with strong contribution to the disease. Unveiling the cellular and systems neuroscience aspects these manipulations produce will certainly advance our understanding of the neurobiological processes likely to be affected in schizophrenia.

Other environmental factors proposed to play a role in the disease have also been modeled in animals, including vitamin D deficiency, pre- or postnatal stress, gestational hypoxia. Although less studied, these models are no less important. If we accept that we do not need to validate models in terms of disease reproducibility and that models are useful tools to test specific questions about consequences of possible pathophysiological scenarios, then all environmental-based models have an important role to play.

Genetic Models

Perhaps the group of animal models that has grown most rapidly is the cluster of genetic manipulations possibly associated with schizophrenia. Although schizophrenia is a disorder with a clear genetic predisposition, the role of genes is complex. Although the common view involves interactions among multiple gene variants, each contributing a very small risk, and environmental factors, recent work has identified a few genetic modifications with high penetrance. These include chromosome deletions such as the 22q11 and other copy number variants (CNVs). As schizophrenia-predisposing gene variations continue to be identified, mouse models expressing such variations are developed. This is a long list that cannot be addressed in its totality. Examples of single-gene mutations with suspected link to the disease include dysbindin (for which knockout mice exist), DISC1 (for which several manipulations also have been used in animals), and neuregulin. Unfortunately, a great deal of effort has been placed on proving the validity of these models. Even if we admit that individual genes may contribute only a small fraction of the risk, studying the neurobiological processes triggered by altering the ERB4 gene or the DISC1 gene is extremely useful. By gaining such basic understanding, we can then link these genes with cellular activity, brain circuit function, and animal behavior in a manner that can illuminate about factors that can be affected in the disease.

Several genetic manipulations have recently been used to test specific hypotheses about the impact of a specific gene variation on neurobiological processes. For example, among the diverse genetic variations that confer risk for major psychiatric disorders stands a truncated DISC1 gene. A Scottish family with a chromosome translocation in which 70% of its members present with schizophrenia or bipolar disorder permitted the DISC1 gene to be identified as one of the truncated genes in the translocation (Millar et al. 2000). Interestingly, the protein encoded by the DISC1 gene proved critical for NMDA synapse development and cortical interneuron function. A mouse overexpressing a truncated DISC1 gene, which acts as a dominant negative, produces several behavioral, neurochemical, and electrophysiological changes that emerge in the adult animal and are shared by several other animal models (Hikida et al.

2007). Another genetic model was produced to test the hypothesis that deficits in NMDA receptors in cortical inhibitory GABA interneurons can selectively produce abnormal behaviors and schizophrenia-relevant endophenotypes. The obligatory NR1 subunit of NMDA receptors was knocked out of PV interneurons in the cortex, resulting in loss of high-frequency oscillations, reduced prepulse inhibition, and altered cognitive functions (Belforte et al. 2010). Again, this is a hypothesis-testing use of an animal model that is not constrained by lack of validity. One could argue that there is no loss of NMDA receptors in schizophrenia, but these mice have been critical to show the impact of altered interneuron function on a number of schizophrenia-relevant phenomena, thereby proving useful to test hypotheses about loss of PV interneuron function.

Finally, there is strong impetus in testing mouse models that recapitulate rare, highly penetrant gene variations. Mice with a microdeletion in chromosome 22 (22q11), similar to what in humans produces a high incidence of schizophrenia, have shown altered PFC–hippocampal synchrony (Sigurdsson et al. 2010), thus providing a link between a gene variation with strong association with the disease and a relevant pathophysiological construct. Several open questions can be addressed with the diverse genetic models available today, such as why mutations can in so many different genes lead to a common pathophysiology or whether there are convergent biochemical/cellular pathways or developmental processes affected by different genetic manipulations. These are examples of possibilities in which the hypothesis-based use of animal models can help move the field forward.

Biomarkers and Endophenotypes

Perhaps the best use of animal models is to test hypotheses related to biological processes that can underlie schizophrenia endophenotypes and ultimately to help identify biomarkers that can be associated with endophenotypes and pathophysiological conditions. For example, the currently popular notion that cortical disinhibition is critical for cognitive deficits requires extensive animal work to be translated in more efficacious treatments. As inhibitory interneuron deficits may be a central tenet of the disinhibition scenario, animal models which test the impact of altered interneurons will be extremely useful in determining a variety of outcomes that can be related to schizophrenia phenomena. Many reports have emerged over recent years of altered cortical oscillations in diverse models that affect cortical interneurons, thus opening the door to establishing clinical neurophysiological readouts of interneuron deficits. More, however, remains to be done. Although there are animal studies using EEG and auditory evoked potentials in a manner similar to what is used in schizophrenia patients, these studies typically employ intracerebral or subdural electrodes. The signal obtained with these electrodes is clearly stronger but may differ greatly from the scalp recordings used in humans. A more human-like recording strategy (i.e., outside of the skull) is required for EEG and related signals to become more easily translatable. Such an approach would allow animal work to unveil neurobiological processes related to human neurophysiological signals and to understand processes that can alter them.

Another theme that is gaining ground in schizophrenia research is the possible role of immune activation, inflammation, and oxidative stress in the disorder. This is an area in which animal model work will be extremely important. Pending questions include whether inflammation and oxidative stress can be expressed selectively in interneurons, providing a link to the disinhibition hypothesis and perhaps information regarding mechanisms that can yield disinhibition. In addition, we need to establish whether inflammation and oxidative stress can yield cognitive and behavioral anomalies. By testing these questions in animal models and learning about biological processes associated with these variables, we may gain information regarding human biomarkers and how they relate to endophenotypes.

A theme that the current research with animal models should incorporate is the role of dopamine. Most recent animal model work has concentrated on cortical GABA and glutamate. Although these are clearly important players with a critical role in cognition, the link between dopamine and positive symptoms cannot be discounted. It is essential that dopamine systems gain more prominence in animal model work. With the emergence of the disinhibition hypothesis, the dopamine hypothesis seems to have taken a backseat. There are many open questions that need to be answered to obtain a better integrative view of GABA, glutamate, and dopamine systems. Can dopamine alterations emerge as a consequence of cortical disinhibition or are they unrelated? Does dopamine play a role in putting disinhibition into evidence? Animal testing of positive symptoms is problematic; arguably, they cannot be reproduced in a rodent. However, if we focus less on the validity of the models and more about using manipulations to test hypotheses related to the role of dopamine in behavior, we may be able to obtain information that can subsequently be used to guide human studies.

Animal Models and Novel Therapeutics

Ultimately, animal modeling should be at the service of novel medication development. Although a large number of targets (e.g., GABAergic, cholinergic, glutamatergic) have been identified using the models described above, we have so far failed to identify useful targets. For example, as cortical disinhibition hypothesis gained support with animal work, it was reasonable to consider developing new compounds that targeted a disinhibited cortex. Although there were some promising leads, such as the initial report of a metabotropic glutamate agonist mGluR2/3 having similar efficacy as olanzapine (Patil et al. 2007), others failed (e.g., Buchanan et al. 2011). Furthermore, subsequent studies with the mGlu2/3 receptor did not provide conclusive data. Many factors played a role in this process: some trials were underpowered or showed a high placebo effect, patients were not selected according to specific biomarkers, etc. As these efforts are costly, some are now cautioning against the use of animal information to drive human trials. I argue that these efforts were conducted too early, with little biological information other than a hypothesis developed based on an array of data. More work needs to be done in animal models to determine whether reducing excess glutamate or increasing GABA-A tone does restore excitation-inhibition balance and, if they do, what are the optimal tools (mGluR agonists vs. allosteric modulators; what GABA-A receptor selectivity works). Furthermore, schizophrenia is a heterogeneous disease; it is more likely part of a continuum of neurobiological processes which spans across other related psychiatric disorders. We need to embrace heterogeneity in animal studies and design hypotheses to illuminate how biological processes (and potential treatment targets) can cause diverse sets of clinical outcomes. On one hand, disinhibition may be a feature of bipolar disorder and autism; on the other, there may be a subset of schizophrenia patients in which cortical disinhibition is a prominent feature and others in which it is not. The same goes for dopamine alterations or any proposed pathophysiological scenario.The field, therefore, needs to identify biomarkers that can be associated with pathophysiological conditions. If we are able to determine EEG signals, evoked potentials, imaging alterations or cognitive tests that have a strong correlation with disinhibition in animal models, we can then use those markers to select patients for trials based on biology. Finally, when considering the cognitive realm, it is possible that any benefit of novel agents may be offset by the deleterious impact of traditional antipsychotics on cognition if the trials were designed with the new drugs as adjuvants. Animal models could be useful in determining whether differing effects can be expected from isolated or adjuvant administration of a particular novel compound. Animal models need to be used differently and more extensively before moving on to the next generation of treatment. This would permit trials to use the most likely to succeed targets, schedules, and patient population.

Conclusion

Animal models of psychiatric disorders are important, and it is crucial that we have a diverse set of tools to test biological processes relevant to these disorders. If we knew what the pathophysiological processes in schizophrenia were, we would only need one or a few models to reproduce it. But we don't. Therefore, we need to avoid the pressure of having the "most valid" model and instead use the models to explore specific hypotheses about the contribution of different factors, from genes to the environment. Animal models can also be better employed to seek correlates of neurobiological processes with readouts that are similar to human biomarkers. To achieve this, we need to replace the notion of validity with usefulness. A useful model would allow us to test the impact of factors that are hypothesized to play a role in the etiology or pathophysiology of schizophrenia. A useful model would also allow us to study neurobiological processes that are affected by any suspected factor without being limited by a perceived lack of validity. If we recognize that schizophrenia cannot be reproduced in animals, then we are free to use animal manipulations to explore biological processes that can have relevance to schizophrenia as well as other psychiatric disorders. The affected neurobiological processes in these studies could then be tested in patients using imaging or other techniques. By de-emphasizing the validation aspect of a model (i.e., the need to mimic the human disease), we can move the field forward by using animal manipulations to test hypotheses about the roles of genes, development, neurotransmitters, or environmental factors.

Why are clinical trials on novel compounds that were designed on the basis on animal data not working? Briefly, concepts developed with animal work were taken to the clinic too early. All of the attempts based on the disinhibition hypothesis were doomed to fail because they were predicated on small pieces of evidence; we did not have a complete understanding about the mechanisms that were yielding to the disinhibition observed in different models. More work with models is needed, for example, to test whether cortical disinhibition is indeed responsible for cognitive deficits and to elucidate the cellular and/or systems mechanisms that may yield disinhibition or any other pathophysiological construct following developmental, genetic, or environmental manipulations. Only with that information will we be able to understand the neurobiological mechanisms of stimulating mGluR receptors or enhancing GABA-A receptor activity. However, we should not throw out the baby with the bathwater. As our gaps in knowledge are being filled, novel compounds can still be tested on those models in which a pathophysiological state relevant to their targets is present. The pharmaceutical industry needs to invest more heavily in testing compounds in animal manipulations which model the pathophysiology intended for the new agent and, most importantly, which identify biomarkers that can be associated with a positive effect of these agents. Only then can a sufficiently powered clinical trial be conclusive in accepting or rejecting a particular target.

How, then, can animal models be better utilized? The answer is simple: by using them to test specific hypotheses related to the flow of etiological/risk factors from pathophysiological processes to behavior and clinical manifestations.

Acknowledgments

I thank Kevin J. Mitchell, André Fenton, Danielle Counotte, Thibaut Sesia, Meriem Gaval, and Hugo Tejeda for very useful comments on an early version of this manuscript.

12

How Can Computational Models Be Better Utilized for Understanding and Treating Schizophrenia?

Daniel Durstewitz and Jeremy K. Seamans

Abstract

This chapter discusses computational neuroscience approaches which could be used to establish mechanistic and causal links between structural, biophysical, and biochemical factors of the underlying neural hardware, the dynamic properties implementing computational operations, and their relationship to cognition and behavior. This process is illustrated using an example relevant to schizophrenia: the bidirectional dopamine regulation of dynamic network regimes in prefrontal cortex and their relation to higher cognitive functions like working memory and flexibility. Thus, dynamic system properties (like attractor states or bifurcations) provide the glue between neuronal hardware and cognitive function. Importantly, they are not mere abstract mathematical concepts, but rather properties which can be derived from experimental measurements. This way computational tools may help gain a mechanistic understanding of how various schizophrenia-related biochemical and genetic changes could be related to the functional and cognitive deficits, and could be used to develop novel treatment options by identifying yet unknown parameter configurations that reinstall "healthy dynamics."

What Is Computational Neuroscience and What Types of Questions Can It Answer?

Computational neuroscience has become a broad field, with major contributions from (theoretical) physics, computer science (informatics), mathematics, biology, and psychology, and has grown tremendously, especially over the last decade. Common to all areas of computational neuroscience is the usage of theoretical, mathematical, and computational methods to address questions about brain function. There are at least two major aspects to this field. The first is the idea that the brain is fundamentally a computational system; that is, a system that tries to compute (in an algorithmic sense) from sensory inputs sensible behavioral outputs, given certain internal states and stored information (Figure 12.1). At this level, the types of questions are:

- What mappings (functions) between input and output patterns does a specific molecular network, a single cell, a local circuit, or a given brain region perform?
- How are these mappings implemented at a biophysical, biochemical, anatomical, etc., level?
- What constraints exist on the types of mappings that can be realized by the neural system under consideration?

For instance, a single neocortical pyramidal cell receives spatiotemporal patterns of synaptic inputs driven by spikes, and transforms these patterns into temporal patterns of axonal output spikes. Are these mappings linear or are they (highly) nonlinear? Do they acknowledge the temporal and/or spatial structure in the inputs, or do they somehow average across temporal and/or spatial aspects? Can they be cast in terms of classification or regression problems? How do the input/output transformations performed by the cell arise from the mathematical operations represented by various voltage-gated ion channels (e.g., fast Na^+ channels impose a kind of binary threshold operation on synaptic inputs, either strongly amplifying them or letting them decay away)? Thus, the computational neuroscientist tries to translate the behavior of a given neural system into a (class of) mathematical function(s), and attempts

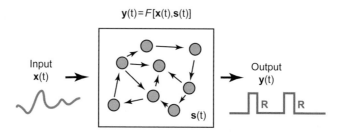

Figure 12.1 Understanding the brain as a computational system. Brain functions may be mathematically characterized as mappings, F, between spatiotemporal patterns of sensory inputs, $\mathbf{x}(t)$, internal states of the system, $\mathbf{s}(t)$, and behavioral outputs, $\mathbf{y}(t)$. The goal is to work out the exact mathematical form of these functions, F, what class of functions a given neural system can compute, and which ones it cannot, and how these are implemented in terms of the underlying component operations and biophysical/biochemical mechanisms. The considered system does not have to be the whole brain at once, of course, but may consist of single brain areas, local networks (e.g., columns), single neurons, or simply subcellular molecular networks.

to work out how this function is implemented in terms of the mathematical operations realized by biophysical and biochemical components.

A second major theme in computational neuroscience is the idea of gaining insights into neural system function through simulation. The brain is a highly complex system consisting of millions to billions of constantly interacting nonlinear feedback loops, organized at many different interacting levels of spatial and temporal resolution (from molecular networks to single cells, to local neuronal ensembles, to brain areas, to systems of interacting brain areas and so forth). To gain mechanistic insight into the workings of this system requires more than intuition based on experimental data. For instance, increasing excitatory input to pyramidal neurons *in vivo* or pharmacologically enhancing NMDA currents might be seen as a means to enhance their spiking activity, but we may paradoxically find that the firing rate actually decreases strongly. Among the various possible reasons, this may be because we overlooked the fact that Na^+ channels severely inactivate at constant levels of higher depolarization, or because this leads to enhanced Ca^{2+} influx which in turn triggers long-lasting Ca^{2+}-dependent K^+ currents, or because we underestimated the feedback inhibition from GABAergic interneurons. Because of the many positive and negative feedback cascades at work in a fully functional neural network (in contrast perhaps to its dissected components in some experimental preparations), we will inevitably run into plenty of apparently paradoxical and counterintuitive effects.

This limitation may become particularly severe once we try to understand how different levels of nervous system description link up with each other; that is, how for instance particular cellular components, like AMPA channels, contribute to neurodynamic phenomena like oscillations, or how neurodynamic phenomena in turn are related to specific behaviors. In our view, current understanding is mostly correlational in terms of how different levels are related to each other, and, in most cases, lacks truly mechanistic insights. We may know, for instance, that turning a specific biophysical knob (say, increasing NMDA conductance) has a specific effect on slow oscillations or behavior in a specific task, but we might not understand why this is the case and how these effects are specifically mediated. It is clear that such a level of mechanistic understanding could, in principle, greatly boost our ability to devise very specific interventions into the system that would alleviate or abolish particular dysfunctions. One of the reasons why simulations of neural systems could be a particularly powerful tool is that, unlike experimental systems, one has full access to every and each single variable: one can both simultaneously monitor all dynamic variables (e.g., the membrane voltages and ionic conductance of all cells) as well as independently manipulate each single variable or parameter of the system. One can even introduce artificial manipulations or constructs into the system (e.g., insert purpose-designed ion channels into subpopulations of cells to test a specific hypothesis, set up arbitrary interneuronal wiring diagrams, or regulate the tone of different modulatory inputs in a very specific way). This maximum degree of control over the simulated system permits us to address research questions that are far beyond the scope of experimental techniques. In experimental setups, even with the most advanced present-day techniques (like optogenetics or multi-tetrode recordings), only a tiny minority of the system's variables are directly observable or simultaneously measurable. Computational models allow us to obtain insights into details of the system dynamics and mechanics, or to derive systematic simplifications of it to improve understanding, which otherwise might not be obtainable through purely experimental work. In the natural science or medical context, it is, of course, clear that computer simulations are only tools to gain insight into experimental data, and that the predictions yielded require experimental verification and approval.

In summary, computational models of neural systems are particularly suited to gain mechanistic and causal insight into the relation between various levels of nervous system description: how specific constellations of biophysical, biochemical, and anatomical factors give rise to specific classes of dynamic phenomena, and how these, in turn, relate to behavior.

Neural Dynamics, Information Processing, and Cognitive Function

A common concept in theoretical neuroscience is that cognitive and computational properties of the brain are implemented in terms of the system dynamics. Mathematically, a neural substrate may be described by a system of dynamic equations (differential equations) which govern the evolution of the system's variables (e.g., the membrane voltages or firing rates of neurons) in time and space. The space spanned by all the dynamic variables of the system is called the *state space* (Figure 12.2a), since a point within this space fully describes the state of all the system's dynamic variables at a given point in time. As time passes, the firing rates, membrane potentials, ionic conductance, etc., of the neurons will change, and so as the system's state wanders through the state space governed by the system's dynamic equations and external influences, it will follow a particular path or *trajectory* through this space (Figure 12.2a). State spaces of complex dynamic systems are filled with diverse geometrical objects that determine the flow of the trajectories, the most important being *attractor states*. To get an idea, consider the electrical circuit for a passive patch of cell membrane governed by the simple first-order differential equation, $C_m \, dV/dt = g_L(E_L - V)$, where V is the membrane potential, C_m the membrane capacitance, g_L the conductance of the so-called leakage channels in the membrane, and E_L the associated leakage reversal potential. If $V < E_L$ (i.e., below the leakage potential), the temporal derivative dV/dt will be positive and V will grow in time until $V = E_L$, where $dV/dt = 0$. If $V > E_L$, dV/dt will be negative and V will decay back to E_L. The state $V = E_L$ is thus an attractor state of this simple system, a state which "attracts" nearby trajectories that will converge to it (or will converge back after a small perturbation). In this simple example,

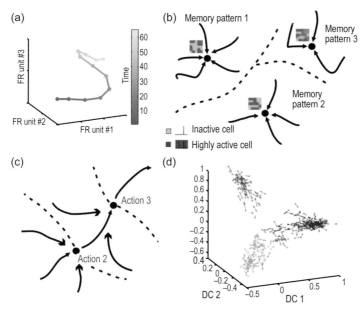

Figure 12.2 Concepts in neural system dynamics. (a) Any (bio-)physical system that can be described by a set of differential equations ("equations of motion") may be represented in a state space that spans all dynamic variables of the system. In the example shown, the space is spanned by the instantaneous spiking rates of all neurons in the system; for the purpose of visualization just three here. A point in this system uniquely characterizes the current state of the system (in this case, the spiking rates of all three neurons). A trajectory in this space is a path laid out by temporally consecutive system states; that is, it describes the evolution of the system's state in time, as illustrated. Modified from Durstewitz and Balaguer-Ballester (2010). (b) State spaces are filled with different geometrical objects which govern the flow of trajectories. One important class of such objects is attractor states or (sets of) points toward which neighboring states converge in time. The domain of convergence is called the attractor's basin of attraction. In this example, several attractors with their associated basins of attraction (delineated by dashed lines) are shown. In neural terms, each attractor state may correspond to a specific, distinct pattern of neuronal firing rates, which in turn may represent a specific memory pattern. Convergence to a specific attractor state would thus correspond to the process of memory retrieval, initiated for instance by external cues. (c) A behavioral action sequence or "memory chain" may be represented neurodynamically as a sequence of transitions among attractor states. (d) Attracting states are not mere theoretical constructs; they are an inherent property of natural systems that can be described by sets of (nonlinear) differential equations, and can be measured experimentally. In this example, attracting states were reconstructed from multiple single-unit recordings *in vivo* from rats performing a working memory and decision-making task. Using dimensionality reduction techniques, a three-dimensional space was obtained from 10–30 simultaneously recorded units. Arrows give the direction of flow at each point in the space, and the vector length indicates the velocity of flow. Activity tends to converge on one of several attracting states, with the system's state changing quickly when it is far away from any attracting state, and slowing down as it approaches the center of a state. Each of the shown states corresponds to a specific cognitive act (e.g., choice, reward, memory period). After Balaguer-Ballester et al. (2011).

the attractor state is just a point ($V = E_L$) in state space (thus called a "fixed point attractor"); however, in general, attractor states can have more complicated geometries like closed orbits (associated with oscillatory activity) or fill up fractal volumes in state space ("chaotic attractors"). This simple example also highlights the point that attractor states or other dynamic properties are not pure mathematical inventions or metaphors for the system's behavior. They do exist in biological (and other physical) systems and are experimentally measurable entities (in this case the stable resting potential of a cell).

A common idea in neurocomputational theory (e.g., Hertz et al. 1991; O'Reilly and Munakata 2000) is that attractor states represent outcomes of cognitive computations and that transitions among attractor states represent computational processes. For instance, a neural attractor state may correspond to a pattern of firing rates across a set of neurons representing a retrieved memory item (Figure 12.2b), and a sequence of transitions among attracting states may implement a sequence of behavioral actions (Figure 12.2c). Neural attracting behavior associated with cognitive processing can, in fact, be extracted from experimental data (Balaguer-Ballester et al. 2011), as shown in Figure 12.2d. In this case, neural population activity patterns are derived from multiple single-unit recordings from the rodent prefrontal cortex and embedded into a lower dimensional state space as illustrated in Figure 12.2d. The convergence of neural ensemble activity to specific attracting states is illustrated by the arrows plotted at each point in space (the so-called flow field) which indicate the direction into which the system will move when in that state, while the vector lengths give the movement velocity at those points. The different states toward which neural activity evolves correspond to different cognitively defined task stages, such as choices the animal makes, rewards it receives, or a delay phase associated with working memory load. Although extracting such dynamic behavior from experimental data may require special mathematical tools (Balaguer-Ballester et al. 2011), this example demonstrates that attractor dynamics is not merely an abstract mathematical concept; it has a clear and experimentally measurable correspondence in neural systems.

How Computational Models Could Enhance Understanding of Schizophrenia

How, then, are the concepts introduced in the last sections relevant to schizophrenia? A wealth of disparate and often seemingly unrelated changes have been linked to schizophrenia: changes in transmitter and neuromodulatory systems (Carlsson and Carlsson 1990b; Benes and Berretta 2001), in molecular pathways (Turner et al. 1997), in cellular morphology, anatomical properties, and synaptic connectivity (Meyer-Lindenberg et al. 2001; Penzes et al. 2011), across a variety of different brain regions (most prominently the prefrontal cortex, hippocampus, and striatum). There are also a number of diverse genes, some of them related to the systems mentioned above (e.g., COMT alleles and dopamine), that seem to increase the risk for schizophrenia significantly (Meyer-Lindenberg and Weinberger 2006). Many of these diverse physiological and structural effects are likely to interact in some nonintuitive way. Thus their net effect on network dynamics and computation is hard to predict. However, these types of alterations in "neuronal hardware" can usually be translated into neural network simulations, either directly through their known biophysical mechanisms, or indirectly through proper data-based phenomenological models. Thus, in principle, computational models could provide a platform to examine systematically the impact of some of the physiological, structural, and gene-related factors, or various combinations of them, on network dynamics and computation. Thereby, such studies would enhance our mechanistic understanding of how changes in biophysical and biochemical parameters associated with schizophrenia could give rise to the dynamic underpinnings of this disease (e.g., changes in 40 Hz oscillations; Uhlhaas and Singer 2010), and through those on neural information processing and cognition. To integrate or "simulate" the impact of comparatively "soft" social and environmental risk factors (Lederbogen et al. 2011; Tost and Meyer-Lindenberg 2012) in neurocomputational models, however, is much harder or even impossible, unless these can either be distilled to very simple differences in sensory stimulation patterns (mimicking an experimental situation in the lab) or translated into specific hardware alterations (which, ultimately, of course, their longer-term effects will come down to, although these changes may sometimes be quite subtle). Likewise, questions of the etiology of the disease may be more difficult to map into a computational framework, as it will be hard to simulate whole etiological trajectories of neural hardware changes, especially if these are partly driven by environmental inputs. Insight may be gained, however, by taking snapshots of the neural hardware at different times during the development of the disease.

To illustrate these ideas, we use alterations in prefrontal cortex (PFC) dopamine modulation as an example, being aware that prefrontal dopamine dysregulation is only one of many possible contributors to the disease. Dopamine inputs to the PFC arise mainly from the midbrain ventral tegmental area and act through two major (metabotropic) receptor classes, abbreviated D1R and D2R. Stimulation of either D1R or D2R triggers a barrage of molecular processes which ultimately result in a number of profound changes in the biophysical parameters of ionic conductances, voltage-gated and synaptic ones, in both pyramidal cells and interneurons (Yang and Seamans 1996; Seamans et al. 2001a, b; Gorelova et al. 2002; Gao and Goldman-Rakic 2003). The first important point to emphasize here is that dopamine is not a classical neurotransmitter. Its cellular actions are neither properly described as being excitatory nor as being inhibitory; in fact, its direct influence on membrane potential is minimal (e.g., Yang and Seamans 1996). Dopamine is a true *neuromodulator* in the sense that it does not directly excite or inhibit a cell, but rather modulates the way it responds to inputs by changing the biophysical properties of various currents. For instance, D1R stimulation enhances both NMDA (Zheng et al. 1999; Seamans et al. 2001a) and $GABA_A$ synaptic currents (Seamans et al. 2001b); at the same time it diminishes synaptic release probability (Seamans et al. 2001a), enhances certain classes of high voltage-activated Ca^{2+} channels while diminishing others (Young and Yang 2004), reduces slow potassium currents (Yang and Seamans 1996), and so on, while D2R stimulation often has effects on cellular parameters and synaptic currents opposite from those of D1R action (e.g., Trantham-Davidson et al. 2004; Zheng et al. 1999). Hence, the cellular and synaptic actions of D1R and D2R stimulation are manifold and often opposing, so that it is hard to make sense of their functional meaning. However, biophysical network simulations offer a framework for interpreting these physiologically measured effects in a functional and information processing context. In a biophysical network simulation of PFC, we observed that most of the physiological effects of D1R stimulation, although apparently quite disparate, seemed to converge onto a common function (Durstewitz et al. 2000; Durstewitz and Seamans 2002). In systems-dynamical terms, this led to a "deepening" and "widening" of PFC attractor states associated with spontaneous or active memory-related firing patterns (Figure 12.3a, b); that is, to an increased "energy barrier" between states. In contrast, D2R stimulation, based on its mostly opposing cellular and synaptic effects, "flattened out" the PFC attractor landscape (i.e., led to a reduction of the "energy barrier" between states). Functionally this means that in a D1R-dominated mode it is much harder to switch between different attractor states (corresponding to memory items, goal states, etc.; see discussed in the previous section). In turn, this may, for instance, improve working memory performance by shielding active memory states against distraction and noise (Durstewitz et al. 2000; Durstewitz and Seamans 2002). Conversely, in a D2R-dominated mode it becomes much easier to switch among attractor states, potentially enhancing functions like set shifting or memory search (Durstewitz and Seamans 2008).

These alterations in network dynamics caused by D1R (or D2R) stimulation are a result of the fact that several of the D1R-induced (D2R-induced) biophysical effects act synergistically on the dynamics, not necessarily the physiology. For instance, in biophysical terms, the D1R-mediated enhancement of NMDA conductance, which increases with membrane voltage, leads to a strengthening of recurrent excitation specifically in currently active memory states, whereas the concurrent enhancement of $GABA_A$ current leads to further suppression of competing states. Thus, although these two effects seem physiologically opposing (one being excitatory, the other inhibitory), dynamically speaking they lead to the same thing: an increased "barrier" between currently active and inactive states. One outcome of this is that D1R effects on neural activity are state-dependent: currently active neurons that are embedded into an active cell assembly may increase their activity further, while rather inactive neurons which fire at a spontaneous level may be further diminished in their activity, or at least show much less of an increase. This has indeed been observed experimentally *in vivo* (Lavin et al. 2005; Vijayraghavan et al. 2007). Perhaps even more surprisingly, however, the dynamic effects conveyed by D1R and D2R stimulation could theoretically happen without any obvious changes in neural firing rates at all: the attractor states shown in Figure 12.3a, b may not change their (R_{PC}, R_{IN}) position in the space of pyramidal cell and interneuronal firing rates, yet the associated basins of attractions may still be deepened, thus

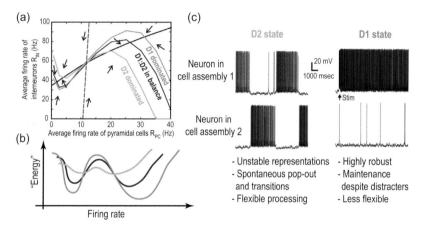

Figure 12.3 Dopamine modulation of PFC attractor landscapes. (a) A cortex-like network of pyramidal cells and interneurons is represented here by a two-dimensional state space projection which is spanned by the average firing rates of the pyramidal (x-axis) vs. interneurons (y-axis). Arrows indicate that activity in this space converges to one of two possible attractor states associated with low vs. high firing rates. These points of convergence are formally given by the intersection of two lines (called "nullclines"): one gives the steady-state firing rates of the pyramidal cells as a function of a fixed average rate of the interneurons (bluish and greenish curves); the other shows the steady-state firing rate of the interneurons as a function of a fixed pyramidal cell rate (black curve). Hence, where these lines intersect, both pyramidal neurons and interneurons are in their steady states yielding a "fixed point." The two regions of convergence for the low- and high-firing rate attractors are separated by the black dashed line. While D1 stimulation leads to a stretching of the pyramidal cell nullcline along the x- and y-axes, D2 stimulation leads to a contraction along both dimensions. (b) Representation of the information from (a) (with corresponding line colors) in terms of an "energy landscape" (schema). Minima of the energy correspond to the fixed point attractors in (a); the state of the system may be envisioned as a ball rolling down into the nearest minimum. The local slopes in this graph depend on the sign and magnitude of the derivatives of the underlying system, as given by the flow field (cf. Figure 12.2d). The graph makes clear that it becomes much harder to switch between different attractor states in the D1-dominated regime as the troughs move apart and the "valleys" become much steeper. Conversely, in the D2-dominated regime, the valleys become so flat and nearby that noise may easily push the system from one state into the other. (c) Network simulation illustrating the fact that the system spontaneously switches or cycles among different attractors (neural representations) in the D2-dominated regime, while robustly maintaining a once elicited attractor in the D1-dominated regime. Figure modified from Durstewitz (2007) and reprinted from Durstewitz and Seamans (2008) with permission from Elsevier.

increasing the "energy barrier" between states. Functionally, this would have the same consequences as discussed above. This has fundamentally important implications for *in vivo* electrophysiology: functionally highly relevant changes in network dynamical properties may not necessarily be picked up by simply measuring changes in neural spiking rates! Instead, in such circumstances, perturbation experiments, now that they are feasible through optogenetic tools (Fenno et al. 2011), may be necessary to reveal the functional implications.

Previously we have summarized the observations above derived from biophysical network simulations fed by *in vitro* electrophysiological measurements under the term "dual-state theory of prefrontal dopamine function" (Durstewitz and Seamans 2008). We have proposed that there is a D1R-dominated state that enhances stability (of, e.g., working memory contents) and a D2R-dominated state that facilitates flexibility (switching among states). These two receptor types are commonly co-localized on prefrontal neurons, but they possess different affinities for dopamine within different affinity states (Durstewitz and Seamans 2008; for a review, see Richfield et al. 1989). Thus, one factor that could regulate the relative balance between D1R and D2R stimulation, and hence shift the network among different modes of computational operation, is simply dopamine concentration. Having established this biophysical and network dynamical framework, one could evaluate the functional consequences of different D1R versus D2R receptor densities in PFC, or changes in their affinities or affinity states. For instance, moving some D2R from low- into high-affinity state, as has been linked with psychosis (Seeman et al. 2006), may lead to a relative dominance of the D2R mode, which could give rise to positive symptoms. If prefrontal attractor basins become permanently very shallow (resulting in very labile attractor states), frequent switching, working memory deficits, fleeing thoughts, and highly associative, incoherent thinking may be the cognitive consequence. Even hallucinations may theoretically be explained in this way, as perception-related attractor states may pop out spontaneously when noise drives the system across one of its shallow "energy valleys." Conversely, an increase in D1R density, which has been proposed in schizophrenia (Weinberger 1987; Abi-Dargham and Moore 2003), may lead to D1R dominance and thus to network dynamical regimes which, in extreme cases, may lead to persistence of a few dominant attractor states, to perseveration, to strongly internally driven dynamics with low sensitivity to external inputs, and consequently to behavioral effects such as flattened affect and lack of social interaction characteristic of negative symptoms.

Regardless of whether dopamine imbalance in PFC contributes to schizophrenic symptomatology in this way or not, this discussion illustrates how network dynamical implications may be derived from biophysical factors, how these may translate into functional and thus ultimately cognitive-behavioral consequences, and how variations in the underlying biophysical/biochemical factors may explain alterations in cognitive processing via their impact on network dynamics. This general approach could be applied to any other set of biophysical, biochemical, or anatomical/morphological factors that have been found to be altered in schizophrenia. For instance, changes in the GABAergic system that have been widely described in schizophrenia (Benes and Berretta 2001) could be implemented in a quite straightforward manner by altering GABAergic cell numbers, synaptic inputs, and/or conductances in the model networks. Likewise, effects of schizophrenia risk genes, like COMT, DRD2, or CACNA1C whose role in regulating specific ion channels or transmitter systems is relatively well understood, would map directly onto corresponding changes in ion channels (e.g., conductance of high voltage-activated Ca^{2+} channels) or transmitter systems (e.g., basal and induced levels of dopamine receptor activation) in the model. Risk genes or other molecular factors, for which the precise biophysical mechanisms of action are less well understood, may still be captured by phenomenological models that represent their effects on excitatory and inhibitory synaptic currents and input/output functions of neurons, provided these have been determined electrophysiologically. One particular advantage of the modeling approach, however, is that some or all of these diverse neurobiological and genetic effects can be studied for their *combined* effects on network dynamics and computation.

How Computational Models May Help to Develop Novel Treatments for Schizophrenia

Finally, we comment briefly on how computational models may be used to develop novel therapies and treatments for schizophrenia. Here the fact may be exploited that in a computational model one can independently access and manipulate each single parameter of the system, or can even introduce novel parameters which do not (yet) have a clear physiological equivalent. The basic idea would be to start with a network configuration that mimics the illness state; that is, one where structural (anatomical, morphological, connectivity) and physiological parameters of the model have been set in accordance with our knowledge about how these neural system characteristics are altered in schizophrenia (see previous section). If our knowledge about these neural hardware changes is sufficiently comprehensive and complete, the resulting model network should exhibit dynamic signatures and computational deficits that are characteristic for schizophrenic patients (e.g., alterations in oscillatory activity, deficits in working memory performance). If our knowledge is still rather incomplete, the model may guide us toward further neurobiological factors that may be altered in schizophrenia, by means of systematically examining in the model which parameter changes could potentially contribute to experimentally measured alterations in functional and behavioral variables. In this latter case, a phenomenological approach (see previous section) may again be an alternative if decent animal models are available for which cellular and synaptic changes can be characterized electrophysiologically without precise knowledge about the underlying biophysical and biochemical mechanisms.

Either way, once we have a reasonable network model of schizophrenia in place, we can systematically scan the model's parameter space for configurations that restore a "healthy" network dynamic in the system—one that is characteristic of a normal operating mode in "control" networks. This can and should be done in close interaction with pharmacologists, who would guide the selection of suitable parameters toward those which are, in principle, pharmacologically accessible (this does not mean that a specific drug already needs to be available, but only that it is conceivable from a biochemical perspective). Returning to the dopamine example, in addition to obvious manipulations like alleviating a hyper-D1 state through D1R antagonists, we may find that a pharmacological cocktail designed to partially block both NMDA and $GABA_A$ receptors would do the same job. We may even find that variations in cellular or synaptic parameters completely unrelated to those affected by dopamine receptor stimulation may lead to the same system's dynamical outcomes. For instance, reducing Na^+ or low voltage-activated (T-type) Ca^{2+} conductance in both pyramidal cells and interneurons (thus reducing their input/output gain) may potentially have similar consequences for prefrontal attractor dynamics as those conveyed by $NMDA/GABA_A$ changes. Hence, we would exploit the fact that there are likely several different biophysical/biochemical routes toward inducing similar modifications in network dynamics and thus function. Another particular advantage of computational approaches is that many different parameter configurations could be probed over a relatively short time period, compared to animal experiments, for their potential to alter dynamical regimes. In addition, they obviously come at much lower cost and could facilitate the testing of potential compounds utilizing fewer animals.

Conclusions

We began with a brief introduction into the field of computational neuroscience, and how computational approaches could be used to establish mechanistic and causal links between structural, biophysical, and biochemical factors of the underlying neural hardware, the dynamic properties implementing computational operations, and their relationship to cognition and behavior. We illustrated this process along a particular example relevant to schizophrenia, namely bidirectional dopamine regulation of dynamic network regimes in PFC and their relation to cognitive functions, such as working memory or set shifting. In this context, we wish to emphasize again that these dynamic system properties provide the glue between neuronal hardware and cognitive function; they are not mere abstract mathematical concepts, but rather properties that can be derived from experimental measurements. These tools might be useful in gaining a mechanistic understanding of how various schizophrenia-related biochemical and genetic changes could be related to the functional and cognitive deficits observed in this disease. In addition, they could be used to derive novel treatment options by identifying yet unknown parameter configurations that reinstall "healthy dynamics."

Computational neuroscience has traditionally focused more on basic neuroscientific questions, such as mechanisms of neural coding or information processing. Most of the research that targets psychiatric conditions is very much in its infancy. Nonetheless, the basic computational tools needed to address questions of direct relevance to schizophrenia are mostly in place and can be utilized to enhance our understanding of schizophrenia, and with it possibly yet undiscovered treatment paths.

Acknowledgments

This work was supported by grants from the German Federal Ministry for Education and Research (BMBF, 01GQ1003B) and the Deutsche Forschungsgemeinschaft (Du 354/6-1 and 7-2).

First column (top to bottom): Kevin Mitchell, Jay Gingrich, Akira Sawa, Patricio O'Donnell, Bita Moghaddam, Joshua Gordon, and Akira Sawa
Second column: Patricio O'Donnell, André Fenton, Bita Moghaddam, Bill Phillips, André Fenton, Wolfgang Kelsch, and Daniel Durstewitz
Third column: Bita Moghaddam, Joshua Gordon, Wolfgang Kelsch, Jay Gingrich, Bill Phillips, Daniel Durstewitz, and Kevin Mitchell

13

How Can Models Be Better Utilized to Enhance Outcome?

A Framework for Advancing the Use of Models in Schizophrenia

Kevin J. Mitchell, Patricio O'Donnell, Daniel Durstewitz,
André A. Fenton, Jay A. Gingrich, Joshua A. Gordon,
Wolfgang Kelsch, Bita Moghaddam,
William A. Phillips, and Akira Sawa

Abstract

The heterogeneity of schizophrenia at the clinical and etiological levels presents a huge obstacle to understanding the biology of this disorder, or even knowing how to conceptualize it. This chapter discusses how animal, cellular, and computational models can be used to explore convergence at the intervening level of pathophysiology. It considers such models as experimental platforms to investigate specific neurobiological hypotheses, in particular to elucidate causal chains of pathogenic events, from initial molecular and cellular disruptions to eventual effects on neural networks and brain systems underlying specific symptom domains. The ultimate goal is to increase understanding of the neurobiological underpinnings of all aspects of the disorder (etiology, pathogenesis, pathophysiology, symptomatology) to a point where we can rationally identify new therapeutic targets or points of intervention to help break the deadlock in the development of treatments for this devastating disorder.

Introduction

What is the point of making an "animal model of schizophrenia"? What are we hoping to accomplish? Is it even possible? What is it that we are really trying to model?

We propose that animal models are best considered as experimental reagents or platforms to investigate the neurobiological underpinnings of schizophrenia. This contrasts with the idea that animal models in some way recapitulate the disorder in its entirety or are mainly useful as a proxy for drug screening. Despite the commonly used shorthand, it is obviously not possible to generate an animal model of schizophrenia, given its etiological and phenomenological heterogeneity, and considering the uniquely human expression of so many of its symptoms. Moreover, if schizophrenia is an open construct, the boundaries and features of which are difficult to delimit even in humans, attempting to generate an animal that recapitulates the disorder as a whole is even more unrealistic.

The approach we propose is generally fairly agnostic about *face* and *predictive validity*, terms which have preoccupied the field for some time. Face validity means that the animal presents with some behavioral phenotypes that resemble particular human symptoms. Predictive validity refers to those phenotypes that can be reversed in the animal model using current antipsychotic medications. While such information is indeed very valuable and reinforces the notion that one is on the right track, face and predictive validity are not good exclusion criteria for saying whether an animal is really a "model of schizophrenia."

The expectation that a particular pathophysiological disturbance will manifest in an overtly similar way in animals and humans is not always justified. On the contrary, one might more reasonably expect a species-specific expression at the behavioral level. Manipulations that do not result in obvious face validity should thus not be rejected as irrelevant to understanding the disease. Similarly, limiting oneself to studying only those phenotypes that are responsive to current medications—especially using them to screen for drugs—inevitably becomes a circular exercise and may explain why no new drugs with novel mechanisms of action have been found using this approach (Carpenter and Koenig 2008; Abbott 2010).

We emphasize a different approach and propose that the term *animal model* be used to refer to an animal that has been manipulated in some way that is either known to be of etiological relevance to schizophrenia or that is thought to recapitulate a phenotype of relevance to some aspect of schizophrenia phenomenology. Different models may be useful for investigating etiology, pathogenesis, pathophysiology, or other aspects of the disease. As such, they represent discovery platforms to test specific hypotheses and elucidate the underlying biology.

In addition to the use of animal models, this research framework importantly includes human cellular models, such as neural cells derived from schizophrenia patient biopsies, for example, and computational models, which can be used to formally describe the interactions within and between levels of biological phenomena and to predict the effects of manipulations of various components. In this chapter, we present a conceptual framework for relating different levels of analysis of experimental models (genetic, molecular, cellular, circuits, systems, behavioral) and for encompassing the heterogeneity that is apparent at each level.

A Heuristic Framework for Schizophrenia Research

The clinical picture of schizophrenia is one of heterogeneity at the level of clinical symptoms (in terms of the particular profile of symptoms portrayed by any individual patient) as well as at the level of etiology, with a large number of distinct risk factors identified. Thus, to think of this heterogeneity while retaining the integrity of the central construct poses a major challenge. At the present time, the degree of heterogeneity at the intermediate level of pathogenic and pathophysiological mechanisms is largely unknown. Our working hypothesis is that there will be some reduction in heterogeneity at the level of pathophysiology, with convergence onto a smaller set of common mechanisms underlying various symptom domains. In this section, we consider how experimental models can be used to approach this question empirically.

Figure 13.1 presents a conceptual framework that encompasses these parameters in animals, based on a similar framework described for humans (see Corvin et al., this volume). A diversity of etiological risk factors (E_1–E_n) may

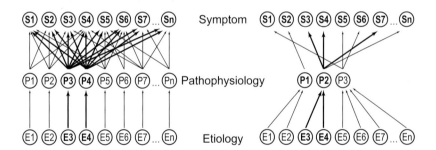

Figure 13.1 Etiology–Pathophysiology–Symptoms (E–P–S) framework. Two alternative scenarios are presented that relate the heterogeneous etiological factors associated with schizophrenia to the heterogeneous clinical symptoms (or behavioral phenotypes in an animal model). The scenario on the left depicts equivalent heterogeneity at the intervening level of pathophysiological mechanisms. Thus, pathogenesis arising from etiological factors E3 and E4 involves distinct pathophysiological mechanisms, P3 and P4. The alternative hypothesis is illustrated on the right, in which the degree of heterogeneity at the pathophysiological level is drastically lower, with phenotypic convergence onto a smaller set of common mechanisms which underlie diverse clinical symptoms. In this case, E3 and E4 induce a common pathophysiological mechanism. Note that the level of pathophysiology itself has multiple hierarchical levels (not shown), with possible convergence from various etiological factors at the level of biochemical pathways, cellular or developmental mechanisms, or emergent neural dynamics in microcircuits and extended brain systems.

impact a range of molecular and cellular processes, leading to the emergence of a spectrum of pathophysiological phenotypes at the level of neural circuits and brain systems (P_1–P_n). These phenotypes may singly, or in combination, lead to the range of clinical symptom domains observed in patients or to the impairment in the animal equivalent of such systems (S_1–S_n). The question is whether there exists for each etiological factor a distinct and unique route of pathogenesis, or if there is instead some convergence onto a smaller set of pathophysiologies. Conceptualizing schizophrenia models within this Etiology–Pathophysiology–Symptoms (E–P–S) framework will be advantageous to elucidate neurobiological mechanisms of relevance to the disorder.

Epilepsy provides a useful exemplar to illustrate how such convergence can emerge (Figure 13.2). There are a large number of Mendelian conditions in which recurrent seizures are one of the clinical symptoms. The genes involved can be roughly subdivided based on the cellular level phenotypes observed or the protein function, including, for example, genes involved in proliferation or

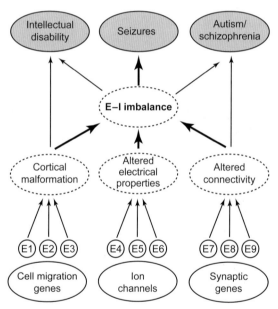

Figure 13.2 Epilepsy as an example of phenotypic convergence. Multiple strong genetic risk factors for epilepsy fall into several categories based on the functions of the encoded proteins (with the obvious potential existence of many more than are depicted). Mutations in genes within each of these groups may converge onto a distinct primary pathology affecting a particular cellular process, such as cortical morphogenesis, ionic flux, or synaptic connectivity. There may be further convergence in the downstream consequences of these changes, which may all lead to an alteration in the excitation-inhibition (E–I) balance in various parts of the brain and a predisposition to seizures. Depending on the pathophysiological mechanism and its penetrance, additional clinical symptoms may also emerge, including ones associated with intellectual disability, autism, schizophrenia, and other psychiatric conditions.

cell migration, which can lead to cortical malformation when mutated, genes affecting synapse formation, metabolic genes, and genes encoding ion channels (Poduri and Lowenstein 2011; Greenberg and Subaran 2011). The heterogeneity of etiological factors can thus be reduced by defining gene function or direct phenotypic effect. Further reduction in heterogeneity is observed at the next level as each of these kinds of disturbance can result in a state of altered excitation-inhibition balance in some part of the brain, resulting in seizures. This is, of course, a superficial level of description—there are certainly distinct ways in which this balance can be disrupted—but it encapsulates a common theme: a type of common pathophysiology that can emerge from diverse primary insults affecting quite different cellular parameters (cytoarchitecture, synaptic connectivity, metabolic flux, or ion channel expression). At the symptoms level, there is also heterogeneity in the type and location of seizures and course of epilepsy. In addition, some genes that predispose to epilepsy also increase risk for other neuropsychiatric disorders, with a number of manifestations other than seizures (including autism, intellectual disability, and psychosis), emphasizing the point that none of these clinical categories is a closed construct.

Populating the E–P–S Framework

We can already begin to populate this framework for schizophrenia at various levels, based on information from diverse sources. At the etiological level, we now know of multiple strong genetic risk factors (Mitchell and Porteous 2011; Sullivan et al. 2012a), in addition to a number of loci with statistical evidence of association from genome-wide association studies (Sullivan et al. 2012a) and a multiplicity of environmental and experiential factors identified from epidemiology (Tandon et al. 2008; McGrath and Susser 2009).

Currently there are at least nine specific recurrent copy number variants for which there is compelling statistical evidence that they predispose to schizophrenia with relatively high penetrance, dramatically increasing risk compared to the general population (Sullivan et al. 2012a). Most of these, however, are also associated with other clinical outcomes, including autism spectrum disorder, epilepsy, and intellectual disability, adding another degree of heterogeneity to the E–P–S framework. Schizophrenia is thus just one possible endpoint caused by mutations in such genes. In addition to these, many other mutations have been identified where the statistical evidence for association with schizophrenia, in particular, is not yet compelling but where the aggregate evidence of some neuropsychiatric manifestation, including schizophrenia in some carriers, is quite strong (e.g., DISC1, SHANK2 and 3, CNTNAP2) (Mitchell 2011a). Regardless of how many cases of schizophrenia will eventually be shown to be associated with such mutations of strong effect, their identification provides an entry point to elucidate the underlying mechanisms experimentally.

As an example of the value of this approach, the identification of the genes underlying Mendelian forms of Alzheimer's disease, including APP, presenilin-1 and presenilin-2, opened an entire field of biological inquiry and ultimately revealed the involvement of these proteins much more generally in this disease (Bertram et al. 2010). We can hope for similar progress in schizophrenia research by following the strong leads we now have in hand. The recent identification of strong etiological risk factors provides an opportunity to follow a proven discovery path in schizophrenia research (Mitchell et al. 2011). It will be especially informative to compare the phenotypes in such models with those observed in well-characterized models generated by pharmacological, anatomical, or environmental manipulations. Such models have proven extremely informative in defining potential pathophysiological mechanisms and relating them to behavioral phenotypes (see O'Donnell, this volume).

At the level of pathophysiological mechanisms, there are also a number of good leads that can be included to help generate testable hypotheses. Pathophysiological mechanisms can be multilayered, with molecular phenotypes yielding synaptic and cellular alterations, which in turn drive circuitry and systems changes. At the molecular level, examples of leads include interleukin-6 and oxidative stress (Behrens and Sejnowski 2009) as well as NMDA receptors (Belforte et al. 2010). At the circuit level, leads include alterations in GABAergic interneurons (Gonzalez-Burgos et al. 2011; Lewis et al. 2005) and dopamine systems (Lisman et al. 2008; Howes and Kapur 2009; Grace 2010). Systems pathophysiological mechanisms currently studied include alterations in cortical or thalamocortical oscillations (Lisman 2012; Uhlhaas and Singer 2012) and hippocampal-prefrontal connectivity (Sigurdsson et al. 2010).

At the level of clinical symptoms and their behavioral correlates in animals, a range of well-established paradigms are available where phenotypes are consistently or at least repeatedly observed across various animal models, including genetic, pharmacological, developmental and others. These include behavioral traits, such as general hyperlocomotion, increased anxiety, and reduced social interactions, as well as task- or challenge-related phenotypes, such as working memory deficits, sensitivity to amphetamine, impaired prepulse inhibition, and others (van den Buuse 2010; Moore 2010; Young et al. 2010). Again, none of these is seen in all models nor should any of them be thought of as an exclusive criterion of the validity of any particular model. Some of them can be related quite directly to human traits, tasks, or psychological constructs, whereas for others a direct parallel is less obvious.

The E–P–S framework includes not just the specific factors at each level but also the known or putative relationships between factors at different levels. These represent the links in the causal chain (or network) from each etiological factor to the clinical manifestation. Any one of those putatively causal arrows represents a specific hypothesis that may be directly testable with the range of reagents and techniques we can now bring to bear in experimental neuroscience. Such hypotheses will be most precisely and tractably defined between adjacent levels of the framework, rather than stretching to test relationships across distant levels, where intervening, unknown complexities may exist.

Identifying Convergent Pathogenic Mechanisms

As stated above, a major goal in the experimental modeling of the effects of schizophrenia risk factors is to identify points and pathways of phenotypic convergence and possibly common pathophysiological states. The identification of such hubs would importantly provide new potential points of therapeutic intervention to reverse or compensate for a particular pathophysiological state that underlies one or more symptoms, or to prevent the emergence of such a state. A key component of such a research program is therefore to provide systematic comparison across multiple models in search of points of convergence at various levels.

Convergence may emerge in some cases at the level of primary cellular mechanisms mediated by the mutated genes. For example, several implicated genes, including *NRXN1* and *CNTNAP2*, play a role in cellular interactions at the synapse (Mitchell 2011a), which may mediate synapse formation and activity-dependent refinement. Members of the SHANK, DLG, DLGAP, and CNTN protein families may act in similar cellular processes, possibly even in the same biochemical pathways (Betancur et al. 2009; Ting et al. 2012). Mutation of other genes, such as DISC1 or CHRNA7, may also have an effect on synapse composition through different molecular pathways (Brandon and Sawa 2011; Lozada et al. 2012). Convergence on particular processes and pathways from analyses of multiple single-mutation models will also highlight potential molecular and cellular phenotypes to assess using human-derived cellular models, where oligogenic effects may be explored (see below).

In other cases, the primary molecular and cellular mechanisms may be very different, but there may be convergence at a higher level of the framework. For example, several models show alterations in gene expression and function of inhibitory interneurons in prefrontal cortex. These include mice expressing dominant-negative DISC1 (Hikida et al. 2007; Shen et al. 2008) as well as amphetamine-sensitized rats (Peleg-Raibstein and Feldon 2008) or rats with prenatal or neonatal manipulations that affect prefrontal cortical and hippocampal development, such as the antimitotic MAM or a neonatal hippocampal lesion (Lodge et al. 2009; O'Donnell 2011). Alterations in inhibitory neuron markers are one of the more consistently observed differences in postmortem studies of human patients and could represent homeostatic responses to reductions in pyramidal neuron activity (Gonzalez-Burgos et al. 2011).

Changes in dopaminergic signaling in striatum and cortex are also observed across many models (Lipina et al. 2010; van den Buuse 2010; Seeman 2011), paralleling consistent observations in human patients, including at prodromal stages (Howes and Kapur 2009; Howes et al. 2012a). Again, such changes could be induced secondarily through reactive mechanisms (Lisman et al. 2008; Grace 2010).

At an even higher level, changes in neural synchrony and oscillations are observed across several models. Neural dynamics at this scale are an emergent property of neuronal ensembles and may be affected by diverse insults. For example, a common pathophysiological state at the level of neuronal populations can emerge due to quite distinct effects at the single neuron level of various psychotomimetic drugs with different modes of action (Wood et al. 2012). Synchrony of neural oscillations may enable communication within and across regions that underlie various aspects of cognition, perception, and behavior. Defects in hippocampal-prefrontal cortex synchrony have been observed in animals modeling the 22q11 deletion (Sigurdsson et al. 2010), in animals that received a neonatal hippocampal lesion (Lee et al. 2012), and in animals subject to maternal immune activation *in utero* (Dickerson et al. 2010, 2012). Such changes correlate with defects in working memory and parallel observations in humans (Meyer-Lindenberg et al. 2001).

Although many details of the causal chains of events remain to be elucidated, these examples illustrate the kinds of explanation that might emerge within this framework and suggest specific and testable hypotheses at multiple levels. Importantly, more selective experimental manipulations in models present the opportunity to move beyond observational approaches and correlations to test causality directly across levels. For example, transgenic animals lacking a particular protein only at some stages or only in some cell types or regions provide tremendously powerful reagents to causally link specific cellular phenotypes to specific pathophysiological outcomes.

Sources of Phenotypic Variability

In considering the relationship between any genotype and an associated phenotype, it is important to consider not just the starting and ending positions, but also the developmental trajectory which connects them. This is especially relevant for the study of schizophrenia, where we know that phenotypic heterogeneity is high among carriers of the same mutation and even between monozygotic twins. How such variable expressivity might manifest in inbred mouse lines is an open question and an important one to keep in mind. Phenotypes may change on different genetic backgrounds, so a profile observed in one strain may not represent a ground truth.

The eventual phenotype may also be affected by environmental risk factors or experience and stress. Animal models provide a powerful platform to test for such effects, especially using animals that may have been sensitized by a "first hit," such as a predisposing genetic mutation (Oliver 2011). Incorporating possible interactions between genetic and environmental risk factors into animal modeling will be an important goal within this framework.

Another important source of variability may be far harder to control or study, however, and that is chance. The processes of neural development are incredibly complex, involving the activities of thousands of different molecular components. These processes are sensitive to what engineers call "noise": random thermal fluctuations at the molecular level which affect gene expression, protein interactions, and other molecular activities on a moment-to-moment basis (Eldar and Elowitz 2010). Such noise can affect the outcome of developmental processes, which can readily be observed at the neuroanatomical level as a probabilistic expression of cellular phenotypes across a population of cells (Raj and van Oudenaarden 2008). When this randomness is played out independently across the brain, it can lead to variation on a macro scale and variation in concomitant physiological and behavioral phenotypes (Mitchell 2007). For example, while the tendency to develop epilepsy is very strongly heritable, the precise type and anatomical focus of seizures are much less so (Corey et al. 2011). These parameters are far more affected by randomness in developmental outcome. One could certainly imagine how a similar scenario played out across other brain circuits could account for some of the variability in presentation in schizophrenia (Woolf 1997; Singh et al. 2004; Mitchell 2007). In animal studies, this variability could be a problem when phenotypes are compared across groups of animals. Alternatively, it could be leveraged by studying individual animals in greater detail, allowing correlation of the severity of defects across levels.

Pleiotropy and Cascading Effects

It is interesting to consider the possible relationships between different phenotypes observed in particular mutants. Co-occurrence of particular behavioral phenotypes could reflect

- a defect in a single underlying neural system on which they both rely,
- the independent expression of a single type of defect in multiple regions of the brain, or
- multiple mechanisms that are independently affected by mutation of the gene (mechanistic pleiotropy).

For example, an alteration in dopamine-mediated signal transduction can influence multiple cognitive functions, such as attention and working memory, stress reactivity, reward and motivational processing, and goal-directed movement (Howes and Kapur 2009; Stephan et al. 2009; Fletcher and Frith 2009).

It is also possible that a single type of defect arises in multiple parts of the brain, with diverse consequences. For example, we can hypothesize that a deficit in visual contour recognition observed in some patients may reflect an alteration in microcircuitry within the primary visual cortex (Butler and Javitt 2005). The identical microcircuit alteration in the prefrontal cortex, however, might alter synchrony of neural ensembles and also functional coherence with the hippocampus, thus causing a defect in working memory.

An alternative scenario is that a particular gene may be involved in quite different cellular processes in different contexts, including in tissues outside of the brain, and pleiotropic effects may arise through very different cellular mechanisms. DISC1 is a prominent example because it interacts with a wide range of proteins in various cellular processes (Soares et al. 2011). This kind of mechanistic pleiotropy is obviously even more likely when the effects of copy number variants are considered, where multiple genes are deleted or duplicated.

Many phenotypic effects will also be very indirect due to cascading effects of the primary cellular pathology. For example, alterations in cell migration or synapse formation will necessarily change future patterns of electrical activity, indirectly altering the activity-dependent refinement of circuitry that occurs at later stages and in other brain areas (Ben-Ari 2008; Ben-Ari and Spitzer 2010). This raises an important point when considering why it is that mutations in so many different genes may lead to quite similar and specific phenotypic outcomes. One possibility is that the convergence represents a property of the developing brain itself, in the way it reacts to a wide range of primary insults (Mitchell 2011b; Lisman 2012). It may not be the crime but rather the cover-up that does the damage.

For example, a lesion to the ventral hippocampus in early postnatal animals alters the development of cortical circuits, resulting in an excitation-inhibition balance that emerges during adolescence and a change in dopaminergic tone in the developing striatum and cortex (O'Donnell 2011). In turn, homeostatic synaptic mechanisms react to this change by altering the levels of dopamine receptors, which is thought to result in subcortical hyperdopaminergic state that may mimic aspects of psychosis. Such a "common pathway" may thus emerge as an active reaction to a range of insults, rather than the endpoint of a passive propagation of cascading effects.

Modeling the Time Course of Schizophrenia

An important and defining aspect of schizophrenia is the typical time course of the emergence and progression of the illness. Although subtle, quantitative differences in behavior can be seen early in life, most patients are typically without significant symptoms until early adolescence. At that time, typically between 12 and 18 years of age, a prodromal phase is often seen, characterized by a decline in social, cognitive, and educational performance. Progression to frank psychosis typically occurs in late adolescence to early adulthood. This time course is a unifying theme which cuts across much of the diagnostic and phenomenological heterogeneity of schizophrenia, pointing to the adolescent period as a crucial factor in the pathogenesis of schizophrenia.

There are a number of distinct and testable hypotheses to explain why schizophrenia typically manifests in adolescence:

1. It reflects a neurodegenerative process with a time course that coincides with adolescence.
2. Changes in some hormone levels directly induce symptoms in a vulnerable brain.
3. There is abnormal development of brain structures that are not fully online at early stages and which cause defects when integrated later.
4. Ongoing developmental processes in the adolescent brain are directly affected by the etiological factors and go awry at that time period.
5. Normal cellular processes of maturation reveal a latent circuit-level deficit due to initial differences.

Given the conservation of physiological changes during adolescence, animal models offer the means to distinguish these hypotheses. In particular, it is important to test whether these processes are aberrant in situations predisposing to schizophrenia and to examine the interaction between processes of maturation and primary phenotypic effects.

Adolescence is characterized by a host of coordinated changes in various structural and neurochemical parameters, including synaptic pruning, ongoing myelination, and changes in the expression of various neurotransmitter receptor subunits (Sturman and Moghaddam 2011). In humans, imaging data revealed that cortical thickness in the prefrontal cortex acquires adult profile late in adolescence (Shaw et al. 2006a), and cortical oscillations exhibit dramatic changes during this period (Uhlhaas et al. 2009). Furthermore, the activation of reward and cognitive systems also matures during adolescence (Galvan 2010; Casey et al. 2010). These processes appear largely conserved across mammalian species and may be driven by a diverse set of neurobiological phenomena which are also known to mature during adolescence. These include changes in the density of prefrontal dopamine innervation (Rosenberg and Lewis 1995) and dopamine receptors (Brenhouse et al. 2008), functional changes in the modulation of excitation-inhibition balance (Tseng and O'Donnell 2007) and processing of salient events by prefrontal cortex (Sturman et al. 2010) and striatum (Sturman and Moghaddam 2012). Thus, when using the E–P–S framework in schizophrenia models, it is critical to consider developmental aspects including those which take place during adolescence.

In addition to explaining the typical age of onset, longitudinal studies of animal models may be used to parallel studies of high-risk, prodromal, first-episode, and chronic schizophrenia patients. The development of powerful small-animal neuroimaging methods offers the means to follow the same individual animals over time with a technique that provides data directly comparable to that from human patients.

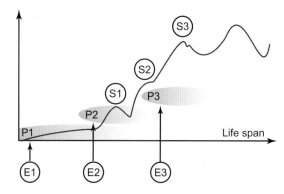

Figure 13.3 The E–P–S framework across the life span. Different etiological factors may come into action at different points in time (e.g., prenatal development, early postnatal critical periods, or during maturational processes of adolescence) and may give rise to interacting or independently acting pathophysiological mechanisms. Distinct pathophysiological sequelae may thus arise at different ages, with the subsequent emergence of clinical symptoms with a specific course. These distinct developmental trajectories, which lead to the emergence of pathophysiological states and behavioral symptoms, can be investigated discretely or collectively in accordingly designed animal models.

Using these approaches we may be able to map the timing of effects of different genes, the emergence of specific pathophysiological phenotypes, and the correlated emergence of behavioral phenotypes relevant to clinical domains (Figure 13.3). Thus, looking across the temporal domain offers another route to dissect the heterogeneity across levels.

Incorporating Computational Models into the Framework

Making sense of a framework that incorporates data across such disparate levels of analysis, from a large number of different models and experimental investigations, requires computational methods and can be greatly informed by computational theories. In particular, understanding the emergent properties of cells, synapses, microcircuits, or brain systems is essential to interpret how changes to specific components yield specific phenotypes. The study of neural dynamics offers a particularly promising tool (see Durstewitz and Seamans, this volume).

Dynamical properties of biophysical/biochemical systems, such as attractor states, oscillations or synchrony, arise from the nonlinear interactions among its many constituent components (e.g., molecules or cells), and may provide specific links between the neural "hardware" and the "software" level (cognition, behavior). For example, the activity of fast-spiking, parvalbumin-positive interneurons is known to drive the synchronous oscillations of local ensembles of pyramidal cells within the gamma frequency range (Cardin et al. 2009). These gamma oscillations and the associated synchrony in the spiking activity of neurons have, in turn, been linked to specific behavioral, cognitive, and perceptual functions by providing a basis for the neural coding of perceptual or mental objects (Uhlhaas and Singer 2012). Thus, through these dynamical mechanisms, genetic or molecular factors which interfere with the normal functioning of fast-spiking interneurons may lead to disorganization of cortical representations, and consequently to some of the symptoms observed in schizophrenia. Durstewitz and Seamans (this volume) describe another example; namely, how alterations in dopaminergic receptors may lead to changes in prefrontal cortical "attractor landscapes" with consequences for behavioral flexibility and information maintenance.

Thus, such a computational and neurodynamical framework may allow prediction of the effect of a mutation in some specific gene on neural dynamics at various scales, which in turn will have specific implications for behavioral and cognitive functions. It is important to note, however, that inferences in the reverse direction are much more difficult; given a particular behavioral difference, there will usually be a number of potential neurodynamical candidate mechanisms compatible with it. Even more limiting, a rather large variety of changes in one or more molecular components may have the same consequences for neural dynamics, making the backward inference from neural dynamics to underlying molecular cause extremely hard, if not impossible.

One current limitation for computational models is that many biologically important parameters or their statistical distributions may still be unknown or not sufficiently described. Filling in these blanks and using them to generate more hypotheses is thus an important research goal that will require an iterative and ongoing dialog between experimental and computational biologists. Ultimately, the detailed computational models of the systems involved will provide a powerful discovery engine for screening *in silico* through large areas of parameter space to identify potential molecular targets for therapeutic intervention.

The Promise of Human Cellular Models

A major difficulty in investigating the cellular correlates of a specific mutation in humans is that the cell types one is most interested in (neurons and glia) are inaccessible. A number of new technologies provide the means to derive neural lineage cells from human patients or carriers of specific mutations (Wilson and Sawa, this volume; Table 13.1). Many different molecular and cellular parameters of these cells can be characterized *in vitro*, including gene expression patterns, morphology, dendrite and axon extension, as well as synaptic connectivity. Derived neural cells can also be injected into animal brains to examine neuronal migration, synaptic integration, and other properties. Comparison of

Table 13.1 Overview of current methods used to derive neurons from human tissue samples. Depending on the experimental question, the scale of the investigation, and other logistical considerations, different methods may be optimal under different circumstances.

Method	Advantages	Disadvantages
Derivation of iPS cells and differentiation into neural cells	• Can be expanded for any number of analyses • Protocols are improving; should allow more standardized, systematic analyses of the effects of different mutations or high-risk genotypes	• Expensive and time consuming
Direct conversion of fibroblasts to neural cells	• Is cheaper and faster	• Does not generate a permanent, expandable bank of stem cells
Olfactory epithelium biopsy to obtain neural cells	• Does not require genetic manipulation of the cells • Does not use differentiation protocols	• Only obtains olfactory neurons

these parameters across various primary mutations may cellreveal phenotypic convergence in some cases at the proximal level of particular biochemical pathways or cellular processes.

The ability to derive such neural cells will provide the unprecedented opportunity to examine the possible molecular and cellular pathology associated with specific patient genotypes in individuals who are also characterized at clinical, neuroimaging, psychological, and genetic levels. Such investigations will be even more powerful when derivations can be integrated with data from similar analyses in animals with the cognate primary genetic lesion.

In addition to characterizing and comparing the cellular effects of identified mutations, derived neural cells provide a platform for discovering and dissecting genetic etiology. Gene expression differences may highlight strongly deleterious mutations in cases where genome sequencing provides a long list of potentially causal candidates, for example. Perhaps more importantly, derived neural cells offer the means to assess the effects of a mutation in the context of the entire genotype of an individual. So far, all of the mutations identified as increasing risk of schizophrenia show incomplete penetrance and highly variable expressivity, manifesting in diverse ways across individual carriers. Genetic modifiers are very likely to have large effects on the phenotypic expression of any particular mutation but could act at very different levels. Comparing cellular phenotypes in cells derived from patients with a specific mutation versus healthy carriers of the same mutation could reveal genetic background effects at the level of a particular molecular or cellular phenotype. Alternatively, it might suggest that they come into play at a higher level, in how the system reacts to earlier developmental differences.

There may also be many cases where multiple mutations are involved in pathogenesis (e.g., Girirajan et al. 2012). Though these may be difficult to identify and probably impossible to model directly in animals, the molecular pathways and cellular processes affected by such high-risk genotypes could still be investigated in cells derived from human patients.

Summary

Our ultimate goal is to increase understanding of the neurobiological underpinnings of all aspects of schizophrenia—etiology, pathogenesis, pathophysiology, and symptomatology—so that new therapeutic targets or points of intervention can be rationally identified to break the deadlock in treatment development for this devastating disorder. However, the heterogeneity of schizophrenia presents a huge obstacle.

To address this, we developed the Etiology–Pathophysiology–Symptoms framework as an overarching heuristic to guide experimental modeling of various aspects of schizophrenia. Rather than trying to overspecify criteria for validity of any particular preparation, experimental assay, or phenotype, the E–P–S framework embraces heterogeneity at etiological, pathophysiological, and clinical levels.

The point of generating models is not to recreate an entire disease state but to test specific hypotheses experimentally. In addition to well-characterized nongenetic models, the growing number of identified high-risk genetic lesions offers a proven discovery pathway to elucidate pathogenic mechanisms and provides explanatory links from molecular and cellular phenotypes to dysfunction of neural networks and brain systems underlying specific symptoms.

Given the very high degree of etiological heterogeneity, the identification of points or pathways of phenotypic convergence at the level of pathophysiology remains a major goal to be achieved. This will require systematic comparison across many different models and integration across levels of analysis. Fortunately, the tools to follow up on strong etiological entry points are now available, especially in terms of our capacity to analyze phenotypes at multiple levels in individuals, both in animals and humans.

Development and Treatment

14

Why Kraepelin Was Right
Schizophrenia as a Cognitive Disorder

René S. Kahn

Abstract

Classification of schizophrenia as a psychotic disorder has greatly inhibited progress in understanding and treating the disorder, as cognitive underperformance lies at the core of schizophrenia. Cognitive function is an important determinant of global functional outcome. This chapter reviews evidence in support of the view that schizophrenia is a cognitive disorder. Risk factors, cognitive decline, and developmental trajectories are discussed, and the consequences of considering schizophrenia a cognitive instead of a psychotic disorder are explored. It is proposed that development of effective treatments needs to focus on the cognitive aspects of the disorder.

Introduction

Schizophrenia is currently classified as a psychotic disorder; be it DSM or ICD, schizophrenia is defined by its psychotic symptoms. This chapter will attempt to show that this emphasis on psychosis is not only a fallacy, it is an error that has greatly contributed to the lack of progress in our understanding of this illness and hence has hampered the development of adequate treatments. Indeed, prognosis of schizophrenia may not have changed substantially since the introduction of chlorpromazine over fifty years ago, and some argue it has not meaningfully improved since the illness was first described (Hegarty et al. 1994). One of the reasons may be that the focus on psychosis has obscured the obvious: schizophrenia is not a psychotic disorder; it is a cognitive illness.

Obviously, this notion is not new. When Kraepelin first delineated the disorder in 1893 he named it "dementia praecox" for a good reason: he considered the illness to be a cognitive one. Indeed, when Kraepelin first described the disorder in the fifth edition of his *Lehrbuch*, his description begins with the observation of slow (occasionally rapid) cognitive decline that he typically found in his patients during adolescence (Kraepelin 1896). In his opinion, the hallmark of the disorder is a decrease in intellectual performance that begins almost a decade before the onset of psychosis and continues for many years thereafter. In fact, only after elaborating on this aspect of the illness for seven pages does he mention the presence of psychotic symptoms.[1] Kraepelin, of course, was not the only preeminent psychiatrist who considered psychosis to be a secondary or associated part of the illness. Bleuler (1911/1950), who coined the term "schizophrenia," viewed delusions and hallucinations as accessory symptoms as well; the core of this illness was determined by disturbance in affect, cognition (associative thinking), social interaction (autism), and volition (ambivalence).

Our current state of knowledge supports Kraepelin's notion of schizophrenia as a cognitive disorder for several reasons. First, low intelligence is a risk factor for schizophrenia. Second, cognitive decline and intellectual underperformance precede the onset of psychosis by many years. Third, decline in cognitive functioning continues after psychosis onset. Fourth, although cognitive underperformance prior to psychosis has not definitively been shown to be specific to schizophrenia, it does distinguish it from the "other" major psychotic illness (bipolar disorder). Finally, cognitive underperformance is an important predictor of general functional outcome in schizophrenia.

Low IQ as a (Genetically Mediated) Risk Factor

Low intelligence and intellectual underperformance have consistently been shown to constitute risk factors for the development of schizophrenia. A recent meta-analysis of 12 studies in population-based cohorts and nested case control studies, which included over 4,000 cases and more than 700,000 controls, found that low IQ increases the risk for developing schizophrenia in a dose-response fashion (with an effect size of 0.43): every point of decrease in IQ raised risk by 3.7% (Khandaker et al. 2011). A separate meta-analysis conducted by Dickson et al. (2012), which partially overlapped with the study by Khandaker et al. (2011), included only those studies which assessed participants aged 16 years or younger; Dickson et al. also found that low IQ increased the risk for schizophrenia, with an effect size of about 0.5. Interestingly, in this meta-analysis, risk was already evident by age 13, that is, many years prior to the onset of psychosis.

A different, but relevant indicator of intellectual underperformance—scholastic achievement—is also related to an increased risk of developing schizophrenia. In a nationwide cohort of Swedish individuals, school performance at age 16 was inversely related to the risk of developing schizophrenia in a dose-response fashion. Children who received the lowest grades had a fourfold risk

---

[1] In subsequent printings of the *Lehrbuch*, Kraepelin greatly expands on the description of this syndrome, eventually separating it into hebephrenic, katatonic, and paranoid subtypes. Even though psychotic features (e.g., hallucinations and delusions) gain prominence in the paranoid and katatonic subtypes, the hallmark of the disorder remains the cognitive decline during adolescence.

of developing the illness. Interestingly, repeating a school year (which occurs in some European countries when grades are insufficient) carried the highest risk: a hazard ratio of 9 (MacCabe et al. 2008).

Some of this risk may be, at least in part, related to the genetic risk of developing schizophrenia. Population-based studies in first-degree family members, as well as data obtained from selected samples and twin studies, all suggest that low IQ is related to the genetic risk of developing the illness (Aukes et al. 2009). In fact, it has been suggested that a substantial portion of the phenotypic correlation between schizophrenia and cognition is caused by shared genetic effects (Toulopoulou et al. 2010).

Cognitive Decline Prior to Onset of Psychosis

Although low IQ is a robust risk factor for schizophrenia, it is unclear whether low IQ is present at birth or is the result of a relative developmental *decline* in IQ that occurs at some point in time prior to the onset of psychosis (or both). Unfortunately, only a few studies have addressed this cardinal issue. One study compared childhood scholastic test performance in Iowa (the Iowa State tests of basic skills and educational development was used) from 70 subjects who later went on to develop schizophrenia with the population average. This scholastic test is administered to all children across the state of Iowa in grades 4, 8, and 11 (corresponding to the ages 9, 13, and 16) to assess five cognitive domains. Although the (prospective) patients did not differ from the State average at ages 9 and 13, they underperformed significantly at age 16 (with an effect size of around .35); this underperformance was most pronounced on language skills (Fuller et al. 2002). These results suggest that intellectual performance declines between the ages of 13 and 16 in individuals who go on to develop schizophrenia.

Retrospectively, we compared a robust and objective measure of high school performance (defined as "doubling" or repeating a grade, a mandatory measure in the Dutch schooling system when grades fall below a certain standard) in a sample of over 80 twins (MZ and DZ) discordant for schizophrenia with that of a matched sample of healthy twins. Not only did the twin who went on to develop schizophrenia underperform his or her unaffected co-twin in 90% of cases, this underperformance was evident at age 13 and preceded the onset of the first psychosis by an average of nine years (Van Oel et al. 2002). In a fully prospective study, Reichenberg et al. (2010) used data from the Dunedin birth cohort, where cognitive performance was tested at ages 7, 9, 11, and 13; final symptomatic follow-up was then conducted at age 32. Not only did the 35 subjects who went on to develop schizophrenia underperform their healthy cohort controls at all measurement points, they started to lag further behind their peers between the ages of 7 to 13 (Reichenberg et al. 2010). In short, these studies not only suggest that children who will go on to develop schizophrenia progressively underperform their healthy peers, but that this (relative) decline in intellectual performance starts early in adolescence, years prior to the onset of psychosis.

Cognitive Function after Onset of Psychosis

Numerous studies have assessed global (or specific aspects of) intellectual functioning in schizophrenia once diagnosis is established, finding IQ to be about 2 standard deviations (SD) lower than age-matched controls (e.g., Keefe and Fenton 2007). Interestingly, only about 20% of the schizophrenia population can be considered to have an unimpaired IQ, defined as being less than 1 SD below the normal mean (Keefe and Fenton 2007). However, even this group may have a lower IQ than would be expected on the basis of the level of education in their respective families. Indeed, as Keefe et al. (2005) show, IQ in schizophrenia patients is lower than would be expected on the basis of the level of their mothers' education. This suggests that IQ may well be lower in all patients, certainly when compared to their (genetic and environmental) intellectual potential. However, the decrease in IQ does not indicate *when* the decline begins. As indicated, some of it occurs prior to the onset of psychosis, but does this decrease in IQ continue once psychosis is established? Since the degree of cognitive impairment of 2 SD in patients is (much) larger than the 0.5 SD observed in individuals prior to psychosis onset (Woodberry et al. 2008), it is highly likely that IQ continues to decline after psychosis sets in.

Although a considerable number of studies have examined intellectual functioning in patients with established schizophrenia over time, many are impossible to interpret. This is because most studies that have assessed IQ over time failed to include a healthy control group, so that effects of practice cannot be ruled out (Goldberg et al. 2007, 2010). In fact, very few studies have included healthy controls in their assessment of IQ change in schizophrenia over time. In a recent meta-analysis (Hedman et al. 2013), which summarizes data from eight studies (including 280 patients and 306 healthy controls), we conclude that IQ increases significantly less in patients over time (0.33 points) than it does in healthy controls (2.1 points), resulting in an effect size for (relative) cognitive decline of .48. Thus, although the number of studies and subjects attests to the lack of well-designed studies that examine cognitive change in schizophrenia, available results suggest that intellectual performance continues to decline after the onset of psychosis in schizophrenia.

Specificity of Cognitive Decline

Although the specificity of cognitive decline in schizophrenia has hardly been studied, it appears that cognitive dysfunction—at the very least prior to the onset of psychosis—distinguishes it from the other major psychotic illness (i.e., according to current classification systems): bipolar disorder. Although the number of studies that examine cognition prior to the onset of bipolar disorder is more limited than those in schizophrenia, a consistent pattern has emerged: low IQ constitutes a risk factor for schizophrenia, but it does not in bipolar illness. Using data from the Israeli Draft Board, where IQ was assessed in adolescents, those who were later hospitalized for bipolar disorder did not differ in intellectual performance from the normal population (Reichenberg et al. 2002). Similarly, data from Swedish (Zammit et al. 2004) and Danish (Sørensen et al. 2012) draft boards suggest that draftees who later go on to be hospitalized for bipolar disorder do not differ significantly in IQ from healthy individuals. In fact, a recent study in over one million Swedish men found that high intelligence carried a 60% increased risk (HR 1.59) for later hospitalization for bipolar disorder, at least in those without comorbidity (Gale et al. 2013). Consistently, in a study examining school grades from all Swedish schoolchildren between 1988 and 1997, individuals with excellent school performance had an almost four times higher risk of developing bipolar illness compared to those with average grades (MacCabe et al. 2010). Clearly, low IQ is not a risk factor for bipolar illness.

The issue of whether a *decline* in cognitive function precedes the onset of bipolar disorder has not been addressed in population-based studies. However, we conducted a study in MZ and DZ twins discordant for bipolar disorder, similar in design to the study mentioned earlier in schizophrenia (Van Oel et al. 2002). In contrast to the discordant schizophrenia twins, in this study we found that the discordant bipolar twin pairs only showed a temporary decline in functioning, and over the longer term did not underperform the healthy control twins. In addition, in contrast to what we found in schizophrenia, the twin who went on to develop bipolar disorder did not do worse in school than his or her unaffected co-twin (Vonk et al. 2012).

Similarly, at illness onset, patients with bipolar disorder, in contrast to those with schizophrenia, do not appear to perform worse than healthy controls. In a study by Zammit et al. (2004), which examined this issue, recent onset bipolar disorder or mania patients performed significantly better than first-episode schizophrenia patients on a broad variety of cognitive tests and only underperformed healthy individuals on two of the sixteen subtests: delayed verbal memory and category fluency. Consistent with the studies reviewed above, the estimate of premorbid intellectual functioning was normal in the bipolar group and decreased in the schizophrenia patients (Zanelli et al. 2010). Finally, a meta-analysis of cognitive functioning in patients with established illness showed that those with bipolar disorder perform significantly better than patients with schizophrenia in almost all cognitive domains, with an effect size of around 0.5. This difference was found for actively ill patients as well as for those in remission (Krabbendam et al. 2005). Taken together, the evidence strongly suggests that low IQ, cognitive underperformance during adolescence as well as at first presentation of psychosis differentiates schizophrenia from bipolar disorder.

Cognitive Dysfunction as Predictor of General Outcome

The central role of cognitive dysfunction in schizophrenia is solidified by ample evidence that cognitive function is an important determinant of global functional outcome. Although some studies suggest this aspect of the disorder is the strongest predictor of outcome (Bowie et al. 2006), others find that cognitive function is independently, but not necessarily predominantly, predictive of outcome (Mohamed et al. 2008). At any rate, cognitive dysfunction in schizophrenia is unaffected by current pharmacotherapy, which for schizophrenia is almost entirely based on the use of antipsychotic medication. These drugs, which essentially have not been pharmacologically altered since the introduction of chlorpromazine over half a century ago, are indeed effective antipsychotics. However, despite many claims to the contrary, none of these compounds have proven effective in improving cognition in schizophrenia to any meaningful degree. Numerous studies have examined the effect of first- and second-generation antipsychotics on cognitive function in schizophrenia. Although several studies claim improvement in some specific subtests, global cognitive change in large comparative studies in first-episode and chronic schizophrenia rarely reaches an effect size of over 0.3. A meta-analysis of first-generation antipsychotics found an effect size of 0.22 (Mishara and Goldberg 2004), whereas more recent studies that directly compare first- and second-generation antipsychotics have not found differential effects of these drugs with effect sizes of around 0.3 (Keefe et al. 2007; Davidson et al. 2009). However, even this small effect is likely to be no more than practice-related: when healthy individuals are included in the trial design, their improvement on the same tests that are administered to the patients is of a similar effect size as that observed in the patients (Keefe et al. 2008, 2011a). Thus, although cognitive dysfunction is central to outcome in schizophrenia, current pharmacological treatment does not appear to ameliorate it. More promising may be efforts which combine cognitive interventions with rehabilitation programs (Zanelli et al. 2010).

Conclusion

Cognitive underperformance is at the heart of schizophrenia. It constitutes a (genetic) risk factor, precedes the onset of psychosis by many years, continues to worsen after psychosis is established, and determines outcome. Underperformance is broad, evident, and relevant, expressed throughout school in the years prior to the onset of the first psychosis. This underperformance at school constitutes one of the highest hazard ratios found for schizophrenia, only surpassed by the risk of having a sibling with the illness. Although low IQ at primary school may already constitute a risk factor for schizophrenia, a (further) decline in global cognitive functioning most likely occurs in early puberty, preceding the onset of psychosis by almost a decade. This decline does not halt once the psychosis develops, but appears to progress even further, unstopped by current (pharmacological) treatment methods. Whether this process of cognitive decline prior to psychosis onset is specific to schizophrenia has not been well studied, but it has not been found in bipolar disorder.

What are the consequences of considering schizophrenia primarily and foremost a cognitive instead of a psychotic disorder? First, cognitive decline prior to the onset of psychosis (in most cases retrospectively established) should be part of the diagnosis. This (under)performance should be particularly evident when compared to the intellectual performance of parents and siblings. Second, treatment of cognitive deficits should be central to any guidelines, and are not at present. Third, the whole concept of schizophrenia as an illness which presents with psychosis should be discarded: *schizophrenia presents with cognitive decline*. Fourth, the age of onset of schizophrenia is probably a decade earlier than we now assume.

As proposed in this chapter, schizophrenia is an illness that starts (at least) in early adolescence, around the age of 12–14 years, and is accompanied by a decline in global cognitive functioning relative to healthy peers. As Kraepelin so aptly stated (Kraepelin 1896): *Je weiter sie aber fortschreiten, desto schwerer wird es ihnen mit ihren Kameraden Schritt zu halten* (the more they [affected individuals] continue, the more difficult it is for them to keep up with their peers). This perspective implies that early recognition and prevention programs which focus on (brief and intermittent) psychotic symptoms happen too late in the disease process and, more importantly, fail to address the core aspect of the illness. Indeed, much of the social damage has already occurred once the psychosis finally manifests itself, in late adolescence or young adulthood: a person may have dropped out of school, lost friends, or failed to reach his full potential. Clearly, we have been focusing on the wrong risk phenotype: being prone to psychosis may not constitute the highest risk of developing schizophrenia, but rather the propensity to cognitive decline or intellectual stagnation during adolescence. Just like psychosis (Verdoux and van Os 2002), this phenotype will most likely be much more prevalent in the general population than we now consider it to be. It may be present in many individuals who will not go on to develop psychosis, let alone schizophrenia. Thus, to understand the genetic and environmental influences that lead to schizophrenia, we need to study the interaction between the (genetically mediated) cognitive underperformance during adolescence, and the environmental (and genetic) factors that determine why some of these individuals will eventually develop schizophrenia.

It may be that the boundaries of the disorder that are characterized by progressive cognitive decline prior to and after the onset of psychosis are narrower than those of the illness now defined as schizophrenia. It may be that even within that (sub)type, various causes will be identified. What is completely clear, however, is that by defining schizophrenia as a psychotic disorder, we have done our patients a disservice. Putting the focus back on cognition may facilitate the search for a cure for the illness that we should, for lack of any better term, have called Kraepelin's Disease.

15

What Will the Next Generation of Psychosocial Treatments Look Like?

Kim T. Mueser

Abstract

This chapter defines the broad range and scope of psychosocial treatments for schizophrenia (also called psychiatric rehabilitation) and discusses how new conceptualizations of recovery from mental illness, which emphasize meaning and purpose in life over a narrow focus on symptom remission, have shaped the nature and delivery of services. Substantial progress in psychosocial treatment has been made over the past several decades; rigorous controlled and replicated research has demonstrated the effectiveness of a variety of interventions, including contingent reinforcement, family intervention, supported employment, cognitive behavioral therapy, cognitive remediation, illness self-management training, and social skills training. Despite this progress, most of these services are not available in routine care. Obstacles to disseminating psychosocial treatments are considered, including insufficient training of professionals prior to entering the workforce and the need for more research on the science of implementation. Recommendations for improving the quality of psychosocial interventions include targeting predictors of response to treatment, evaluating the critical components or mechanisms underlying effective programs, improved precision of goal setting and monitoring outcomes to facilitate individual tailoring, and training of clinicians across the range of effective treatments to maximize the creative use of clinical expertise.

Introduction

The past four decades have witnessed an unprecedented growth in the development and evaluation of psychosocial treatments of schizophrenia. By the late 1960s, research on the token economy demonstrated that grossly impaired psychosocial functioning in long-term psychiatric inpatients could be improved through contingent reinforcement (Ayllon and Azrin 1968; Paul and Lentz 1977), and in the 1970s, the rudiments of systematic skills training approaches to rehabilitation for schizophrenia were refined, followed by a growing array of curricula (Liberman et al. 1986). In the 1980s, several models of family intervention were developed and shown to be effective (Falloon et al. 1985; Leff et al. 1985), and in the 1990s, supported employment was standardized and shown to improve competitive work outcomes in controlled research (Drake et al. 1996, 1999).

Despite these advances, concerns have been raised that little progress has been made in the treatment of schizophrenia over the past century (Insel 2009). What accounts for the apparent discrepancy in perspectives on the outcome of schizophrenia, and what are the implications for the future of psychosocial treatment? This chapter seeks to address these issues and to suggest promising directions for the development of more effective psychosocial interventions. Following a discussion of the definition and scope of psychosocial treatment, the current state of knowledge of psychosocial treatment for schizophrenia is reviewed, including empirical support for interventions and issues related to implementation and dissemination. The chapter concludes with suggestions for improving the effectiveness of psychosocial treatment.

Definition and Scope of Psychosocial Treatment

In this chapter, the terms *psychosocial treatment* and *psychiatric rehabilitation* are considered to be interchangeable. Treatment and rehabilitation are sometimes distinguished from one another, with the former defined as focusing on the management of symptoms and the latter addressing the restoration of functioning. This distinction, however, serves little practical use in schizophrenia because both symptoms and impaired functioning are central to the diagnosis of the disorder, both cause significant distress, but they are semi-independent of one another (Strauss and Carpenter 1972). While some interventions focus more on symptoms and others on functioning, treatment needs to address both—often in an integrated fashion.

Psychiatric rehabilitation is broadly aimed at maximizing the ability of individuals with a mental illness to function as effectively and with the greatest satisfaction as possible, in the least restrictive living environment, and with a minimum amount of professional intervention (Anthony et al. 2002; Corrigan et al. 2008). Psychosocial treatment typically focuses on either teaching more effective skills to improve functioning or more adaptive coping with symptoms, or providing environmental supports, prompts, or contingencies to facilitate optimal functioning. The word *psychosocial* distinguishes this type of intervention from other treatment methods that are more biological in nature, such as pharmacological treatment or electroconvulsive therapy.

Psychosocial treatment comprises a broad range and diversity of methods that go far beyond the traditional notions of "talk therapy." The variety of different psychiatric rehabilitation approaches can be described by considering the domains targeted by interventions, different treatment modalities, and the settings in which treatments may be provided (Liberman and Mueser 1989; Spaulding et al. 2003). A wide range of potential domains or levels are promising targets for psychosocial interventions, including improved functioning in the areas of *cognition* (e.g., cognitive remediation), *social cognition* (e.g., social cognition training), *interpersonal relationships* (e.g., social skills training), *work* or *school* (e.g., supported employment), *socio-environmental adjustment* (e.g., family psychoeducation, contingency management), *symptom severity* and *relapses* (e.g., cognitive behavioral therapy for psychosis, training in illness self-management), and *psychophysiological arousal* (e.g., stress management training). Psychosocial treatment can be provided in a variety of different formats, including naturalistic settings, in individual, group, and family modalities as well as in a similarly wide range of settings, including acute and long-stay inpatient hospitals, residential milieu, community mental health centers, in the client's home, or in other community settings (e.g., workplace, store, or on public transportation).

Rehabilitation and Recovery

It has been over one hundred years since Kraepelin (1919/1971) presented the bold thesis that schizophrenia is a single disease, and the validity of his claim continues to be hotly debated. Although the limitations of Kraepelin's single disease model have frequently been noted (Spaulding et al. 2003; Bentall 1993), no alternative conceptualization of the disorder, including proposed subgroups of diseases, has thus far been shown to be sufficiently more useful or compelling to result in a paradigm shift to a new model. There is, however, abundant evidence that both biological and environmental factors play a role in the development of schizophrenia (van Os and Kapur 2009), and a range of subtle neurodevelopmental, cognitive, and social problems have been established to precede the onset of the disorder for some people by many years (MacCabe et al. 2013). Whether the heterogeneity of schizophrenia reflects multiple causal factors which converge on a final common pathway or separate disease states, by the time schizophrenia is fully manifested, the broad range of impairments across the different levels of systematic functioning have gained a degree of functional independence from one another (Strauss and Carpenter 1972, 1977). Thus, until specific causes of the characteristic impairments of schizophrenia are known and can be targeted for treatment, there is a need for interventions to address problems at different levels of systemic function.

Treatments range from interventions which address more molecular levels of systemic functioning, such as medications for neurophysiological dysfunction, to those which target more molar levels of functioning, such as psychiatric rehabilitation to improve psychosocial adjustment. While psychosocial treatment can viewed as an interim solution that future advances in the understanding and treatment of schizophrenia may obviate, these interventions are important because they address levels of systemic functioning that have been beyond the reach of pharmacology (e.g., role functioning, social relationships). Indeed, the importance of psychiatric rehabilitation has been highlighted in recent years by an active movement of individuals with major mental illness who have questioned the emphasis of treatment providers on symptoms and associated impairments, arguing that attention to psychosocial functioning and psychological adjustment should be the true focus of treatment.

For over twenty years, the *recovery movement*, spearheaded by individuals with a serious mental illness who have received psychiatric treatment (referred to as *consumers* in the United States or *service users* in Great Britain), has evolved to be a major force in changing how schizophrenia and other major mental illnesses are understood and treated (Silverstein and Bellack 2008). The impetus for this movement came from the objections they raised to the pessimistic messages they were often given about the long-term outcome of schizophrenia and other disorders, pointing out that they were both "spirit-breaking" and inaccurate in light longitudinal research, which showed symptom remission and functional improvement in significant proportions of people (Deegan 1990; Harding et al. 1987). Consumers also called for a reduction in the use of coercive, often "retraumatizing" interventions (e.g., involuntary hospitalization, use of seclusion and restraint, forced medication) (Brase-Smith 1995; Jennings 1994), and a shift from traditional hierarchical medical decision making, based on the assumption that "the doctor knows best," to a more collaborative approach that respects an individual's preferences and the right to determine their own treatment priorities (Chamberlin 1997a; McLean 1995). Perhaps most importantly, this movement challenged the notion that recovery from mental illness can only be defined in medical terms; it was argued that recovery should be defined in more nuanced and personally meaningful ways to empower consumers and give them hope for the future (e.g., Frese 2008; Ralph 2000).

New conceptualizations of recovery focus on personal growth and the establishment of meaning and sense of purpose in life, despite having a mental illness (Anthony 1993). The desire for more personally meaningful definitions of recovery other than symptom remission frequently evokes the need to improve different areas of psychosocial functioning. For example, in the United States, the President's New Freedom Commission on Mental Health (2003:6) defines recovery as "…the process in which people are able to live, work, learn, and participate fully in their communities." The consumer/recovery movement has underscored the importance of improving psychosocial functioning as a treatment priority over a narrow focus on symptom management and relapse prevention.

The focus on functional outcomes need not be to the exclusion of treatment that targets characteristic symptoms and cognitive impairments. However, it does suggest that attention to these areas should be driven primarily by difficulties in making progress toward established functional goals or attention to high levels of psychological distress. The emphasis of psychosocial treatment on functioning also underscores the fact that, given the relative independence of different levels of systemic dysfunction (Spaulding et al. 2003), targeting one area of systemic dysfunction (e.g., cognitive impairment) will not necessarily lead to benefits in other areas, such as role functioning, unless they are also explicitly targeted for treatment (a summary of research on cognitive remediation is provided below).

Progress in the Psychosocial Treatment of Schizophrenia

Following both the discovery of antipsychotic medications in the 1950s and the deinstitutionalization movement, which began in the 1960s spurred on by economic forces as well as an improved capacity for clinical management (Johnson 1990), an increasing variety of psychosocial interventions for schizophrenia and other serious mental illnesses were developed and empirically evaluated. Meta-analyses show that adding broadly defined psychosocial intervention to pharmacological treatment has a global impact on improving outcome in schizophrenia compared to medication alone, and those individuals with the most severe impairment tend to experience the greatest benefit (Mojtabai et al. 1998). Furthermore, specific approaches to psychosocial treatment are routinely included in treatment guidelines for schizophrenia, such as the PORT recommendations in the United States (Dixon et al. 2010) and the NICE guidelines in the United Kingdom (National Collaborating Centre for Mental Health 2009).

Defining Empirically Supported Psychosocial Treatments

There is no clear consensus on how to categorize specific approaches to psychiatric rehabilitation and the evidence evaluating them. Should the validation of a practice, for example, be highly specific to certain brands or types of programs, such as Beck et al.'s (1979) versus Lewinsohn's (1974) treatment approaches to depression, Linehan's (1993) dialectical behavior therapy for borderline personality disorder, Falloon's approach to behavioral family management for schizophrenia (Falloon et al. 1984), or Hogarty's cognitive enhancement therapy for schizophrenia (Hogarty et al. 2004)? Alternatively, should validation be based on a scientific understanding of the active ingredients of an intervention?

The active ingredient approach would appear more legitimate, but many validated treatment approaches justifiably incorporate procedures that are not validated. Furthermore, the process of dismantling an effective intervention to identify its critical components and better understand its mechanisms of action is complex, time consuming, and costly, and thus not practical for most complex psychosocial interventions for schizophrenia. On the other hand, narrowly focusing on the validation of specific, "brand-like" interventions without attention to their core ingredients can be misleading when "new" treatments are developed based on methods of established interventions but containing additional unique but unproven components, and such an intervention is touted as both new and effective (Lohr et al. 1999).

In practice, the evaluation of empirical support for psychosocial treatments for schizophrenia has been based on a combination of the theory which guides the intervention and the methods employed, the treatment modality, and the targeted domains of functioning, with less attention paid to the specific "brand" of program. For example, although there are many different approaches or "brands" of cognitive behavioral therapy for psychosis (Beck et al. 2009; Kingdon and Turkington 2004), they are typically grouped together for reviews of research; the basis for this lies in the fact that they share a common general theory of the relationships between thinking, feeling, and behavior and employ a common set of therapeutic techniques to enhance coping and evaluate thoughts and beliefs associated with psychotic symptoms. In contrast, family interventions for schizophrenia tend to be grouped together based on a combination of their shared use of the family treatment modality, their focus on reducing family stress, and their agreement on a common set of principles underlying intervention (e.g., collaborative relationship with mental health professionals, provision of information to the family about mental illness and treatment, inclusion of the client in family sessions), despite differences between programs in theoretical orientation and therapeutic techniques employed (Anderson et al. 1986; Barrowclough and Tarrier 1992).

Empirically Supported Psychosocial Treatments for Schizophrenia

There is some variability in the specific criteria used by different organizations or reviewers to determine whether a psychosocial treatment is empirically supported (Chambless and Ollendick 2001; Drake et al. 2005; Herbert 2000). Generally, empirically supported psychosocial interventions for schizophrenia are identified using similar criteria to those employed for defining evidence-based medicine (Sackett et al. 1997). For schizophrenia, several standardized interventions have been shown to improve broadly accepted, important outcomes (e.g., symptoms, social functioning, work, or school) in multiple randomized controlled trials (RCTs) conducted by independent research teams (e.g., Drake et al. 2001; Ganju 2003). More than thirty years ago, the token economy program, which involves systematically modifying environmental contingencies in individuals living in inpatient settings to reinforce more adaptive behaviors (Ayllon and Azrin 1968), was the first psychosocial treatment approach shown to be effective for people with serious mental illnesses, facilitating discharge from long-stay hospitals into the community (Paul and Lentz 1977). Following empirical validation of the token economy, at least six other psychiatric rehabilitation programs have demonstrated strong empirical support (Mueser et al. 2013a), as described briefly below.

Family Intervention

Research in the 1970s demonstrated that high levels of family stress (i.e., "expressed emotion") were predictive of increased risk of relapse and hospitalization in recently discharged people with schizophrenia who were living, or in close contact, with their relatives (Brown et al. 1972). This finding led to the development and empirical validation of five different family intervention programs aimed at improving family coping, lowering overall stress in the family, and reducing client risk of relapse (Anderson et al. 1986; Barrowclough and Tarrier 1992; Falloon et al. 1984; Kuipers et al. 2002; McFarlane 2002). These programs differ in their specific targets for treatment and methods to achieve them: some train families in stress management or communication skills (Barrowclough and Tarrier 1992; Falloon et al. 1984), whereas others increase family support through multifamily groups (Kuipers et al. 2002; McFarlane 2002). Despite differences in theoretical orientation and specific therapeutic methods used in these family intervention programs, they share a common set of features which may, in part, explain some of the similar beneficial effects found across the programs (Lucksted et al. 2012; Pitschel-Walz et al. 2001). These common features include

- long-term (minimum nine months) family intervention provided by mental health professionals,
- emphasis on creating a collaborative relationship with family, and avoiding blame and pathologizing of relatives' behavior,
- inclusion of the client in some or all family work,
- provision of information about schizophrenia and its treatment, and
- focus on reducing family stress.

Most research on family intervention programs has targeted clients with a recent symptom relapse or hospitalization. Research findings (Pharoah et al. 2010) include 53 RCTs conducted throughout the world; family intervention reduces relapses and hospitalizations over a period of one to two years and may facilitate treatment adherence. In addition, clients in families who receive family intervention show modest improvements in psychosocial functioning, and relatives experience some reduction in stress and tension.

Supported Employment

Traditional vocational rehabilitation approaches for schizophrenia have typically focused on extended training programs to prepare clients to enter the workforce, with some including work experiences in sheltered or other protective settings. Research has shown, however, that these "train-place" approaches fail to improve vocational outcomes for people with a serious mental illness (Bond 1992). Based on the "train-place" philosophy, supported employment focuses on helping clients obtain competitive work, and then provides the training and support necessary to succeed at these jobs. The most thoroughly standardized and evaluated supported employment program is the *individual placement and support program* (Becker and Drake 2003), which is defined by the following characteristics:

- Desire for competitive work is the only inclusion criterion for participation in the program (e.g., clients are not excluded because of symptoms or cognitive impairments).
- Focus is on rapid search for competitive jobs in integrated community settings and no required prevocational training.
- Client preferences are respected regarding the type of desired employment and disclosure about psychiatric disorder to prospective employers.
- Follow-along supports are provided after job attainment.
- Vocational and clinical services are integrated.
- Counseling is provided to inform clients about special incentives for work and impact of work on disability benefits.

Research on supported employment indicates that out of 25 RCTs, 15 used the individual placement and support model (Bond et al. 2012). Most studies were conducted in the United States, with some in Europe. Over the one- to two-year study period, supported employment was found to be superior to other vocational programs in terms of competitive work outcomes (e.g., proportion who worked, hours and weeks worked, wages earned). The effect sizes ranged from .58 to .67. However, the impact of supported employment on work after vocational supports are removed is unclear, given the limited research that has addressed this question.

Cognitive Behavioral Therapy for Psychosis

The persistence of hallucinations and delusions in a significant proportion of people with schizophrenia (Lindenmayer 2000) and the distress and functional impairment associated with these symptoms (Racenstein et al. 2002) led to the adaptation of techniques from cognitive behavioral therapy that were used to treat depression and anxiety to psychotic symptoms. Multiple programs for cognitive behavioral therapy for psychosis have been developed (e.g., Beck et al. 2009; Chadwick 2006; Kingdon and Turkington 2004) and generally share the following features:

- "normalization" of psychotic symptoms to reduce embarrassment and stigma,

- identification of individual and situational factors that influence severity of symptoms,
- development of a shared formulation of symptoms,
- the teaching of more effective coping strategies,
- evaluating thoughts and beliefs related to psychotic symptoms, and
- conducting behavioral experiments to obtain more information about psychosis-related beliefs.

Research on cognitive behavioral therapy for psychosis indicates (Granholm et al. 2009; Wykes et al. 2008) that over 33 RCTs were conducted, mostly in United Kingdom. Effect sizes ranged from .35 to .44. Significant effects were found for psychotic symptoms, and there was some indication of effects on negative, mood, and anxiety symptoms, as well as psychosocial functioning. The latter outcomes, however, need to be replicated by more rigorous trials.

Social Skills Training

Impaired social and self-care functioning are hallmarks of schizophrenia that often precede the onset of the disorder and are also associated with a worse course of illness. To systematically teach more effective interpersonal and self-care skills, social skills training approaches were developed based on the principles of social learning (Bandura 1969), with the broader aim of improving social and community functioning. Social skills are usually taught in groups according to the following sequence (Bellack et al. 2004a; Liberman et al. 1989):

- Establish a rationale for learning a skill, and break complex skills into component parts.
- Model (demonstrate) a skill in role play.
- Engage each client in practicing the skill in role play (one at a time).
- Elicit and provide positive reinforcement about client's performance, followed by suggestions for improved performance.
- Engage client in 1–3 more role plays, followed by positive and corrective feedback.
- Collaboratively develop home assignment for each client to practice skills on their own.
- Program generalization of skills through *in vivo* community practice and/or involvement of natural supports in the client's environment (e.g., family, residential workers).

Skills training programs have been developed to address specific areas of impaired functioning, such as independent living (e.g., personal hygiene, use of public transportation), occupational functioning (e.g., job interviewing, responding to feedback on the job), interpersonal relationships and leisure (e.g., conversation skills, making friends, exploring leisure activities), psychiatric and physical illness self-management (e.g., taking medication, discussing medication side effects with prescribers, developing relapse prevention plans), and coping with social situations that involve alcohol or drugs (e.g., resisting offers to use substances). Although summaries of research on skills training tend to focus on broad outcomes that are evaluated across multiple applications of skills training (e.g., social and community functioning), the implementation of skills training programs usually targets very specific domains, with research supporting its effects across different areas. In addition, skills training methods are often incorporated into other, previously described practices: teaching communication skills in some family intervention programs (Falloon et al. 1984), improving work-related skills in supported employment (Mueser et al. 2005; Wallace et al. 1999), and teaching coping skills in cognitive behavioral therapy (Tarrier et al. 1993) for psychosis and illness self-management (Gingerich and Mueser 2011). Research on social skills training indicates (Kurtz and Mueser 2008; Pfammatter et al. 2006) that over 25 RCTs were conducted. Significant effects were found on learning specific social skill-related information and behavioral competencies, improving social and community functioning, and improving negative symptoms. Effect sizes ranged from .39 to .77. Skills training programs also exert smaller but significant effects on reducing symptoms and relapses, possibly through improved social support and interpersonal coping, leading to reduced sensitivity to stress, in line with the stress-vulnerability model of schizophrenia (Liberman et al. 1986; Nuechterlein and Dawson 1984).

Cognitive Remediation

The cognitive impairment characteristic of schizophrenia (Heaton et al. 1994) and the association between cognitive and psychosocial functioning (Green 2006) makes cognitive functioning an obvious treatment target for the disorder. Cognitive remediation is defined by an intervention that directly focuses on improving attention, memory, psychomotor speed, and executive functions, or reducing the effects of cognitive impairment on psychosocial functioning. Over the past 30 years, a wide variety of cognitive remediation approaches have been developed that range in focus from elemental cognition to complex social cognition and problem solving (e.g., McGurk et al. 2007), with some packaged modalities incorporating the entire range and even interfacing with social skills training (Brenner et al. 1994; Hogarty et al. 2004). Common elements of programs include (a) practice of cognitive exercises on computer or paper and pencil tests and (b) use of self-monitoring and errorless learning. Some programs provide coaching on cognitive strategies to improve cognitive performance during practice tasks, some programs teach coping or compensatory strategies to reduce the impact of impaired cognitive functioning on psychosocial functioning, and some do both.

Significant progress in research on cognitive remediation has occurred, especially over the past decade. Major findings from two meta-analyses (McGurk et al. 2007; Wykes et al. 2011) indicate that over 40 RCTs were conducted, mostly in Europe or the United States. Significant effects were found on cognitive functioning (effect sizes were .41 to .45) and psychosocial functioning (effect sizes were .36 to .42), with weaker effects on symptoms (effect sizes were .18 to .28). The impact of cognitive remediation on psychosocial functioning is moderated by the provision of adjunctive or integrated psychiatric rehabilitation; cognitive remediation improves functional outcomes when it is added to (or integrated with) psychiatric rehabilitation (compared to psychiatric rehabilitation alone) but not to usual services (compared to usual services alone). In addition, impact is dependent on whether strategic training was provided; when strategy training is provided in the context of rehabilitation, effect size is doubled.

Training in Illness Self-Management

The recovery/consumer movement emphasized the importance of actively involving clients in their own treatment, involving them in collaborative decision making with treatment providers, and empowering them to determine their own treatment goals (Farkas 2007). Illness self-management programs are aimed at providing clients with the information and skills needed to manage their illness in collaboration with others (e.g., reducing symptoms, preventing relapses). A variety of programs have been developed and evaluated, including individual and group formats, with durations ranging from several months to over a year (Gingerich and Mueser 2011; Hogarty 2002; Kopelowicz et al. 1998). Common components of illness self-management training programs include

- education about serious mental illness and its treatment,
- teaching strategies for improving medication adherence,
- training in coping skills to manage persistent symptoms,
- developing a relapse prevention plan, and
- social skills training to strengthen social supports.

Research has been conducted on both the individual components of illness self-management and comprehensive programs which target the broad range of skills. The benefits of teaching illness self-management, including psychoeducation, behavioral tailoring to incorporate medication adherence into the client's personal routine, coping skills training, and relapse prevention have been shown in over forty controlled studies (Lincoln et al. 2007; Mueser et al. 2002). Three RCTs of the illness management and recovery program (Gingerich and Mueser 2011), which incorporate the aforementioned strategies, have shown significant improvement on outcomes related to self-management. Some evidence indicates that consumer-provided training in illness self-management is effective (e.g., Wellness Recovery and Action Plan, or WRAP) (Cook et al. 2012).

Other Programs

There is growing evidence for the effects of other psychiatric rehabilitation approaches for serious mental illness: integrated treatment of co-occurring psychiatric and substance use disorders (Drake et al. 2008; Barrowclough et al. 2010), training in social cognition (Kurtz and Richardson 2012), and modifying the living environments of individuals with severe cognitive impairment to prompt self-care behaviors and sustain community living (Velligan et al. 2002, 2006). Although not a rehabilitation approach per se, alternative methods for delivering pharmacological and psychosocial treatment to clients with serious mental illness living in the community who do not access available services on their own (i.e., assertive community treatment) (Stein and Santos 1998) have been developed and shown to be effective, primarily in the United States, where access to psychiatric services for this population is most problematic, but also in Australia (Coldwell and Bender 2007; Nelson et al. 2007; Rosen et al. 2007).

In summary, the preponderance of evidence across multiple studies (more than 20 RCTs for most interventions) indicates that potent psychosocial interventions have been developed, with most producing effect sizes in the moderate range (Cohen 1992). New interventions continue to be developed, and along with them a growing body of evidence supporting them. The progress that has been made in psychiatric rehabilitation for schizophrenia provides realistic hope for improving the quality of lives of people with this disorder, if access to these effective services can be assured.

Poor Adoption of Effective Psychosocial Treatments

Despite steady advances in the development and validation of effective psychosocial treatments for schizophrenia, the gap between science and implementation of effective interventions has continued to widen in some countries. The problem of poor access to empirically supported psychosocial treatment has been well known for many years (e.g., Drake and Essock 2009; Lehman and Steinwachs 1998; Resnick et al. 2005). As awareness of this problem has grown, repeated calls have been issued to increase access to effective practices (Drake et al. 2001; Institute of Medicine 2001; President's New Freedom Commission on Mental Health 2003), although the success of these calls has not been readily apparent. This situation seems to differ from other countries where national guidance and audit define the types of treatments that should be available. In more coherent health systems, successful implementation is difficult but not impossible. However, even where mandatory guidance and training are available, difficulties remain.

What accounts for the failure to disseminate effective psychosocial treatments for schizophrenia? There is no single answer to this question, as there are likely multiple obstacles to dissemination which vary as a function of country and its economic wealth, the geographic setting (e.g., urban vs. rural), and the intervention itself. Below, two different explanations for the poor dissemination of empirically supported interventions for schizophrenia are offered: (a) policy failures in training mental health professionals despite adequate tools for implementing effective treatments; (b) more research is needed to inform the process of implementing and adapting effective interventions in the context of routine service delivery.

Training Obstacles

A clear prerequisite to the dissemination of empirically supported psychosocial treatments for schizophrenia is the availability of suitable resource materials for different practices and established methods for implementing those practices in routine treatment settings. As previously described, empirically supported psychosocial treatments are standardized in manuals to guide clinicians in providing the intervention. Fidelity scales have also been developed to evaluate whether psychosocial interventions are provided with good adherence to the defining principles of each treatment approach (Bond et al. 2000; Teague et al. 2012). In addition, standard training methods have been developed to facilitate the implementation of different treatment programs. One research project has evaluated the effectiveness of combining these three resources into a cohesive package for implementing empirically supported psychosocial treatments for severe mental illness, as described below.

The National Implementing Evidence-Based Practices Project

This study was aimed at evaluating whether empirically validated practices could be implemented and sustained in routine mental health treatment settings (Bond et al. 2009; McHugo et al. 2007; Mueser et al. 2003). For each of five different psychosocial treatment approaches (supported employment, family intervention, illness management and recovery, integrated treatment for co-occurring disorders, and assertive community treatment), a standardized "toolkit" was created and included the following components:

- practitioner's manual,
- information brochures for different stakeholders, including clients, family members, practitioners, supervisors, and policy makers,
- instruments for evaluating outcomes,
- implementation tips,
- a 15–20 minute introductory video to the practice, and
- a 1–3 hour training video for the practice.

For each practice, a standardized two-day training to be delivered by experts was developed and access to expert consultation was facilitated for a two-year period.

A total of 53 publicly funded mental health centers participated in this study by receiving training and implementing two of the five interventions, which were chosen by each agency. Routine fidelity evaluations were conducted at baseline and at six-month intervals for two years. Findings indicate that all five programs were implemented with acceptable levels of adherence to the program models over the first 6–12 months of the project, and that acceptable fidelity levels were maintained up to the two-year assessment point. Results show that empirically supported psychosocial interventions for schizophrenia can be successfully implemented in routine treatment settings and suggest that other factors may be at least partly responsible for the failure of such treatments to be more widely implemented and disseminated.

Policy Implications for Training Mental Health Professionals

To ensure the provision of empirically supported psychosocial treatment for schizophrenia, four basic requirements are needed:

1. Standardized manuals for the intervention and methods for monitoring the quality of its delivery.
2. Practitioners who are trained in the treatment models.
3. Sufficient resources to support the provision of the treatment, including, if necessary, the training of practitioners.
4. Guidelines or incentives that prioritize the delivery of the treatment over less empirically supported interventions.

Standardized manuals for different empirically supported treatments exist, as do fidelity scales for evaluating implementation quality. In the National Implementing Evidence-Based Practices Project (McHugo et al. 2007), the cost of training practitioners was borne by the research project, not the agency; the voluntary engagement of agencies in the project probably ensured some level of motivation or incentive to implement the chosen practices as faithfully as possible to the fidelity criteria. This suggests that policy implications in the training of mental health professionals in psychosocial treatment deserve special scrutiny.

Training of Mental Health Professionals

In the United States, individuals who enter the mental health profession from fields such as clinical psychology, social work, nursing, occupational therapy, and even psychiatry receive little training in specific empirically supported psychosocial treatments for schizophrenia or even in programs with an evidence base over two decades old, such as family intervention (Dixon et al. 2001). For example, clinical psychologists graduating from Ph.D. programs in the United States, accredited by the American Psychological Association (APA), are not required to demonstrate competence in either the assessment or treatment of people with schizophrenia, nor does any state require such competence for the licensing of psychologists. Consequently, there are limited training opportunities for working with people with schizophrenia in clinical psychology Ph.D. programs, and many programs lack any faculty expertise in the treatment of serious mental illness (Reddy et al. 2010). The problem is not appreciably different in the other fields of social work, occupational therapy, or nursing. This shifts much of the cost of training in psychosocial treatments for schizophrenia from professional schools to the healthcare system, which leads to financial strain on limited resources. As proposed for clinical psychology (Mueser et al. 2013b), requiring competency in the psychosocial treatment of serious mental illness from students, who obtain advanced degrees and are licensed in the mental health profession, could reduce the burden of training clinicians on the healthcare system.

The relative lack of training in empirically supported psychosocial treatments for schizophrenia in professional programs in the United States for disciplines such as clinical psychology could also reflect the absence of generally understood roles or niches for the special skills of each profession within the psychiatric rehabilitation community and postgraduate career disincentives to working with this population. This is in marked contrast to the situation in Great Britain and Europe, where more defined roles for clinical psychologists and other mental health professionals in the treatment of serious mental illness have been established.

Still, even concerted training efforts may be insufficient to foster the broad uptake of some psychosocial interventions for schizophrenia. The most notable case example of this is family intervention for schizophrenia. Despite the development and empirical validation of several family interventions for schizophrenia, as reviewed here, and programs for training clinicians and disseminating the practice (Tarrier et al. 1999), access to these interventions remains problematic in both the United States and Great Britain. A wide variety of factors have been identified that influence the implementation of family intervention programs, including clinicians' attitudes about the effectiveness of family programs, organizational issues, clinicians' specific profession, and the willingness of relatives to engage in services (Fadden 1997; McCreadie et al. 1991; McFarlane et al. 2001; Wright 1997). These issues have been insufficiently addressed in research on the implementation of family intervention and other empirically supported programs.

Research on Implementation and Dissemination

The problem of poor access in routine care to effective psychosocial treatments for schizophrenia is not unique to the disorder, but is also present across the broader range of mental health and preventive interventions, where numerous other relatively complex interventions also enjoy a strong evidence base

(Glasgow et al. 2003; Institute of Medicine 2006). The challenge of making effective treatments widely accessible has led to a growing interest in the rapidly evolving field of implementation science, which has been described as the translational step between the development and empirical validation of interventions and the integration of these services into systems of care (Proctor et al. 2009). Although still in its infancy, a better understanding of the processes and factors relevant to the implementation of psychosocial treatments may be necessary to close the gap between science and practice.

The terms diffusion, dissemination, and implementation should be distinguished. *Diffusion* refers to the spread and uptake of new practices into systems of care (Rogers 2003), with *diffusion research* being the study of factors critical to the adoption of empirically supported interventions by providers of treatment for a specific population (Proctor et al. 2009). A wide range of contextual factors has been identified as influencing the spread of new practices, such as norms and attitudes about particular health conditions, organizational structure and process, resources, policies and incentives, networks and linkages within the organization, and media and other change agents (Mendel et al. 2008). Whereas diffusion may be a passive process, *dissemination* refers to active, targeted efforts to persuade key stakeholder groups to adopt a specific intervention, and the distribution of related information and materials designed to promote its successful adoption (Greenhalgh et al. 2004). Finally, *implementation* is the use of specific strategies aimed at introducing an empirically supported intervention within a specific treatment setting (Proctor et al. 2009).

Just as successful treatments for schizophrenia are evaluated by changes in client outcomes (such as symptoms and relapses, cognitive functioning, and psychosocial functioning), implementation efforts require attention to a different set of outcomes (Glisson and Schoenwald 2005; Proctor et al. 2009). Primary outcomes of interest to implementation research include:

- Feasibility: Is it possible for the intervention to be incorporated into routine services in an agency, including organizational structures and costs?
- Fidelity: Can the intervention be provided with good adherence to the defining elements of the practice?
- Penetration: What proportion of the targeted population in the setting receives the treatment?
- Acceptability: Can clients be engaged in the intervention and complete it, and are they satisfied with it?
- Sustainability: Can the intervention be maintained over the long-term?

The successful implementation of a practice is not sufficient to ensure that it will continue to be provided. A host of factors (e.g., organizational leadership, change in funding priorities) can lead to the dismantling of new practices (Massatti et al. 2008).

The issue of evaluating fidelity to the program model is a thorny one, as modifications to the original model may be undertaken to ensure the feasibility of implementing the practice into routine treatment, and to achieve acceptable levels of penetration (Aarons et al. 2011). Successive modifications of an intervention lead to the inevitable question: When has the implemented practice deviated significantly from the original practice, and is the deviation better or worse? A related issue is that high levels of program standardization are required to attain the necessary precision to evaluate an intervention's impact on the targeted outcomes and to establish its efficacy and effectiveness. For example, most cognitive remediation programs specify a core curriculum of cognitive skills and exercises as well as a set number or range of hours or sessions during which the training is to be completed (McGurk et al. 2005; Wykes and Reeder 2005). Similarly, family intervention programs also recommend curricula in terms of information and skills to be taught to families, and specific timeframes in which the teaching is to be accomplished (Barrowclough and Tarrier 1992; Falloon et al. 1984). However, in routine practice, there is a need to provide clinicians with practical guidance about how to tailor the intervention to the client's (or family's) needs (e.g., targeted teaching of curricula), how to determine whether the client has benefited, and when the intervention should be abbreviated or extended from the standard parameters originally developed for it. Implementation research is needed to address how (a) to modify and develop practical clinical guidelines for providing empirically supported interventions for schizophrenia in routine practice settings, (b) to evaluate whether adapted programs continue to improve targeted client outcomes, and (c) to alter fidelity criteria accordingly.

Models for guiding the evaluation of implementation efforts are still in the developmental stage and tend to be more descriptive than theoretical in nature (Aarons et al. 2011; Atkins 2009). Proctor et al. (2009) have proposed a heuristic framework for implementation research which provides a classification of the multiple levels that implementation strategies may target and that need to be assessed to evaluate performance improvement; this includes the larger system and environment (e.g., reimbursement, regulatory policies), the organization (e.g., structure, leadership), groups or teams (e.g., cooperation, coordination, sharing of knowledge), and the individual (e.g., knowledge, skill, expertise). The importance of recognizing the broad scope of change agents and factors which may be critical to successful implementation has led to interdisciplinary approaches that are more broadly inclusive and collaborative in involving multiple stakeholders (e.g., clients, clinicians, family members) over the full range of treatment development, implementation, and dissemination (Gonzales et al. 2002; Wells et al. 2004). The recent growth in use of participatory action research approaches to mental health (Knightbridge et al. 2006), including psychosocial treatments for serious mental illness (Cook et al. 2010), is an example of this trend, which has long-term potential to improve access to effective psychiatric rehabilitation for schizophrenia.

Increasing the Effectiveness of Psychosocial Treatments

While great progress has been made in psychosocial treatments for schizophrenia, even the most potent interventions fail to help a significant proportion of individuals. For many who do benefit, improvement is only modest. For example, although the majority of people with serious mental illness who receive supported employment obtain some competitive work over a period of 1.5–2 years, many clients work little or not at all, and those who do work often have brief job tenures marked with unsuccessful job endings (Bond et al. 2008; Mueser et al. 2004). In addition, the current armamentarium of psychiatric rehabilitation approaches targets a limited range of domains of functioning and consequences of schizophrenia. There is a need for effective interventions that address needs in other areas, such as psychological well-being, the effects of stigma and self-stigma, physical health, close personal relationships, and parenting skills. Several approaches to improving the effectiveness of psychosocial interventions and research on treatment are described below.

Targeting Factors Related to Change in Effective Psychosocial Treatments

Research aimed at understanding the individual who fails to benefit from empirically supported psychosocial interventions for schizophrenia, and the mechanisms underlying effective treatments, has the potential to lead to more effective interventions.

Who Benefits from Psychiatric Rehabilitation?

The identification of illness-related predictors of response to psychosocial treatment can lead to interventions which directly target those areas. Two examples illustrate the utility of research that has identified impaired cognitive functioning as a predictor of attenuated response to psychiatric rehabilitation. First, the severity of cognitive impairment is associated with poorer vocational functioning in schizophrenia and less benefit from a range of vocational rehabilitation approaches, including supported employment (McGurk and Mueser 2004). To address this issue, several research teams have developed cognitive remediation programs aimed at improving cognitive functioning and employment outcomes, in the context of vocational rehabilitation programs. RCTs of these combined intervention programs suggest that the addition of cognitive remediation is associated with greater improvements in both cognitive abilities and employment rates compared to vocational rehabilitation alone (e.g., Lindenmayer et al. 2008; McGurk et al. 2009; Vauth et al. 2005).

Second, more impaired cognitive functioning is associated with reduced acquisition of skills in social skills training (Mueser et al. 1991; Smith et al. 1999). Several approaches have been developed to address this problem and improve the ability of clients with more impaired cognitive functioning to learn social skills. Silverstein et al. (2009b) developed an attention training program for clients whose inattention prevents them from learning in social skills training groups and showed that the incorporation of this training into skills training groups led to better skills acquisition than social skills training alone. Brenner and colleagues (Brenner et al. 1994; Roder et al. 2011b) developed Integrated Psychological Therapy (IPT) to target systematically impairments in cognitive functioning, social cognition, and social skills, with the initial focus on practicing basic cognitive processes thought to be critical to learning more complex social cognition and interpersonal skills. Research has shown that the IPT program improves social functioning in schizophrenia (Roder et al. 2011a).

Research on How Psychosocial Treatments Work

Empirically supported psychosocial interventions for schizophrenia tend to be complex and multifaceted, creating a challenge for any concerted effort aimed at disentangling the mechanisms or critical ingredients responsible for treatment effects. Nevertheless, there are common or defining elements and goals of each intervention which may serve as beginning hypotheses. For example, the core features of family intervention programs are developing a therapeutic alliance between the family and treatment team, the provision of information to families about schizophrenia and its treatment, and the reduction of stress in the family. Critical elements of social skills training include systematic teaching (i.e., shaping) and practice of social skills in simulated situations, encouragement to try skills in social situations, and facilitated practice of skills in real-world settings. Although many different approaches to cognitive remediation have been developed, all seek to increase cognitive performance. The most effective approach for improving functioning includes a combination of practice and teaching strategies on cognitive exercises, which aids the transfer to psychosocial functioning. Key elements of illness self-management programs include providing information about mental illness and its treatment, teaching strategies to improve medication adherence, developing relapse prevention programs, and teaching coping strategies for persistent symptoms. Defining components of cognitive behavioral therapy for psychosis include the development or enhancement of coping strategies as well as cognitive restructuring aimed at helping people evaluate the evidence which supports upsetting thoughts and beliefs. Critical features of supported employment include rapid job search for competitive work, provision of practical supports in finding and keeping jobs, and attention to client preferences.

Evaluating whether an intervention succeeds in modifying the immediate targets of treatment (e.g., improved knowledge about schizophrenia in family intervention and illness self-management training programs), and the association between changes in those targets and functional outcomes, could serve to identify which targets are most critical. Similarly, determining which elements of an intervention are necessary for improving outcomes, and which are not, could shed light on the critical mechanisms that underlie benefit from treatment. For example, there is some evidence supporting many of the individual components used to define supported employment, such as rapid job search and follow-along supports (Bond 2004; Bond and Kukla 2011). Of course, different targets may be critical for different individuals under different circumstances, further complicating the process but underscoring the importance of efforts to understand how interventions work.

Increasing Specificity and Routine Monitoring of Targeted Outcomes

Psychosocial treatment methods for schizophrenia developed over the last several decades have the potential to incorporate considerable detail in assessment, treatment planning, decision making, and outcome evaluation of interventions (Spaulding et al. 2003). Such detail is necessary for the personalization and tailoring of treatment to the individual. These methods represent a convergence of several lines of work, including the case formulation approach to cognitive behavioral therapy (Beck et al. 1979), social learning theory (Bandura 1969), functional assessment and analysis of behavior (Bijou and Peterson 1971), problem-solving models of clinical practice (D'Zurilla and Goldfried 1971), and new conceptualizations of recovery (Anthony 1993).

The essence of person-oriented treatment is that individuals are actively involved in setting their own treatment goals, identifying the outcomes they most want to change and the treatment strategies they want to use, and actively working together to implement interventions. It should be noted that the individual's perspective, attitudes, values, and beliefs are incorporated into a biosystemic understanding of mental illness, as part of the individual's psychosocial functioning (Spaulding et al. 2003). To personalize psychosocial treatment and to make it as effective as possible, several issues related to the targeted outcomes need to be addressed, including

- a comprehensive assessment leading to the identification of specific goals or outcomes,
- access to reliable and sensitive measures of these outcomes, and
- the routine monitoring of progress toward desired outcomes.

Despite advances in the technology of treatment, the ability of clinicians to provide personalized treatment for schizophrenia is limited by overemphasis on symptoms and deficits, goals which focus on treatment adherence rather than desired outcomes, poorly specified goals, and lack of time and effort invested in monitoring progress toward goals. The scope of goal setting needs to address a broad range of psychosocial needs and desires beyond (or often instead of) coping with symptoms, such as:

- emotional well-being (e.g., experience of positive emotions, hope),
- role functioning (e.g., work, school, parenting, homemaker),
- social relationships (e.g., family, friends, intimacy),
- leisure activities,
- self-care and independent living (e.g., money management, grooming/hygiene, shopping/food preparation),
- physical health (e.g., diet, exercise, smoking, management of physical illnesses such as diabetes),
- creative expression, and
- community inclusion and involvement.

In addition, it is important to seek to understand the individual's perspective or experience of mental illness, such as demoralization (Birchwood et al. 1993), self-defeating thinking (Grant and Beck 2009), stigma, and self-stigma (Drapalski et al. 2013). Assessing these perspectives can be informative about other areas of impaired functioning (e.g., unemployment, social isolation) and may serve as potential targets for treatment. Some research suggests that negative attitudes or unhelpful beliefs can be fruitfully targeted, such as defeatist thinking (Granholm et al. 2009) and self-stigma (Lucksted et al. 2011; Yanos et al. 2012), which may contribute to improved well-being, personal growth, and better functioning (Roe and Chopra 2003).

Initial engagement and treatment focused on identifying and pursuing client-centered goals is critical to developing a therapeutic relationship (Tryson and Winograd 2001) and instilling motivation to learn illness self-management (Corrigan et al. 2001). Goals need to be described with sufficient specificity to permit reliable measurement and assessed frequently enough to provide useful information about whether progress is being made. The process of setting and making progress toward personal goals is important to psychological well-being and growth (Elliot et al. 1997; Sheldon et al. 2002). To ensure that clients are able to reap the full benefits of goal setting and attainment, large goals need to be broken down into smaller ones, and periodic monitoring of progress needs to be conducted to reinforce effort and to maintain the therapeutic alliance. Routine monitoring of targeted outcomes also enables the clinician and client to evaluate whether the strategies they are using to achieve the desired goals are working and should be continued, or whether alternative approaches should be considered.

Clinicians often have difficulty specifying goals with clients, and infrequently monitor progress toward goals, sometimes only when required by an agency (e.g., treatment planning conducted every six months). This makes it difficult for clinicians to "keep their eye on the ball" of the goal and to customize their work to the individual client accordingly. Assessment tools are needed which tap the broad range of needs that clients may have, are easy to administer and score, and for which targeted outcomes can be routinely assessed (e.g., monthly) to gauge progress. For the assessment and treatment of serious mental illness, increasing innovation in the use of e-technology may hold promise for the development of more comprehensive assessment packages that can be used both to identify needs treatment goals and to track progress routinely over time (Ben-Zeev et al. 2012).

The notion that rigorous clinical evaluation and systematic treatment are necessary to evaluate outcomes rigorously on a case-by-case basis, even for interventions that have been validated in controlled trials, is increasingly referred to as *practice-based medicine* (Horn and Gassaway 2007; Horn et al. 2010). There is a growing interest in this approach in the broader field of psychotherapy, with the development of commercialized clinical decision supports systems designed to facilitate this practice (Bickman 2008; Chorpita and Daleiden 2009; Lambert 2005), and a few applications for treatment of serious mental illness (Chinman et al. 2004; Iyer et al. 2005; Paul 1986). The recognition of clinical practice has the promise to inform research and treatment development and has led to initiatives aimed at enhancing the capacity of clinicians to use research methods to evaluate the effects of their interventions (Sulivan et al. 2005), and calls for *practice research networks* to facilitate the integration of research methods into clinical practice (Borkovec 2004).

Improving Outcomes by Thorough Training in Empirically Supported Interventions

The primary value of empirically supported interventions is often thought of in terms of having a standardized treatment that improves a targeted outcome. This is important, but as previously reviewed, these interventions do not work for everybody, and there are many domains of functioning for which effective practices have yet to be identified. Providing clinicians with a solid foundation in empirically supported psychosocial treatments may, however, yield additional benefits to clients, both in terms of better outcomes in domains which are the focus of established practices as well as other domains not previously targeted by those practices.

Empirically supported psychosocial treatments are standardized in manuals, and learning them usually involves a combination of reading, attending lectures, observing how other skilled practitioners use the model, practicing skills (both in role plays and with actual clients), and receiving regular supervision on efforts to implement the program. The structure imposed by each specific practice, and the process of learning a practice through observing how experts model it, combined with repeated opportunities to practice specific skills and receive supervisory feedback (i.e., shaping), provides an ideal platform for teaching core clinical competencies that extend beyond the specific practice itself. The repeated practice and honing of skills incorporated into an empirically supported program, to the point of overlearning, may enable clinicians to develop expert performance capabilities in the practice (Ericsson and Charness

1994). Clinicians may then be able to use this expertise in new, creative ways to improve their ability to address other problems and goals.

For example, training in family interventions involves learning how to reach out and engage significant others (along with the client) in a collaborative relationship with the treatment team, demonstrating concern and empathy for the challenges faced by relatives and significant other people, evaluating the concerns and priorities of family members, providing information, enlisting family support, and teaching stress reduction strategies. While family intervention programs were originally developed to reduce the risk of relapse and hospitalization following a recent relapse, the clinical skills involved in working with families have many other potential applications:

- The clinician working with a young mother with schizophrenia who had difficulty caring for her infant enlisted the help and support of her parents and sister to reduce the burden of caring for her child, to help her improve her parenting skills, and to facilitate the management of her mental illness.
- Two leaders of a social skills training program in a residential setting serving severely ill, formerly institutionalized clients with schizophrenia engaged and regularly met with frontline residential staff to educate them about serious mental illness, obtain their perspectives on problematic social situations experienced by the clients, review skills targeted in the skills training program, and collaboratively work out plans to help clients practice and use skills in appropriate situations in the residence (Bellack et al. 2004a).
- The supported employment specialist of a client who frequently missed appointments, and whose family was not supportive of him getting a job, reached out and engaged the family, and identified their chief worry about work-related stress causing a relapse. The specialist addressed these concerns by explaining to the family that involvement in meaningful and structured activities (such as work) can actually *protect* people with schizophrenia from stress, and then reviewed with them the client's current strategies for preventing relapses. This information allayed the family concerns, ensured their support, and led to several job leads from friends of the family.

Similar examples could be provided for other empirically supported treatments. For example, cognitive restructuring, a core skill learned by clinicians who provide cognitive behavioral therapy for psychosis, can be used to:

- address self-defeating thinking in a client who has had difficulty getting work ("I'll never get a job") and shows minimal follow-through on job search activities when participating in supported employment;

- reduce the severity of symptoms such as ideas of reference and thought broadcasting that interfere with a client practicing social skills in community situations; and
- help the parents with a son who has schizophrenia, who are participating in family intervention together, cope with and move past their pervasive feeling of loss and despair about his mental illness by reframing their understanding of what recovery means, and appreciating his potential to live a meaningful and rewarding life, despite having this disorder.

Thus, investment in training clinicians in multiple effective psychosocial practices could have synergistic benefits as many of the practices involve complementary skill sets. Expertise across different practices could increase competence at each of the established practices.

Summary and Conclusions

Psychosocial treatment (or psychiatric rehabilitation) of schizophrenia encompasses a great variety of interventions aimed at improving functioning across multiple domains, including cognitive functioning, social relationships, independent living skills, work or school, socioenvironmental adjustment, symptoms, and well-being. Psychosocial treatments can be provided in a variety of different modalities (e.g., individual, group, family) and may involve environmental modifications such as the provision of practical supports or contingent reinforcement of adaptive behavior. Over the past two decades, the recovery movement (led by persons with a mental illness, referred to as "consumers" in the United States or "service users" in Great Britain) has successfully challenged traditional medical definitions of recovery from mental illness as the remission of all symptoms and relapses, and has argued for a new definition of recovery that emphasizes living a personally meaningful life, including quality of social relationships, independent living, role functioning, and well-being. The recovery movement has also underscored the importance of a client participating actively in their own treatment, including the setting of goals and participation in shared decision making about treatment options.

Significant progress has been made in the development and validation of effective psychosocial treatments for schizophrenia, including contingent reinforcement (i.e., token economy), supported employment, cognitive behavioral therapy, social skills training, cognitive remediation, family intervention, and training in illness self-management. Most of these interventions, however, have not been routinely implemented in standard clinical practice, including programs for which there is an evidence base for well over a decade. Research shows that empirically supported psychosocial interventions can be implemented in routine treatment settings with good fidelity to the models. However, major obstacles to dissemination remain. One obstacle is the relative lack of training in empirically supported interventions for serious mental illness in graduate schools for mental health professionals (e.g., clinical psychologists, social workers). A second challenge is the need for more attention to understanding the processes involved in implementing an effective practice into routine care, such as organizational and financing factors, clinicians' attitudes and skills, the involvement of clients and other stakeholders, and methods for adapting a practice to the treatment setting while maintaining fidelity to the original model. Implementation science is still in its infancy but stands to play an increasingly important role in the field as further psychosocial treatments are empirically validated but wait to become integrated into usual practice.

Future psychosocial interventions for schizophrenia will benefit from further research aimed at understanding the predictors of benefit from empirically supported interventions and the critical components of effective treatment programs. Such research has the potential to target factors that limit response to treatment (e.g., impaired cognitive functioning) and to enhance the most important elements of intervention. There is a need to facilitate accurate and sensitive measurement of goals and outcomes so that clinicians can routinely monitor progress and tailor their interventions accordingly, thereby leading to more personalized treatment. The importance of individual tailoring of treatments also suggests that a linear process of research informing practice should not be assumed, and that practice has much to contribute to research. Thus, practice-based treatment, and establishing practice research networks, has the potential to improve treatment. Finally, the training of clinicians in empirically supported psychosocial interventions may reap benefits above and beyond their ability to provide any one practice. Training clinicians across the spectrum of effective treatments may lead to superior clinical competencies that improve their ability to tailor treatment to the personal needs of the client, and to target other areas for which established effective practices do not yet exist.

16

Creative Solutions to Overcoming Barriers in Treatment Utilization

An International Perspective

Wulf Rössler

Abstract

Despite the noteworthy changes in the provision of mental health services in all industrialized countries, there are still considerable deficits in the treatment and care of persons with mental illness. This is particularly true for those with serious illnesses such as schizophrenia as one- to two-thirds of all severe cases of mental disorders go untreated.

This chapter addresses the determinants of help-seeking behavior and methodological issues related to the assessment of needs for care. Help-seeking behavior is affected by (a) prior personal experiences in looking for assistance, (b) the social environment and the influence of significant others, and (c) the overall disease and treatment concepts of the individual. Several approaches taken toward reducing the proportion of untreated persons in need of help are discussed, ranging from a general political level to more specific health care policy. Finally, strategies are discussed for improving the way in which mental health professionals can shape their personal relationships with patients and learn to respect their ideas about the causes of their disorder so as to include them in all treatment decisions. In mental health care, the relationship between patient and therapist is one of the most important treatment factors, serving as a reliable predictor of outcome, regardless of diagnosis, setting, or type of therapy used.

Introduction

Remarkable changes have been made in how mental health services are provided in all industrialized countries where large state-run mental hospitals have been downsized in favor of outpatient services within communities. This process of deinstitutionalization has not only brought a shift of financial resources from inpatient to outpatient care, it has also been accompanied by a significant increase in those resources (Gustavsson et al. 2011). Furthermore, major developments have occurred in the pharmacological, psychotherapeutic, and psychosocial treatment of mentally ill persons.

Several surveys of the general population have revealed that a high prevalence of mental disorders can create an enormous burden for disease management (Rössler 2006). It also has become clear that many persons with such disorders do not use or do not receive mental health services tailored to their objective treatment needs (Alonso et al. 2004, 2007; Bijl et al. 2003; Kessler et al. 2005; Saldivia et al. 2004). For example, Bijl et al. (2003) analyzed the prevalence rates and treatment estimates from Canada, Chile, Germany, the Netherlands, and the United States—all countries with considerable variance in their mental health treatment settings. Estimates for 12-month prevalence (i.e., the proportion of a population under investigation that has experienced a mental disorder during the past year) range between 17.0% (Chile) and 29.1% (United States). Treatment rates vary significantly across countries, from a low of 7.0% in Canada to a high of 20.3% in Germany, with a U.S. rate of 10.9%. Prevalence rates and overall treatment rates show no direct relationship. Although the probability of receiving treatment is strongly related to the severity of the disorder, between one- to two-thirds of all persons with serious cases obtain no treatment in a given year (Bijl et al. 2003). Undertreatment of serious cases is most pronounced among young, poorly educated males.

European countries show considerable differences in their use of drug and psychotherapy treatments, which are difficult to explain. For example, in Great Britain, almost 40% of all treated patients are given medication, but the proportion of patients receiving psychotherapy is comparatively low (about 12%). By contrast, just over 15% of patients in Denmark with mental health problems receive medication. Overall, the proportion of people who undergo psychotherapy because of mental problems is much smaller (10–20%) than for those who get medication. The fact that nonspecialists rarely provide medication and psychotherapy demonstrates the severe undertreatment of such disorders in primary care (OECD 2012).

These high rates of unmet needs have provoked critical discussion over the structure of psychiatric and psychosocial services. Potential reasons for mentally ill persons not receiving mental health services according to their needs can range from the individual (e.g., a patient's subjective perception of the illness, a caregivers' influence, questions of demand and supply, socioeconomic factors, or the impact of society, such as that pertaining to stigmas associated with mental illnesses) to questions about the appropriateness of professional services that are offered.

In this chapter, this topic is approached from several perspectives. We begin with a general discussion of what is meant by the concept of "unmet needs." Thereafter we focus on those needs for severely mentally ill patients, in general, and for persons with schizophrenia, in particular.

The Concept of Needs

The concept of needs is intuitively quite appealing. It implies that we have a checklist available for the objective criteria of "need." The starting point for such a need evaluation is the prevalence rate of mental illness in the general population.

Over the last few decades, highly structured research interviews were developed that allow a reliable assessment of mental symptoms and, consecutively, the identification of "cases" in large population samples. Over time, modifications to these assessment instruments have revealed their sensitivity to seemingly small changes and, likewise, their limitations when defining needs for care and treatment (Regier et al. 1998). Unfortunately, due to differences in the construction of these instruments, some of the best-known general population surveys have produced quite different rates for individual disorders (Andrade et al. 2003; Andrews et al. 2001; Bijl and Ravelli 2000; Jenkins et al. 1997; Kessler et al. 2003; Regier et al. 1993). Regier et al. (1998) compared two large-scale surveys conducted in North America at approximately the same time: the Epidemiological Catchment Area Study (ECA) and the National Comorbidity Survey (NCS). For ECA, they calculated selected prevalence rates of 4.1%, 4.2%, 9.9%, 1.1%, and 1.6% for diagnoses of alcohol dependency, major depression, anxiety disorder, panic disorder, and social phobia, respectively; in NCS, prevalence rates were 7.4%, 10.1%, 15.3%, 2.2%, and 7.4%. Similar differences were found when lifetime prevalence rates were compared.

It is difficult to interpret such diverging figures and to identify the magnitude of the population in need, if we do not assume that those values indicate true differences. It is much more likely that the difficulties in making a reliable case assessment contribute the most to those differences (Cooper and Singh 2000). However, not only do these contrasting data raise concerns about the comparability of different studies, assessments which reveal consistently high rates also invite serious questions about the clinical significance of all of these disorders. Thus far, this (epidemiological) discussion has made clear that the recourse to symptoms is not sufficient when defining a group of persons in need (Wittchen 2000). DSM-IV included clinical significance criterion, which requires that symptoms cause "clinically significant distress or impairment in social, occupational, or other important areas of functioning." This rule attempts to minimize false-positive diagnoses in situations where the symptom criteria do not necessarily indicate pathology on pragmatic grounds. This argument has been particularly emphasized in the discussion over the revision of the DSM-5 (Regier et al. 2013).

Because most human behavior is located along a continuum, no clear cutoff point exists to separate good health from illness and, as such, define a point where the need for treatment exactly begins. Instead, categorical classification systems do not represent natural illness entities but, rather, constitute agreed-upon definitions for designating a (certain) mental illness on pragmatic grounds.

This continuum approach has been widely accepted for affective disorders (Angst et al. 2003). Comparable emerging debate now concerns psychotic disorders. Within a general population, van Os et al. (2009) have calculated a rate of about 5% for psychotic symptoms that are below the threshold of a psychotic disorder (see also Rössler et al. 2013)—a percentage which is five times higher than that reported for full-blown schizophrenia.

From a professional perspective, a significant proportion of persons with subclinical psychosis display mental symptoms which are, to a varying degree, accompanied by functional disability. Using data gathered over a period of 30 years from a community cohort in the Canton of Zurich, we have concluded that those symptoms are associated with significant dysfunction in social roles (Rössler et al. 2007). These symptoms are of clinical importance because their presence may increase the risk for comorbid mental disorders (Rössler et al. 2011), including those related to substance use (Rössler et al. 2012). In terms of schizophrenia spectrum disorders, the reference population is obviously much larger than has been commonly assumed. Because persons affected at a threshold below that of a psychosis diagnosis are subjectively distressed and in need of help, it is quite likely that physicians "upgrade" the symptomatology to a respective psychiatric diagnosis, which then officially allows them to pursue psychiatric treatment.

Determinants of Help-Seeking Behavior

Help-seeking behavior by an individual is affected by prior personal experiences in looking for assistance, by the social environment and the influence of significant others, and by the overall disease and treatment concepts of that individual. To evaluate these concepts as a whole, Lauber et al. (2000) conducted a representative population survey in Switzerland with a detailed focus on lay opinions about mental disorders and their treatment options (Lauber et al. 2001). Respondents were presented with two vignettes based on DSM III-R: one described depression; the other, schizophrenia. Respondents were then asked for their impressions about what might be helpful in treating those disorders. Respondents could choose from a list of health services and professions, while also selecting various individual treatment measures. Overall, 68% chose "psychologist" first, followed by "family doctor" (57%) and "psychiatrist" (51%). After "psychotherapy" (42%), "inpatient treatment" or individual treatment measures (e.g., "medication" or "electroconvulsion therapy") were recommended by less than 20%. Within a similar magnitude, "homeopathy" (19%) or "natural remedies" (20%) were also proposed.

The answers differed with respect to the two vignettes. A larger portion of the respondents recommended consulting a "psychiatrist" for schizophrenia than for depression. Furthermore, drug proposals varied, with the public distinguishing between antidepressants and antipsychotics. "Psychiatric hospitalization" and "psychotherapy" were considered to be more helpful for schizophrenic persons than for depressive individuals. However, antipsychotics, "psychiatric hospitalization," and "psychotherapy" were considered more harmful for depressive than for schizophrenic individuals.

With regard to the different diagnoses, it was crucial when suggesting a treatment to know if the disorder was perceived as a "life crisis" or a "disease." Those respondents who considered the person to be in a life crisis preferred nonmedical interventions such as "social workers," "telephone counseling," "naturopaths," and "homeopathy." They opposed standard psychiatric therapy that included "psychiatrists," "psychopharmacology," and "psychiatric hospitalization." However, if the described person was perceived to be mentally ill, respondents recommended significantly more traditional psychiatric intervention strategies ("psychiatrist," "psychotherapy," and "psychopharmacology"). Furthermore, treatment strategies viewed as an alternative to traditional medicine (e.g., "naturopaths" or "vitamins") that were also used "to deal with the situation alone," were viewed as harmful.

Such concepts must necessarily have an impact on help-seeking behavior. First, it is striking that laypeople have confidence in individual persons and not treatment measures. Above all, family physicians (and, in Switzerland, also psychiatrists) appear trustworthy; their treatment methods, however, are seen as significantly less reliable. The fact that the professional group "psychologist" and the method of treatment "psychotherapy" received the most nominations makes clear that most people want nonstigmatizing professional help. Pharmacological treatment obviously is considered in opposition to such a concept because it implies that those affected will lose control over their lives when using medication. This is also suggested by the general linguistic usage concerning medication: "chemical straitjacket" or the "tranquilization" of a person.

Against this attitudinal background, an affected individual—often in accord with their family or other important caregivers—then decides in a second step whether and what help he or she wants to utilize. Because financial barriers in Central European health systems do not play a major role in this decision-making process, so-called "convenience factors" shape a person's help-seeking behavior. One well-analyzed factor is the distance or travel time to a required institution. Approximately 40 years ago, a German Expert Commission assumed that a service user would accept a travel time of one hour. However, as several of our own analyses have shown, this threshold is far too high for psychiatric patients. In fact, travel time of just a half hour reduces the number of willing users by 50%. This applies to both outpatient (Rössler et al. 1991) and inpatient care (Meise et al. 1996). Thus, the decision to engage a health service is caught in a delicate balance, as demonstrated by the short-term decline in the utilization of psychiatric emergency services in Scotland during the 1990 Football World Cup (Masterton and Mander 1990).

Socioeconomic Factors

Other factors that influence individual help-seeking behavior include those grounded in the social and environmental living conditions of potential users. The environmental perspective refers to the social characteristics of a geographical region, whereas the social perspective describes the individual psychosocial characteristics of those persons affected. Analyses about how socioeconomic factors affect the onset and course of mental disorders, as well as the utilization of (mental) health services, have a long tradition in social-epidemiological research. Since the pioneering work of Faris and Dunham (1939), the association between social factors and the development of mental disorders has been discussed controversially. Nevertheless, the relationships between indicators of deprivation (mostly socioeconomic indicators of difficult living conditions combined with a lack of social support) and help-seeking behavior are well documented and, in health planning, widely accepted (Folwell 1995; Gaebel et al. 2012; Lancet Global Mental Health Group et al. 2007).

Stakeholders

Unlike most other medical disciplines, agreement has been scarce in the field of psychiatry among patients, caregivers, and professionals over the causes of mental disorders and how they should be treated.

Professionals

By law, in all European and Anglo-Saxon countries, the task of an objective needs assessment lies with a physician. In general, that assessment must precede the making of any decision over a treatment, which must then be labeled "necessary," "sufficient," and "appropriate." If no treatment type can address a need appropriately, then one deems this to be a "no (objective) need."

Objectives defined by professionals, however, do not necessarily correspond to those of patients. Physicians often seek to relieve a symptom, a goal they see as a necessary and sufficient prerequisite for a better quality of life. For patients, however, functional integration into family, work, and society is of utmost importance (Eichenberger and Rössler 2000).

To make things worse, psychiatrists are not the natural allies of persons with a mental illness. At best, psychiatrists hold the same opinion about them as does the general public (Lauber et al. 2004a). If we assess the attitudes of psychiatrists who work in institutional settings, we find that they confirm even more stereotypes about patients (especially with respect to patients with schizophrenia) than does the general public or even other professional groups who are involved in such treatment and care (Nordt et al. 2006).

To ensure that psychiatric treatment is effective, it is crucial to improve cooperation between patients and psychiatrists. We already know that a better therapeutic relationship between the two parties is associated with closer adherence to medication among patients with schizophrenia (McCabe et al. 2012). In mental health care, the relationship between patient and therapist is one of the most essential treatment factors, serving as a reliable predictor of treatment outcome, regardless of diagnosis, setting, or type of therapy used. Any perceived loss of autonomy will accompany a more negative dynamic between patient and clinician (Theodoridou et al. 2012).

This relationship is not necessarily restricted to the dyadic situation between a patient and one therapist, nor does it focus exclusively on clinical outcomes. The therapeutic relationship is quite often extended to an entire team or several members of that team. Thus, a good relationship improves quality of life for the patient in general and vocational outcomes in particular (Catty et al. 2010, 2011).

Patients/Users/Caregivers

Concepts of disease and treatment diverge between patients and professionals and can result in significant tensions. This is especially true for families with a member who suffers from schizophrenia. To address this, various psychoeducational programs have been developed to enhance understanding of schizophrenia, to provide information about various treatment options, and to enrich their coping strategies to deal with crises more successfully. The overall objective has been to reduce a families' burden (Dixon et al. 1999).

From their perspective, caregivers for schizophrenic patients prefer early and prompt (inpatient) treatment over a sufficient length of time, because a significant proportion of caregivers experience physical violence prior to an acute exacerbation of the disease (Lauber et al. 2003). By contrast, patients often try to avoid inpatient treatment or opt for minimum care which takes the least amount of time. Caregivers often require detailed information about the course of treatment, which may not be disclosed to them without the consent of the patient. In the case of (premature) discharge from inpatient treatment, family members desire the broadest possible support for their caring responsibilities.

Some of this may explain the different attitudes of patients and caregivers toward medications. Caregivers prefer, whenever possible, to use depot medication to gain control, whereas patients themselves want maximum autonomy concerning their treatment (Jaeger 2010). For reasons of convenience, depot medication might be, for patients, a reasonable alternative to oral medication. However, on the whole, the degree to which an intervention is acceptable to a patient cannot be ignored (Perkins 2001).

Above all, patients seek to gain self-esteem. The concept of empowerment also entails being able to exercise control over a treatment through self-determination and participation (Scott et al. 1999). These dimensions are strongly correlated with quality of life (Rogers et al. 1997). Although these considerations are part of the British health policy (Lelliot et al. 2001), they are not applied consistently when the primacy of evidence-based medicine is postulated during the selection process of various treatment approaches.

The concept of empowerment, self-determination, and active participation in treatment has led to a new view of the course of serious mental illnesses, particularly schizophrenia. While professionals have for many decades held very pessimistic opinions about the course of schizophrenia, it is quite clear today that there is a remarkable heterogeneity of outcomes for persons with that disease. Even if many of the affected do not return to premorbid functioning, most afflicted persons have a good chance for symptom remission, independent living, vocational integration, intimate relationships, etc. Many services are now available that have adopted such a recovery-oriented view for patients (Farkas 2007).

Public

The general public holds quite specific ideas about mental disorders and the objectives of psychiatric treatment. Their attitudes are determined by their discomfort against the mentally ill, which then leads to demands for utmost security. In the above-mentioned representative population survey in Switzerland, more than 70% of respondents favored compulsory treatment in the case of mental illness (Lauber et al. 2002). Between 60% and 75% said that a driver's license should be revoked in such cases, while 26% to 39% proposed that a pregnant woman should consider an abortion if she has ever suffered from a severe mental disorder. In addition, between 19% and 34% of the overall population has recommended that the right to vote be withdrawn in the event of a mental disorder (Lauber et al. 2000).

These wide-ranging expectations for psychiatric treatment and care make it clear that, in terms of resource allocation, psychiatry cannot have top priority. However, where the security needs of the population are very high, the public believes that considerable investments can be made (e.g., in the case of drug addicts or sex offenders) to ensure that such persons are excluded or marginalized from general societal life. All of this makes clear how urgently we need more public education about schizophrenia and other serious mental illnesses to reduce their attendant stigma and discrimination.

Cost-Benefit Ratio

Accompanying these tense relationships between patients, medical staff, and the population at large is another dimension: cost-benefit ratios. For efficient use of resources, it is essential to choose treatment methods with the best ratio. For example, when assessing schizophrenia treatments, we can either increase the vulnerability threshold (drugs or psychosocial means) or reduce environmental stresses (e.g., via sheltered work and housing). Drugs are highly effective in preventing relapses and can also be applied in psychosocial treatments, albeit to a lesser extent. Mojtabai et al. (1998) have demonstrated in a meta-analysis that a combination of medication and psychosocial treatment delivers the best results. From a health economics perspective, however, the gap is wide among costs associated with these different approaches. Out of the overall cost for treating schizophrenia, sheltered living and working environments account for approximately 40% of the annual totals, while the expense of drugs amounts to only about 6% (Salize and Rössler 1996; Salize et al. 2009). Although the costs of psychotropic medication for treating schizophrenia have tended to remain at that level, a recent cost analysis of six European countries (Salize et al. 2009) has demonstrated a 12-fold difference in the total treatment costs for schizophrenia between the "cheapest" and "most expensive" country. That difference is mostly due to the costs of sheltered living and working in those places.

Overcoming Barriers in Treatment Utilization

Each year, one- to two-thirds of all serious cases of mental disorders go untreated (Bijl et al. 2003). Several significant factors that influence help-seeking behavior, such as structural deterrents to service utilization or attitudinal barriers by the general public, could be addressed at a political level.

Financial barriers do not play a compelling role in precluding utilization in Central Europe: almost 100% of the population is covered by health insurance, and the public generally has free and equal access to all health services. In other parts of the world, including the United States, objective problems arise in providing and, consequently, taking advantage of mental health services.

A remarkable attitudinal change has occurred in some portions of Europe, especially toward affective disorders, for which there has been a major shift in de-stigmatization. This development can be attributed to the fact that the general public acknowledges today that affective symptoms are part of our everyday life, whereas psychotic symptoms seem to refer only to a very small proportion of the population. The emerging topic of subclinical psychosis has allowed professionals to convey to citizens the sense that psychosis is much more widespread in the general population than has historically been thought: (subclinical) psychosis is something that may apply to one's own family and is not something that usually happens only to others.

From a health policy perspective, it seems advisable to broaden the concept of psychosis, because resource allocation in health care follows the idea that financial resources will be preferably placed where we expect the highest health gains for the general population. An enlarged concept increases the proportion of the general population who might be affected and, likewise, improves the probability that those financial resources will be invested toward caring for psychotic disorders.

To reduce the stigma attached to mental illness, we must demonstrate the advantages of a life-crisis model concerning the onset of mental disorders when compared to a disease model. The general public perceives environmental stressors to be a significant source that impacts the onset of those disorders. Media coverage of "burnout" has contributed to that process because it reflects these lay opinions. However, the appeal of a concept such as burnout does not (yet) extend to schizophrenia, which has been conceptualized more as a disease (Lauber et al. 2002). By applying the ideas of subclinical psychosis, where environmental factors are more prevalent (Rössler et al. 2007), the life-crisis model could be promoted in the same manner, thereby leading to greater (societal) acceptance of persons with schizophrenia.

The existence of burnout allows mental health workers as well to take a different view of their own profession. Several studies have identified stressors that are unique to the psychiatric field. These challenges range from the stigma of the profession, to particularly demanding relationships with patients and difficult interactions with other mental health professionals as part of multidisciplinary teams, to personal threats from violent patients. Other sources of stress are a lack of positive feedback, low pay, and a poor work environment. Finally, patient suicide is a major stressor, upon which a majority of mental health workers report posttraumatic stress symptoms (Rössler 2012).

In clinical practice, professionals could encourage better help-seeking behavior by improving the way in which they shape their personal relationships with patients, respect their ideas about the causes of their disorder, and include them in all treatment decisions. In mental health care, the relationship between patient and therapist is one of the most important treatment factors, serving as a reliable predictor of outcome, regardless of diagnosis, setting, or type of therapy used (Theodoridou et al. 2012). Professionals should also reconsider their attitudes toward those with mental illness. Rather than revering the deficit model of schizophrenia, hope and optimism should be offered, knowledge about the illness and relevant services should be provided, and the empowerment of our patients should be supported, as these domains have been identified as important to a recovery orientation (Resnick et al. 2004). Lending hope and optimism will also reduce the self-stigma of the affected person (i.e., blaming oneself for the disorder) because increased self-stigma is associated with a decreased willingness or ability to seek help (Rüsch et al. 2009).

In fact, the true scandal lies in the treatment of the severely mentally ill, particularly individuals with schizophrenia. Based on current knowledge, only a minority of these patients receives appropriate treatment and care. Already in 1998, Lehman stated that significant gaps exist between scientific knowledge about the efficacy of treatments and the availability of those treatments in routine practice (Lehman 1998). The Schizophrenia Patient Outcomes Research Team (PORT) provided, in 2009, a comprehensive summary of current evidence-based psychosocial treatment interventions for persons with schizophrenia (Dixon et al. 2010). PORT produced eight treatment recommendations: assertive community treatment, supported employment, cognitive behavioral therapy, family-based services, token economy, skills training, psychosocial interventions for alcohol and substance use disorders, and psychosocial interventions for weight management. Only a few have been implemented in routine clinical practice settings. This raises serious concerns about access to care, as well as the appropriateness and quality of care that is offered—aspects which lie mainly within our realm of responsibility.

First column (top to bottom): Vera Morgan, René Kahn, Kim Mueser,
Sharmili Sritharan, Wulf Rössler, William Spaulding, and Andreas Meyer-Lindenberg
Second column: Richard Keefe, Wulf Rössler, Anil K. Malhotra,
Andreas Meyer-Lindenberg, Til Wykes, Richard Keefe, and Sharmili Sritharan
Third column: Til Wykes, William Spaulding, Vera Morgan, René Kahn,
Karoly Nikolich, Anil Malhotra, and Kim Mueser

17

What Is Necessary to Enhance Development and Utilization of Treatment?

Vera A. Morgan, Richard Keefe, René S. Kahn,
Anil K. Malhotra, Andreas Meyer-Lindenberg,
Kim T. Mueser, Karoly Nikolich, Wulf Rössler,
William Spaulding, Sharmili Sritharan, and Til Wykes

Abstract

This chapter is framed by four perspectives. The first views schizophrenia as a heterogeneous disorder that may present itself very differently across individuals. Heterogeneity has important implications for treatment approaches: for treatments to be optimally effective, they need to be tailored to the individual. The second perspective integrates what are often considered separate and divergent approaches to treatment in schizophrenia, bringing together, in both complementary and synergistic combination, the biological and the psychosocial. Within this perspective, clinical treatment models capable of being personalized to heterogeneous, individual profiles are proposed. The third perspective is a practical one that examines how the two extreme ends of the treatment continuum, treatment development and service delivery, can be optimized to ensure enhanced outcomes for people with schizophrenia. Finally, treatment is viewed from the perspective of the whole person. This not only has implications for mental and physical well-being and quality of life in people with schizophrenia, but also takes into consideration the social context in which these individuals are placed. Overall, the approach offered, with its integration across multiple domains, emphasizes the potential for improved recovery rates and hope for prevention and cure in this devastating disorder.

Schizophrenia: Cure and Recovery

A Cure for Schizophrenia

This is a very exciting time for schizophrenia research. The journal *Nature* has declared the decade beginning in 2010 to be the "decade for psychiatric disorders" in the hope that the neuroscience tools and paradigms developed in the "decade of the brain" can now be applied to identify new and better treatments for mental illness.

Nowhere is this more urgent than for schizophrenia. The number of mechanistically novel pharmacological treatments for schizophrenia has been disappointingly low. Government-sponsored large-scale naturalistic trials have suggested that there is little difference in efficacy between newer antipsychotic agents compared to older drugs. Most patients with schizophrenia still do not marry, have a severely compromised educational and job trajectory, and die on average 15 years earlier than the general population. The illness definition is currently based on a combination of psychopathological features observed by patients and their caregivers as well as duration criteria. To be useful for treatment, our understanding of the individual processes lumped together under the label "schizophrenia" will have to be refined for individualization to be possible.

The most useful entry points into this process are the clear biological risk factors for the illness: genetic and environmental factors. The pursuit of risk mechanisms is therefore a major strategy to individualize treatment and to find new treatments (Meyer-Lindenberg 2010a) because truly novel targets for molecular therapies can only emerge from the detailed understanding of the molecular mechanisms of the illness; genes with their products, if they have been found through hypothesis-free searches, are pointers to such novel mechanisms. Ever better designed, better characterized, and larger multinational studies are necessary and are being pursued to identify new genetic and environmental risk factors. Longitudinal studies which focus on the period around adolescence are underway in several countries and are expected to give us a better understanding of the way these risk factors interact with brain development. Another potentially paradigm-changing advance is that, through techniques such as induced pluripotent stem cells, we are now able to generate neurons from a person with a known disease history and genetic makeup and study the metabolism and activity in these cells across their development, giving us access to the "target tissue" of psychiatry in a way that seemed impossible even a few years ago (Brennand et al. 2011).

The pursuit of these new targets necessitates, in principle, the use of the entire armamentarium of modern neuroscience. For the first time in history, psychiatrists truly need and can use techniques from whole brain genome sequencing and epigenetics to expression mapping, proteomics, and lipidomics to pursue their goals. One critical and specific task for psychiatric neuroscience is to integrate this information with an understanding on the neural systems level (e.g., in the technique of "imaging genetics") so as to bridge the gap between cellular–molecular mechanisms and disturbed behavior.

An understanding of the mechanisms underlying schizophrenia is also critical for the generation of better animal models for use in the identification of new drug molecular candidates. Schizophrenia affects human-specific faculties such as language and higher cognition. Clearly, these features cannot be modeled in animals. Animal models currently used for schizophrenia have been directly derived from the profile of the currently used antipsychotic agents that are related to dopaminergic blockade. There is little evidence that these behavioral features are a good model for schizophrenia. A better understanding of mechanisms could be decisive in designing a new generation of animal models that are more predictive for efficacy through delineating neural systems that are implicated in schizophrenia.

Sometimes, however, "the best experimental animal is the human." This is especially true in psychiatric drug development, where the success rate in predicting which new medications are effective is disappointing. A new generation of applying systems-level neuroscience in early drug trials in humans will constitute a revival and focusing of experimental medicine in psychiatry. This concept, which has been extraordinarily fruitful in bringing about advances in oncology and hematology, is ripe for application for schizophrenia.

It is entirely possible that, in the end, "the answer" about schizophrenia is sufficiently complex as to require the study of the multiple risk pathways that combine in a given person to push him or her over the threshold to develop the illness. For schizophrenia research, computational approaches and especially computational neuroscience will be tremendously important to be able to quantify the effects that perturbations on genetic and environmental levels have on systems-level function. A comprehensive characterization of the neural risk architecture of schizophrenia through these various approaches, and their integration, provides a crucial translational research strategy for advancing new treatments for the illness.

Recovery in Schizophrenia

Over the past twenty years, the recovery movement has evolved to become a driving force in changing how major mental illnesses, including schizophrenia, are understood and treated (Silverstein and Bellack 2008). "Consumers" of mental health services (also called "service users") have protested against the pessimistic messages they have been given about the long-term outcome of serious mental illness, pointing to longitudinal research that shows symptom remission and functional improvement in significant proportions of people with schizophrenia (Davidson et al. 2005; Deegan 1991). Consumers have also argued for the reduction of coercive interventions and a change from hierarchical decision making to more collaborative approaches that respect their individual preferences and their need to determine their own treatment priorities (Chamberlin 1997b; McLean 1995). Perhaps the most significant impact of this movement has been its challenge of traditional medical perspectives on recovery from mental illness that have emphasized remission of symptoms and associated impairments, in favor of more nuanced and personally meaningful definitions. For example, *remission* from schizophrenia has been defined in the medical community in terms of meeting distinct thresholds of sustained improvement in symptomatic, cognitive, and functional domains of the disorder (Andreasen et al. 2005). Although there is less agreement about how *recovery* from schizophrenia should be defined, it has been broadly conceptualized as encompassing remission of symptoms and functional impairments, while also extending to improved quality of life (Leucht and Lasser 2006).

New conceptualizations of recovery focus on personal growth, and establishing meaning and sense of purpose in life, despite having a mental illness (Anthony 1993). The desire for a more personally meaningful definition of recovery than symptom remission frequently evokes different areas of psychosocial functioning. For example, the President's New Freedom Commission on Mental Health (2003:6) defines recovery as "…the process in which people are able to live, work, learn, and participate fully in their communities." Thus, according to the recovery movement, improvements in psychosocial functioning are a greater treatment priority than symptom management or remission.

Is Prevention Feasible?

Prevention of illness is preferable to cure, but what is the disease target of the prevention? Is it psychosis generally or schizophrenia specifically? The two are not the same, and the focus of the intervention may differ, depending on the disease target. Most studies of risk intervention prior to illness onset focus on psychotic-like experiences. However, to date, there has been minimal success in identifying which young people will convert to psychosis within high-risk and prodromal samples (i.e., help-seeking groups whose at-risk status is determined in the clinic, generally on the basis of psychotic symptoms). Moreover, it remains unresolved whether these interventions prevent schizophrenia or ameliorate its course. Psychotic symptoms may be too far along the illness trajectory to be a viable target for the prevention of schizophrenia. It is likely that, by the time first episode cases are manifest, a critical point for primary prevention has been missed. In this regard, several studies have demonstrated that psychosis is preceded by cognitive and social dysfunction by almost a decade, suggesting that prevention may need to start years earlier, targeting cognition and social function, rather than the more common target of psychotic-like experiences. This does not reject the at-risk state and prodrome as targets for secondary prevention, which we examine in some detail later.

Defining Prevention

We use the term *primary prevention* to refer to broad public health interventions that reduce incidence of illness or comparable problems in the general population, for example, nutritional programs, after-school activity programs, general health education. We use this to refer to interventions in the *pre-prodromal* period in schizophrenia.

We use the term *secondary prevention* to refer to focused interventions that target subpopulations identified as being at risk for developing illnesses or comparable problems, for the purpose of preventing the actual onset. We use this to refer to interventions in the *at-risk* or *prodromal* period in schizophrenia.

We use the term *tertiary prevention* to refer to focused interventions that target subpopulations *after* the onset of illnesses or comparable problems, for the purpose of minimizing morbidity or chronicity. We use this to refer to interventions *after illness onset* in schizophrenia.

A Target for Primary Prevention in Schizophrenia: Relative Decline in Cognition in Early Adolescence

As described by Kahn (this volume), a decline in cognition relative to peers (developmental lag) in late childhood/early adolescence may be the strongest indicator of early manifestations of the illness. We distinguish cognitive decline from enduring cognitive deficit. While both forms of cognitive deficit may increase the risk of schizophrenia, our focus here is on cognitive decline as it is likely to be more reflective of eventual psychosis, and may be most amenable to primary prevention. We note also that cognitive decline is distinguished from normal variability in IQ that has been observed in adolescence (Ramsden et al. 2011).

Our proposal, in its current state, is not a model of clinical intervention, but a research strategy with the potential to lead to primary prevention. The strategy involves identifying, on the basis of school grades or similar measures, children in late childhood/early adolescence living in the general community, who exhibit cognitive decline relative to their peers. These children would be the target of school-based interventions. It is important to note that schizophrenia would neither be a necessary nor sole endpoint in this proposed strategy, as the intervention is likely to have an impact on a range of disorders. By taking this approach, however, we may learn how to predict and prevent schizophrenia on the basis of relative cognitive decline.

What Is This Thing Called Schizophrenia?

The proposed strategy outlined above considers cognitive decline as a closed construct within an open construct. If a putative illness, in this case schizophrenia, is an open construct, its exact features and parameters are indistinct or unknown, and the distinction between primary, secondary, and tertiary prevention is unclear. For example, if cognitive decline is understood to be an early expression of schizophrenia, interventions directed at it are, by definition, after onset, thus constituting tertiary prevention aimed at arresting further cognitive decline and/or further progress of the illness. On the other hand, if cognitive decline is understood to be a risk factor or part of the prodrome, intervention is understood to be primary or secondary prevention. However, these are semantic distinctions. The value of identifying and responding to cognitive decline has obvious importance, regardless of whether it is a risk factor, a prodrome, or an early expression of the actual illness.

Specificity to Schizophrenia

It appears that cognitive dysfunction—at least prior to psychosis onset—distinguishes schizophrenia from bipolar disorder. A consistent pattern emerging from population-based studies worldwide is that low IQ constitutes a risk factor for schizophrenia, but not for bipolar disorder or depression (Reichenberg et al. 2002; Zammit et al. 2004; Sørensen et al. 2012). Moreover, one study found that children with excellent school performance had almost four times the risk of developing bipolar illness compared to children with average grades (MacCabe et al. 2010). A number of studies have reported a decline in cognitive function prior to the onset of schizophrenia (Fuller et al. 2002; Reichenberg et al. 2010). Whether a decline in cognitive function precedes the onset of bipolar disorder has not been addressed in population-based studies. However, a study of monozygotic and dizygotic twins discordant for bipolar disorder (Van Oel et al. 2002) found, in contrast to findings from a similar study of discordant schizophrenia twins, that the twin who went on to develop bipolar disorder, compared to the unaffected co-twin, did not do worse at school and only showed a temporary decline in functioning, with no long-term underperformance (Vonk et al. 2012). Taken together, evidence strongly suggests that low IQ and cognitive underperformance during adolescence and at first presentation of psychosis differentiates schizophrenia from bipolar disorder.

A Research Strategy

This strategy *aims* to identify children and adolescents in the general population who are cognitively at risk of poor future outcomes. Identification of risk should be based on cognitive decline. The assessment of outcome should not be restricted to schizophrenia: schizophrenia is a relatively rare disorder, while cognitive decline in adolescence relates to a broader risk than schizophrenia alone. Since abnormal cognitive development during adolescence may be related to other areas of dysfunction, assessment of other developmental abnormalities in this group is warranted, in particular, abnormalities in social development and the regulation of emotion. This will permit examination of the relationship between cognitive decline and other impairments, and an exploration of underlying mechanisms. It will also help to distinguish cognitive decline from normal variability in IQ that occurs during adolescence, and may have implications for our understanding the nature of schizophrenia and its classification. At present, the direction of the relationship between cognitive decline and these other abnormalities is open. Thus, we propose that this be an area for further research. Consideration should be given to the inclusion of other assessments of at-risk status, such as the assessment of psychotic-like experiences.

The *optimal design* is the naturalistic, longitudinal study of children in early adolescence, aged 10–14 years, who are followed up prospectively over time with multiple assessments so as to permit an examination of distal outcomes for these children, including but not restricted to schizophrenia. However, a major drawback of the optimal design is the length of time required before results from such studies are available to inform intervention strategies.

An *alternative approach* to conventional longitudinal cohort studies is to use population registers to establish an "electronic" cohort. This is a particularly effective approach in jurisdictions such as Sweden, Denmark, and Western Australia, where there are networks of longstanding, whole-of-population administrative databases, including educational testing and psychiatric case registers, with linkage on the individual across registers under prescribed conditions (Morgan et al. 2011). Establishing a cohort for study based on register data offers some advantages:

- Longitudinal data collected over an extended period means that one can examine outcomes that are distal from the exposures of interest.
- Data are prospectively collected, eliminating recall bias.
- Examination of genetic influences and gene–environment interactions are possible if the registers are multigenerational and genealogies can be established.
- The size of the databases ensures sufficient power for most statistical purposes.

In Western Australia, current analysis of register data over the life course for children who develop psychotic illness includes, among others, data on familial liability, obstetric complications, intellectual disability, childhood abuse, school assessments, and mental health (Morgan et al. 2011).

The *interim design*, therefore, is based on the efficient use of extant administrative registers containing educational testing data. Employing a more interactive approach, testing data can be monitored over time to identify decline in performance, with warning thresholds set. However, work would need to be completed that would establish the best thresholds based upon empirical data. Where possible, linking nationwide standardized educational tests to data on psychiatric case registers will be particularly informative at an early stage of study, and could be used to generate hypotheses for the optimally designed longitudinal studies.

The *proposed outcome* is a research strategy that will inform intervention programs for a broad range of children experiencing cognitive decline.

Brain development is quite variable and plastic during adolescence, with brain changes directly related to IQ (Brans et al. 2010; Schnack et al., submitted). Thus, interventions that improve plasticity, such as physical activity (Pajonk et al. 2010) or cognitive interventions, could be beneficial in ameliorating or optimizing brain development during this vulnerable stage in brain maturation. Given the young age of the children, between 10 and 14 years, they are likely to be an especially good target for cognitive remediation.

Are There Other Targets for Primary Intervention?

Given the likely involvement of many genes and environmental risk factors of small effect (such as infection, nutrition, urbanicity and social adversity, and the breadth and complexity of these factors), there is a paucity of clear specific targets for primary prevention of schizophrenia. Meyer-Lindenberg and Tost observe that "the scientific analysis of social environmental risk mechanisms highlights components of modifiable disease risk on the environmental level that provide entry points into both treatment, and, in some cases, prevention. Although many societal stressors such as social inequality are difficult to address, factors such as social components of urbanization may be modified through social policy, thereby enabling a truly preventative approach toward the enormous worldwide burden of mental illness" (Meyer-Lindenberg and Tost 2012:667).

Some of these targets lend themselves to population-based risk prevention programs along the lines of Head Start.[1] Even though the number of cases of schizophrenia prevented would be relatively low, the advantage of these programs is their benefit for many children at risk of wide-ranging adverse outcomes, encompassing psychiatric, educational, and social outcomes.

Alternatively, schizophrenia researchers could build on the mild cognitive impairment (MCI) model of dementia to identify targets for primary prevention. The MCI model identifies people at risk for dementia many years prior to the first signs of dementia through MRI scans of hippocampal volume and PET scans of the accumulation of beta-amyloid. Not everyone with MCI develops full dementia, but those with MCI are more likely to develop it than others. In schizophrenia, selection for testing could be based on the identification of evident and progressive deterioration in school performance and the assessment tools would include cognitive testing and structural MRI scans. Again, population-based selection and intervention reduces the risk of stigma attaching to people with schizophrenia and provides benefits to all who are identified.

---

[1] A comprehensive U.S. health and education intervention aimed at children from low-income families; http://www.acf.hhs.gov/programs/ohs

The At-Risk State and the Prodrome as Targets for Secondary Prevention: When Do We Start, Who Do We Treat, and With What Do We Treat?

The target for secondary prevention is the at-risk state or prodromal phase in schizophrenia, before the onset of frank psychosis. While a number of factors may influence outcome in schizophrenia, evidence of a relationship between longer duration of untreated psychosis and poorer outcomes suggests the importance of the early initiation of interventions, with recent data suggesting that the benefits of early treatment persist over time (Hegelstad et al. 2012). However, determining the critical point for early intervention to halt schizophrenia, or at least reduce its progress, is fraught with difficulty. People at risk of schizophrenia may be identified on the basis of familial risk factors or because they meet other risk criteria such as the presence of subthreshold, attenuated forms of positive psychotic symptoms, or experience a marked decline in cognitive or other functioning. To date, however, there are no accurate disease markers to indicate who among these asymptomatic or only mildly symptomatic individuals will go on to develop the disease. While early intervention is optimal for those who require it, misidentification carries the risk of stigma, potential exposure to unnecessary interventions, and other unintended consequences. As a result, treatment generally begins with the first psychotic episode, when a person is symptomatic to the point of meeting criteria for some form of psychotic disorder. This then is tertiary prevention, applied after the onset of illness according to prevailing diagnostic classifications. This may be the first entry point for pharmacological intervention in jurisdictions such as the United States, where the prodrome is not sufficient for the prescription of antipsychotic medication and a diagnosis meeting DSM criteria is required.

Population-Based Mental Health Promotion

An individual's health, both physical and mental, is influenced by multiple factors. Some of these, such as sex and ethnicity, cannot be modified. Others, however, can: lifestyle risk factors, such as smoking, alcohol consumption, and poor nutrition. Health is also influenced by economic and employment status which, in turn, interact with lifestyle risk factors. In this context, the impact of poverty is notable. These latter factors not only determine an individual's health status but also determine their access to health care.

Evidence-based mental health promotion in the community provides an opportunity to address lifestyle risk factors at the population level and complements other approaches to risk reduction and illness prevention. Rather than focus on those at highest risk, in an area where there are often no clear risk factor thresholds to separate those at risk of mental illness from those not at risk, a general population approach is able to capture the many more individuals in the community at moderate risk, thereby improving the risk profile of the entire population. This approach to mental health promotion is in keeping with the population-based approaches to primary and secondary prevention outlined above.

For example, a model of secondary intervention in the at-risk state or prodromal phase might follow the primary prevention model for cognitive decline described earlier. This could employ a targeted public health approach similar to the approach in Head Start (see above), with the intervention implemented in a normalized way to communities that include high-risk individuals. Although a "mental health for everyone" approach is possible, it runs the risk of missing the very subgroup that it aims to cover. Some interventions which lend themselves to broad implementation include physical activity programs, cognitive remediation, and nutrient supplementation (e.g., omega-3 fatty acids).

Focus of Treatment in Schizophrenia

As a DSM-5 or ICD-10 disorder, schizophrenia is defined in terms of its characteristic symptoms (e.g., positive, negative, and disorganized symptoms) and impaired psychosocial functioning. Commonly associated features include other dimensions of psychopathology such as substance abuse (Mueser et al. 2000), cognitive impairment (Heaton et al. 1994), poor physical health, and premature mortality (Brown et al. 2010). Thus treatments for schizophrenia have targeted multiple domains, ranging from basic brain and cognitive functioning to psychopathology, psychosocial functioning, and physical health.

Antipsychotic medication, the most common treatment in schizophrenia, targets the positive symptoms of the disorder. The use of antipsychotic medication to manage acute symptoms and reduce hospitalization is a priority. However, these medications do not effectively reduce other deficits in schizophrenia, including negative symptoms and cognitive and psychosocial dysfunction. Moreover, through their weight-gain side-effect profile, they contribute to poor physical health in people with schizophrenia.

Although it is commonly accepted that different treatments are required to span the broad range of affected domains, there is a need to focus greater attention on the integration of treatments across domains for three practical reasons. First, different life domains impaired in schizophrenia are moderately interrelated and affect one another (Strauss and Carpenter 1972). For example, reduced cognitive functioning is strongly associated with more impaired psychosocial functioning (Green 1996), whereas a combination of psychopathology (e.g., suicide, substance abuse), unhealthy lifestyle (such as smoking, poor diet and a sedentary lifestyle), and poor management of physical illnesses can all contribute to premature mortality (Druss et al. 2001; Gale et al. 2012; Inskip et al. 1998; Kotov et al. 2010). Second, treatments targeting one domain can interact with other domains, requiring monitoring, coordination and, optimally, integration. For example, the metabolic effects of antipsychotic medications can contribute to weight gain and diabetes (Meyer et al. 2008), pointing to the need for lifestyle interventions aimed at increasing activity level and weight loss (Faulkner et al. 2003; Gorczynski and Faulkner 2010). In another example, cognitive remediation has been found to be most effective at improving functional outcomes when it is paired with psychosocial rehabilitation (Wykes et al. 2011). Third, client motivation to work on one affected area of functioning may be most effectively harnessed by exploring how improvements in that area may be beneficial to the individual's personal goals in another area, suggesting a need for integration across different areas of treatment. For instance, interventions based on the principles of motivational interviewing (Miller and Rollnick 2002) have been used to instill motivation to reduce medication nonadherence and substance abuse in order to help clients achieve personally valued outcomes such as more independent living, work, and improved social relationships (Barrowclough et al. 2010).

Treating Proximal or Distal Outcomes?

Although poor psychosocial functioning in schizophrenia has often been assumed to be the longer-term by-product (a distal consequence) of the more direct (proximal) effects of the disorder (such as cognitive impairments and symptoms such as psychosis), an alternative possibility is that it is more fundamental to the disorder. In other words, reduced capacity to meet social norms with respect to self-care, role functioning, and social relationships could be as proximal a consequence of schizophrenia, or even more so, as the florid psychotic symptoms or characteristic cognitive impairments often thought to be the primary cause of such impaired functioning. Problems in social and school functioning antedate the onset of psychotic symptoms in schizophrenia by many years (see Kahn, in this volume), and could reflect impairments in social drive and stamina that are associated with reduced cognitive performance, but not explained by it. This conceptualization is similar to how negative or deficit symptoms have been hypothesized to be central features of schizophrenia (Andreasen 1982; Carpenter et al. 1988), and Huber's concept of basic symptoms as reflecting core deficits in resilience, drive, and activity (Gross and Huber 2010; Schultze-Lutter 2009). The implications of this possibility is that there may be as much to learn about the nature of schizophrenia from attempts to improve psychosocial functioning as treatment efforts targeting symptoms or impaired cognitive functioning.

A Biosystemic Perspective on Treatment

According to a traditional view of etiology and treatment (Figure 17.1), disease is a linear cascade that emanates from a unitary source. Treatment is only palliative when it targets the source of the cascade, invoking the allopathic

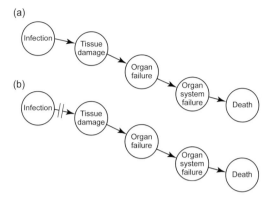

Figure 17.1 A traditional view of etiology and treatment (after Spaulding et al. 2003). (a) In catastrophic disorders, casual cascades are the rule. (b) Catastrophic diseases are effectively treated by disrupting a cascade at a key point.

ideal of a "magic bullet" or ideal therapeutic agent proximal to the cause, that prevents distal consequences.

However, some illnesses, such as diabetes, are systemic (Figure 17.2). A systemic illness has no distinct origin: an ill system is in a state of negative

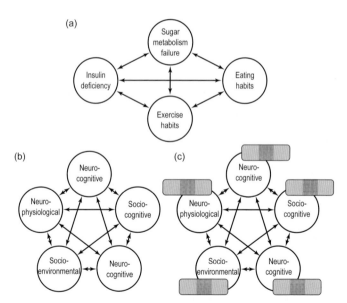

Figure 17.2 A systemic view of etiology and treatment (after Spaulding et al. 2003). (a) In systemic disorders (e.g., diabetes, psychiatric disorders), stable dysregulation (homeorhesis) is the rule. (b) Systemic disorders (e.g., psychiatric disorders) often involve multiple levels of functioning and impairment. (c) Systemic disorders (e.g., psychiatric disorders) must often be addressed at multiple levels simultaneously.

homeorhesis; functional decline is gradual as consequences of impairments radiate throughout the system. Schizophrenia is a systemic condition; impairments interact reciprocally across multiple levels of organismic functioning, from neurophysiological to environmental (Figure 17.2). Treatment at any given point is usually insufficient, even though benefits are also distributed throughout the system. Success is determined not by proximity to a causal origin, but by the multiplicity of interventions at all accessible points. The allopathic imagery of a high-potency bullet gives way to the less appealing but more realistic imagery of multiple low-potency Band-Aids. The idea of being proximal or distal to the source of illness loses its meaning.

Managing Physical Health Outcomes

It is well established that physical morbidity, especially cardiometabolic disease, and mortality are elevated in people with schizophrenia. In schizophrenia, life expectancy is reduced by 18.7 years for men and 16.3 years for women, compared to the general population, with diseases of the circulatory system impacting on life expectancy more than death from external causes (Laursen 2011). A recent national representative survey found 55% of people with schizophrenia aged 18–64 years met the criteria for metabolic syndrome (Morgan et al. 2012). Poor physical health is associated with weight gain as a result of antipsychotic medication use, lifestyle risk factors including high rates of smoking and alcohol consumption, poor nutrition, and low levels of physical activity; recent evidence also points to underlying genetic vulnerability to metabolic disturbance for some (van Winkel et al. 2010). Critically, people with schizophrenia are less likely than the general population to receive appropriate interventions for their physical health conditions, further increasing rates of morbidity and associated mortality (Lawrence et al. 2003).

Improving access to appropriate physical health care is a matter for clinical intervention. From a service perspective, it is essential to address fragmented service delivery across mental and physical health domains. In addition, consideration among medical and mental health practitioners must be given to attitudinal factors that lead to the neglect of the physical health needs of their patients. This includes stigma, which may lead to an under-recognition of physical health issues in people with severe mental illness, as well as a belief that lifestyle change is not possible for this group (Lawn 2012). Regular screening for metabolic syndrome and prescription of medication for those with disease or who are at risk is part of frontline management of physical health for these people. As important is the need to motivate people with schizophrenia to modify lifestyle risk factors. Excess rates of lifestyle risk factors are well documented, and intervention studies support the effectiveness of lifestyle interventions that focus on physical activity and nutrition (Verhaeghe et al. 2011). Nonetheless, little is known about how best to promote the uptake of lifestyle changes in these individuals and to help them self-manage their physical health. Moreover, the quality of existing guidelines is variable; De Hert et al.'s review (2011) identified only four quality guidelines out of 18 published between 2000 and 2010. Many are poorly evaluated, and their implementation is suboptimal (De Hert et al. 2012). Moreover, the guidelines tend to focus on how to measure (screen and monitor) risk, rather than how to modify risk.

As risk may be associated with poor psychosocial function, cognitive remediation may improve functioning and indirectly increase motivation to maintain or regain good health. Increasing their physical health awareness may be another strategy: a person struck by severe schizophrenia in their twenties may miss out on important health promotion messages that their peers are internalizing at the same age.

Benefits of lifestyle risk management extend to mental health. There are interactions between physical health and brain pathology. Physical exercise improves hippocampal neurogenesis (Erickson et al. 2011) while diet impacts on neural growth (Stangl and Thuret 2009). There is a growing literature on the association between physical activity and mental well-being. At the same time, clinicians need to consider the impact of pathology on risk behaviors. Abnormalities in the brain reward system lead to increased risk of smoking addiction in people with schizophrenia, making modification of the reward system a target for intervention. The differential impact of specific antipsychotic medications on smoking behavior needs to be factored into prescribing practices (Montoya and Vocci 2007). In the meantime, much more needs to be understood about genotypes associated with excessive weight gain.

Suicide

In addition to high rates of mortality due to physical morbidity, rates of suicide are also high in this population. It is estimated that 5–13% of people diagnosed with schizophrenia die as a result of suicide (Pompili et al. 2007). Suicide rates peak within a short time of discharge from hospital, making this a critical period for screening and intervention (Lawrence et al. 2001). The range of risk factors include youth, being male, substance abuse, hopelessness, social isolation, deteriorating health after a high level of premorbid functioning, fear of further deterioration, recent loss or rejection, limited external support, and family stress or instability, as well as the experience of either excessive treatment dependence or loss of faith in treatment (Pompili et al. 2007). Nonetheless, what is known about suicide risk is yet to be integrated into effective risk assessment guidelines that enable clinicians to monitor suicide risk in people with schizophrenia and intervene in a timely fashion.

Issues in the Development of Successful Treatments

Over the past two to four years, drug discovery and development for novel therapeutic agents to treat schizophrenia suffered a major setback when several major pharmaceutical companies abruptly abandoned efforts on schizophrenia drug discovery (Abbott 2010; Miller 2010; Nutt and Goodwin 2011). Their reasoning was that (a) CNS drugs take the longest time from discovery to approval, (b) CNS drugs have one of the highest failure rates, (c) neuropsychiatric diseases are heterogeneous, making it difficult to target treatment to the right patient groups, and (d) animal and tissue culture models have shown poor translation into human efficacy (Kaitin and DiMasi 2011; Kaitin and Milne 2011). Economic pressures, patent expirations, uncertainties in the changing health care political environment, and regulatory challenges played an important role in these decisions.

What can the schizophrenia community do to help reinvigorate such vitally important efforts that impact on a large segment of society? How can the process of drug development be restructured to help reenergize the involvement of the pharmaceutical and biotechnology industries?

We believe that there are near-term opportunities for building on what we know today. Ongoing drug development programs with a variety of experimental therapeutic agents have shown positive results.[2] Several programs which focus on negative symptoms and cognitive impairment are at advanced stages of drug development.[3] Given that there are no approved drug treatments for these fundamental components of schizophrenia, support and promotion of these programs is of great importance to the treatment of people with schizophrenia. Surrogate endpoints such as neuroimaging, genetic background, and other biomarkers have the potential to be of great value in refining treatment signal detection. The involvement of regulators will help forward research in this area so that, as this clinical science develops, there is agreement and full acceptance of these endpoints. Meta-analytic techniques have been applied in depression studies (Kirsch et al. 2008) and may also be used to determine the potential benefit of surrogate endpoints in schizophrenia trials.

There are ongoing clinical studies using device-type interventions, such as computer-based cognitive exercises for cognitive remediation. In addition, deep brain stimulation as well as transcranial magnetic stimulation have been

---

[2] Dana Hilt, Herbert Meltzer, Maria Gawry, Susan Ward, Nancy Dgetluck, Chaya Bhuvaneswaran, Gerhard Koenig, Michael Palfreyman. EVP-6124, an Alpha-7 Nicotinic Partial Agonist, produces Positive Effects on Cognition, Clinical Function, and Negative Symptoms in Patients with Chronic Schizophrenia on Stable Antipsychotic Therapy. Presented at the annual meeting of the American College of Neuropsychopharmacology, Kona, Hawaii, December, 2011.

[3] Daniel Umbricht. Effects of the Glycine Transporter Typ1 Inhibitor RG1678 in Schizophrenic Patients with Predominant Negative Symptoms. Presented at the annual meeting of the American College of Neuropsychopharmacology, Miami, December, 2011.

used for the treatment of drug-resistant depression (Cusin and Dougherty 2012; Downar and Daskalakis 2013). These interventions have demonstrated medium effect sizes across a large range of studies and methods (Wykes et al. 2011).

In the mid to long term, better, more effective translation is needed. Delineation of pathomechanisms, next-generation low-cost sequencing, and many other recent techniques will provide high-resolution guidance for genetics to help determine predisposition, susceptibility, and vulnerability genes. Better and more appropriate animal models are needed. In this light, it is worth mentioning that, in addition to existing pharmacological and genetic models, circuit-modulatory models are being developed using optogenetics (Deisseroth 2012). This unique ability to generate switchable phenotypes and symptoms that approximate human symptoms has created a new ability to screen for and test existing drugs. Induced pluripotent stem cells and other cell-based models (Brennand et al. 2011; Dolmetsch and Geschwind 2011) are making important contributions to the understanding of pathomechanisms. We need to accept that we are treating symptoms and this is a viable option when disease-modifying treatments have been elusive.

The Clinical Trial: Learning from Failure

As discussed by Mitchell et al. (this volume), the relevance of current animal models for treatment development has been questioned, although innovative work is underway or has been proposed. One of the greatest gaps in drug development research has been progress from phase 1 studies of healthy humans to phase 2a and 2b, where the efficacy of novel compounds is tested. Results at these early stages are often proprietary and, in some cases, are not made available. Only phase 3 data are made available and, since many drug development programs are abandoned before this stage, the information streams that can facilitate drug development are weak.

Research to enable a "go/no-go" decision at the early phase of drug development will reduce costs and enable a greater number of compounds to be studied, thereby increasing the possibility of bringing effective drugs to market. Several initiatives to meet these goals are underway, including the NIMH "Fast-Fail" Trials (Yan 2013).

The methodologies of phase 3 trials in patients with schizophrenia are a source of constant refinement. Work on several specific issues is underway by international groups of experts (e.g., the International Society for CNS Clinical Trials and Methodology). In multisite trials, the greater the number of sites, the greater the risk of a negative results (Mallinckrodt et al. 2011). Normally, this challenge is met by increasing sample sizes. However, more rational approaches are needed as well as work that addresses site heterogeneity and improves inter-site reliability. Personalized medicine approaches using genetic and other biomarkers, described later in this chapter, may reduce heterogeneity across patients.

Neuroimaging techniques offer the potential to increase signal intensity, allowing for smaller samples in early phase treatment studies. However, few studies have empirically demonstrated that these technological advances surpass conventional clinical tools, such as rating scales and cognitive performance measures. Further, there are few standard activation paradigms that can compare results across trials. Currently, work is in progress to map systematically brain activation responses to cognitive activation tasks in controls, providing normative data that can be used in patient populations on standard protocols.

The cost of phase 2 clinical trials may be reduced by shortening the length of trials. Recent work suggests that treatment efficacy can be established earlier in the course of a clinical trial than is traditionally accepted. For example, the period of greatest sensitivity of antipsychotic efficacy may occur in the first two weeks of treatment, and the traditional endpoint of antipsychotic trials of 8–12 weeks may actually reduce the effect size of new treatments due to patient drop out (Agid et al. 2003; Kapur et al. 2005).

A Conundrum: Sample Homogeneity versus Heterogeneity

A number of schizophrenia trials failed because of challenging patient cohorts, geographic and cultural differences, and inclusion and exclusion criteria that are too broad or poorly defined. In the clinical trial, two principles drive the need for sample homogeneity: (a) an ethical imperative, with safety given precedence over efficacy and (b) the scientific need for homogeneity to maximize signal detection, particularly in trials involving biomarkers, to allow precise estimation of treatment effects. Unfortunately, the need for homogeneity can result in narrowly constrained clinical trial samples that are not representative of the heterogeneous community of treatment users. Thus, a major challenge in schizophrenia clinical trials methodology is to identify patient cohorts that are refined enough to permit the detection of a true treatment signal, yet broad enough to enable treatment efficacy to be generalized to the schizophrenia patient population at large.

New Models and Alternative Approaches

Numerous innovative strategies for testing the safety and efficacy of new treatments for schizophrenia are currently under development. Given the current conservatism in clinical trials, driven in part by the fiscal challenges in industry and academia alike, many of these innovations remain untested. We review a few of these here.

Due partially to the professional and disciplinary separation of investigators from pharmacologic and behavioral traditions as well as the cost of such studies, very few studies have examined the synergistic or complementary effects of behavioral treatments and drug treatments. However, such studies hold much promise and have been effective in other psychiatric conditions (Barr et al. 2008). Analogous to the obvious need for physical exercise when an individual takes steroids to increase muscle mass (Keefe et al. 2011b), behavioral interventions may be included as a platform for all patients receiving a new drug when compared to placebo to enhance the potency of the compound, especially in cognitive paradigms that may require an enriched environment before pharmacologic treatments can become effective.

A less costly approach would be to examine retrospectively the relative efficacy of different treatment designs using existing databases, some of which have tremendous statistical power to address important questions. A recent stroke study with a very large sample used a naturalistic, retrospective design that relied on hospital records and patient retrospective recall. In addition, the South London Case Register Interactive Search system collects data that will allow retrospective study.

Other innovations that have not been sufficiently utilized are virtual reality outcomes and interventions, and adaptive trial designs. Several different research groups are focusing on the use of virtual reality environments as interventions for the treatment of symptoms (Freeman 2008) and cognitive deficits (Spieker et al. 2012), or as outcome measures in clinical trials (Harvey and Keefe 2012). Additional validation work, however, needs to be done on these methods before they will be accepted into later phase trials. One approach would be to include such measures as exploratory outcomes in studies using conventional outcomes as primary endpoints. With regard to adaptive designs, despite encouragement from regulatory agencies, such as the U.S. Food and Drug Administration (Wang et al. 2011), pharmaceutical companies have been hesitant to utilize these approaches.

Other Bases for Therapeutic Intervention in Schizophrenia

On the level of differential psychopathology, there appear to be few new leads. Elaborate systems, such as the Kleist–Leonhardt classification, have not been shown to have therapeutic relevance, and other subdivisions, such as the concept of brief reactive psychosis, may have merit but are well treated with current approaches. On the level of neuropsychology or psychosocial function, no commonly used test or scale has been shown to have clear differential therapeutic relevance; although poor cognition predicts worse treatment response, moving this forward into differential treatment would require a new generation of cognitive interventions.

There may be more promise for the concept of dimensional psychopathology. In particular, diagnoses-transcending dimensions such as depressive symptoms may prompt appropriate adjunctive treatment including antidepressant prescription. Extending the dimensional concept to neuropsychology leads to the National Institute of Mental Health's concept of the Research Domain Criteria (RdOC) (Cuthbert and Insel 2010). Going further, it has been proposed that the most useful defining entities of such dimensions may be neural systems (Buckholtz and Meyer-Lindenberg 2012).

Further down the translational chain, there are few genetic or genomic markers for treatment response available. However, important advances are being made in genetic predictors of side effects, such as metabolic syndrome or tardive dyskinesia, that may be important for therapy. The therapeutic predictive value of environmental risk factors, such as childhood abuse or urbanicity, is almost unexplored and should be investigated. The same goes for epigenetic markers from blood or cerebrospinal fluid. Diagnostic blood markers through proteomics are being marketed (rules-based medicine), but it is currently unclear whether these will be useful to guide therapy under real-world conditions.

New tools may offer novel avenues to enhance treatment prediction. Momentary assessment technologies may give a more comprehensive view of hour-to-hour, day-to-day fluctuations in mood, symptoms, salience processing, and stress. Virtual reality techniques may allow better assessment of social function under controlled circumstances. Investigation of social media activity may show changes or abnormalities in web-based interactions. Eye movements can be tracked naturalistically using a new generation of glasses or using lasers from a distance.

Given the neurodevelopmental nature of the illness, measuring performance during the second decade of life may better define early intervention points, for example, by using and linking school performance and testing data as well as data on social interactions, where these are readily available.

In moving the field forward, a dialectical process between defining intervention points and new therapies is expected. Without differentially effective therapeutics, early intervention has little consequence. Conversely, by understanding the processes early in the illness, especially the pre-psychotic state, new treatment targets are expected to come into focus.

A Framework for Treatment Development

Treatments for schizophrenia cover psychosocial treatments and other therapies, as well as biological treatments. While discussion in this section has focused primarily on new developments within biological treatments, throughout this chapter we stress the key role played by psychosocial interventions and the importance of their integration with biological approaches to treatment.

Finally, it is important to recognize that treatment development does not take place in isolation from other areas of schizophrenia research. These areas are multidisciplinary and include work on animal models, neuroscience research, preventive research, and services research (Figure 17.3). Integration across modalities is an essential ingredient of the development of new paradigms. In addition, to ensure its effectiveness, research needs to link into practice to

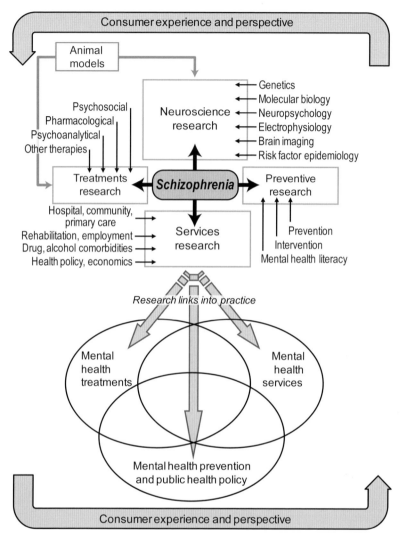

Figure 17.3 A multidimensional, translational treatment model (after Morgan et al. 2006).

inform mental health policy development and service delivery. Critically, we note that the consumer experience and perspective is the fundamental context in which both research and practice occur (Morgan et al. 2006).

Personalized Treatment: Tailoring Treatment to Individual Needs

To explore personalized treatment, we begin with two definitional approaches: (a) a bottom-up approach which starts with the identification of predictors of response that may apply to any patient; and (b) a top-down approach which begins with the patient and assesses what is available for tailoring treatment to that patient's specific needs. Whereas the first approach is generally associated with a broad, biological perspective and the second with an individualized, psychosocial perspective, these associations are not immutable. Ideally, the two approaches converge. These two approaches are then framed by a patient's perspective set within a collaborative model of care.

Personalized Medicine from a Broad, Biological Perspective

The conventional definition of personalized medicine encompasses the idea of predicting a priori response to antipsychotic drugs. Predictors of response include demographic factors (age, age of onset), clinical factors (neurocognitive function), or neurobiological factors, including brain imaging phenotypes or molecular genetic sequence variation. Phenotypes for examination commonly utilize broad clinical response parameters based on drug efficacy but have also focused on side effects of treatment, such as clozapine-induced agranulocytosis.

To date, much of the neurobiological prediction of drug response research has utilized the pharmacogenetics approach. Advantages of this approach are the relative ease of access to DNA, the immutability of genotype, and the increasing ability to interrogate entire genomes in a cost-effective manner. Work on pharmacogenetics of efficacy has, however, not led to clinically actionable results, although this may be secondary to methodological issues pertaining to these studies. For example, most studies have utilized convenience samples derived from clinical trials designed to compare antipsychotics. This serves to diminish power due to multiple treatment arms and incomplete DNA collection, and phenotypes under examination may not be maximally informative for identifying the effects of subtle DNA sequence variation. Finally, it should be recognized that the genetic architecture of clinical treatment response may be complex, and perhaps no simpler than that of disease susceptibility, which has hampered genetic studies aimed at identification of common variation.

Studies of side effects have been more successful, in part because of the increased power due to the decreased measurement error of these phenotypes. Recently, the human leukocyte antigen system has been linked to clozapine-induced agranulocytosis and weight gain with a melanocortin receptor genotype. These studies require follow-up, and clinical trials based on these genotypes are being planned. At the same time, we need better studies of side-effect *burden* if we are to improve patient adherence to medication. Many studies are limited to measuring frequency and severity, without assessing more subjective side effects such as dysphoria or considering the trade-off of side-effect burden against symptom reduction.

New study designs may need to be considered (Malhotra et al. 2012), including the study of alternative phenotypes. For example, brain imaging measures utilized shortly after drug administration may provide novel information more closely linked to the sites of gene action. Focus on more homogeneous patient populations, use of earlier phase patients, and minimization of prior drug treatment may also provide benefit.

Personalized Medicine from an Individualized, Psychosocial Perspective

The idea of personalized medicine converges with related ideas about individualizing treatment that evolved in psychiatric rehabilitation. Broadening the focus from psychopharmacotherapy to include psychosocial treatment addresses multiple levels of functioning, including neurophysiological, neuropsychological, sociocognitive, behavioral, and socioenvironmental processes. These are no longer seen as competing paradigms: they reflect distinct levels of analysis and action within a unified biosystemic understanding of mental illness.

In severe mental illness, measurable impairments are observed at all of these levels and are a source of heterogeneity in schizophrenia, occurring in constellations that differ from one person to the next. While effective pharmacological treatment may resolve acute psychosis or eliminate symptoms, impairments remain in cognition, self-care, interpersonal effectiveness, and social role performance. There is a rapidly growing armamentarium of psychosocial treatment modalities to address specific deficits across levels of functioning. For example, modalities like cognitive remediation act primarily at the neuropsychological level. Cognitive behavioral therapy and related behavioral treatments act at sociocognitive and behavioral levels. Behavioral family therapy extends to the socioenvironmental level. To treat individual constellations of impairments optimally, we must select and apply the respective treatments that correspond to those individual constellations. As with personalized medicine, this creates an assessment burden that may limit complete individualization of multimodal treatment regimens. However, for more severe illness, resulting in more pervasive distribution of impairments across levels of functioning and more severe disability, comprehensive and integrated treatments are necessary. The level of severity and/or pervasiveness at which this type of individualization becomes cost-effective can, in principle, be empirically determined, but as assessment and treatment technologies advance, one would expect the threshold level to become lower.

Coordinated treatment of multiple, functionally independent but interrelated impairments across levels of functioning is problematic in diagnosis-driven clinical decision making and is better accommodated by broader problem-solving approaches. These are familiar to clinicians and have even been canonized in American medical records standards as Problem-Oriented Medical Information Systems (PROMIS). A similar approach, termed case formulation, has evolved within the methodology of cognitive behavioral therapy. With some further formalization and structure, a clinical problem-solving approach can effectively guide multimodal treatment of heterogeneous conditions like schizophrenia.

In clinical problem solving, the unit of analysis is not diagnosis but "problem type." Problem types are jointly defined by types of impairment at the various levels of functioning and by technologies currently available to measure and treat those impairments. For example, psychopharmacology and the neurophysiological impairments it treats is usefully categorized under a "CNS dysregulation" problem type. Measureable deficits in interpersonal functioning that can be effectively reduced by social skills training are usefully categorized under a "social skills deficit" problem type. Failure to recognize one's illness and therefore to adhere to needed treatments can be effectively addressed with psychoeducational interventions, and is usefully categorized as an "illness management deficit" problem type. A problem type reflects what we know from science about the links between the problem and its effective solution, and thus provides a logical justification for treatment selections. An integrated, individualized treatment regimen is achieved with a complete inventory of the person's problem types and a plan for systematically addressing them with evidence-based treatments.

Contextual factors (e.g., patient's perspective, neuropsychological status, developmental and environmental factors) as well as the nature of the treatments drive clinical decisions about how to prioritize or sequence treatment, and determine what configuration of pharmacological and psychosocial approaches works best on a case-by-case basis. For example, the time frame in which the effects of drug treatment can be evaluated is much shorter than that for evaluating skill-training interventions. It is more practical to determine what interpersonal deficits persist after effective drug treatment than what psychotic symptoms persist following social skills training. Similarly, drug treatment of an acute CNS dysregulation may be prerequisite to social skills training in some cases whereas in others, the potential for medication nonadherence must be addressed either before or in the context of providing pharmacotherapy. In the foreseeable future, advances in clinical psychopharmacology may provide additional reasons to prioritize, sequence, coordinate, and integrate treatment of problem types across levels of functioning, for example, the possibility that short-term effects of oxytocin might improve engagement in psychosocial treatments, or that cognitive therapy is necessary to consolidate the effects of deep brain stimulation.

A crucial element in this approach is for treatment to proceed as a quasi-experimental hypothetico-deductive process, wherein the effects of specific interventions are reliably evaluated and reevaluated in iterative cycles. This is how we determine that, for any one person, specific additional problems remain to be treated after maximum pharmacotherapeutic benefit has been achieved. No scientific breakthrough is likely to change this clinical reality.

Personalized Medicine from a Patient's Perspective Set within a Collaborative Model of Care

An additional perspective on personalized treatment is that of the patient. This perspective is framed in a collaborative model of care where the patient is an active participant involved in shared decision making in the treatment process. The relationship between patient and therapist in mental health care is one of the most important factors in successful treatment. It is a reliable predictor of treatment outcome, regardless of diagnosis, setting, or type of therapy. Shared decision making builds on a trustful therapeutic relationship, incorporating concepts such as recovery, empowerment, and self-esteem. It takes into account not only a person's individual circumstances, but also their preferences for outcomes they most value. For practitioners, personalized treatment within a collaborative framework requires that they:

- Determine the problems to be treated.
- Identify the available treatments.
- Recognize individual differences in a patient's psychological, biological, and social makeup.
- Consider the patient's desires.

It is important to note that in a biosystemic view of human functioning, the patient's perspective, beliefs, attitudes, and values are important elements, shaped by social cognition. Sometimes addressing and influencing those elements is beneficial to the patient's recovery. For example, a belief that no better life is possible and all effort will be punished is a frequently encountered perspective; changing that perspective may promote recovery. We have effective methods for facilitating such changes (e.g., cognitive behavioral therapy and motivational interviewing). We can therefore identify self-defeating beliefs as a specific problem type in an integrated and individualized treatment plan, treat it with cognitive behavioral therapy and related methods, and measure the success of the treatment.

However, there is danger in treating perspectives as targets for change. For example, a person's perspective on adherence to treatment may be detrimental to symptomatic recovery, but people often make different choices across treatment options. Historically, we have erred much more on the side of not respecting patients' perspectives than on missing treatment opportunities. Practitioners need to know and understand their patients' perspectives, and some of these perspectives may also be a target of treatment. To this end, respect for patient decision making needs to rest on a foundation of mutual understanding of the nature of the disorder, to the extent that this is possible, and the patients' goals for treatment. In the end, how and where we draw the distinction between perspectives to treat versus perspectives to respect will be settled by social consensus, not science. The recovery movement has contributed importantly to the broader discourse on the roles of perspective, attitudes, and values in psychiatric treatment and rehabilitation, but a working consensus will require participation of all quarters of the mental health community.

Optimizing Service Delivery

Although current treatments for schizophrenia are still far from optimal, there is a reasonable evidence base to inform the clinician as to what works and when. These evidence-based interventions, however, are not implemented systematically, if at all. Pharmacological treatment of positive symptoms is the basis of most treatment regimes, yet patient adherence to treatment is generally poor, and the prescription of depot medication and clozapine is suboptimal. The use of nonpharmacological treatments targeting other aspects of the illness is low, with these other treatment modalities poorly integrated into mainstream treatment regimes. What is needed is a strategic way of refocusing or fine tuning treatment goals, bearing the individual patient in mind. This includes careful management of the balance between pharmacological and psychotherapeutic interventions, with a view to optimizing treatment response across modalities, increasing quality of life, and preventing psychosis relapse.

In terms of optimizing delivery, there are two issues: one for research funders and one for health care providers. First, we need to consider what are the service configurations that will provide the most benefit—this is a research question. The second concerns the translation of current findings into practice. This is not an easy issue to address as it is not clear how individual and effective therapies may best be implemented into "normal" services. Dissemination science or implementation science takes into account social and organizational psychology approaches to drive the implementation. Its outputs can act as a template for translation of therapies into clinical practice.

The arguments made for change will differ according to different groups but perhaps the overarching argument needs to be made in terms of net benefit. This can be defined in different ways. For example, it may be defined as cost savings for some service providers, a political argument necessary for some potential investors. For others, it may need to be defined as a benefit in cost utility; that is, it achieves a desired outcome such as people attending services, which increases the immediate cost in the hope that improvements in functional or other outcomes will result in the future. Net benefit is also a way of costing the outcome. For example, people are more content, have more friends, and rely less on family support.

Impediments to the Implementation of Evidence-Based Treatments

The evidence base supporting the efficacy of specific psychosocial interventions for schizophrenia has grown steadily over the past several decades.

Despite this growth, there continues to be a significant lag, possibly growing, between the science of treatment and the implementation of empirically supported interventions. At least three broad factors can be identified, each of which contributes to this gap in knowledge and implementation.

First, practical knowledge about how to implement and sustain individual practices in routine treatment settings is needed. For example, the successful adoption of an intervention requires attention to the science of implementation. This involves consideration of factors traditionally studied under the purview of organizational psychology. These include structural aspects of a human service organization, distribution and sharing of power and decision making, openness to change and innovation, and access to critical resources such as training and consultation expertise. Attention to these issues is critical both to understanding how to implement a specific practice and to appreciating organizational characteristics and needs in ways that will facilitate broad-scale adoption of the practice. Variables which may have an impact on implementation are: training, supervision, and collaboration among service providers; the attitudes, beliefs, and practices of treatment staff; time allotted to staff to provide services; and the skill of those individuals responsible for overseeing and supervising the practice.

Second, and related to the first point, psychiatric rehabilitation programs based on social learning in institutional and hospital settings are generally understood to have features that are inconsistent or even incompatible with the conventional "medical model" of administrative and clinical practices that predominate in such settings (Paul and Lentz 1977; Liberman 1979; Silverstein et al. 2006a; Tarasenko et al. 2012). These features include:

- The need to supervise direct care staff closely, to ensure high fidelity to procedure manuals which require behavioral responses by staff that are sometimes counterintuitive and/or contradictory to conventional nursing or medical practice.
- Administrative control over direct care staff by a program director, who is directly accountable for treatment fidelity and outcome, rather than indirect control of direct care staff by an administrative supervisor, such as a director of nursing, who is not directly accountable for program operation.
- Psychiatric staff who have a "consultant" role with focused responsibility for pharmacological dimensions of treatment, rather than superordinate authority and accountability for all patient care.
- Individualized treatment prescribed and directed according to a treatment plan that is constructed by an interdisciplinary treatment team rather than through physicians' orders.

These incompatibilities may also apply in community-based service systems outside institutions and hospitals.

Finally, training and mental health policy issues need to be addressed to improve the uptake of these evidence-based practices. All mental health professionals should understand and be competent to provide high-quality, effective treatments for people with schizophrenia. This should happen as part of basic training, further supplemented by continuing practice development as novel and effective treatments emerge. Mental health service providers and purchasers need to incorporate sufficient incentives and accountability for providing such efficacious treatments, not only ensuring they are carried out as prescribed in the evidence but also checking to see that outcomes are in the expected direction and at the expected level.

A Business Model of Service Delivery

For the many reasons discussed above, the effective and individualized use of pharmacological and psychosocial therapeutic options has been limited, despite their availability. One means of overcoming problems with current mental health service delivery is to operate a free market model for some of its components. This type of business model, operating across public and private sectors, would approach patients as "purchasers" and mental health services as "products." In this context, shared decision making and patient education would increase the market for mental health services. Increased demand, in turn, could motivate private companies to invest in treatment development, the training of specialized professionals, and the establishment of quality standards, thereby optimizing quality, diversity, and costs of services—especially psychosocial therapies which are more expensive in the short term than pharmacological interventions, and more difficult to standardize. At the same time, investors may be motivated to increase their market hold and consumer retention through investments in innovative services which target patients' special needs. One example would be the establishment of e-health systems using smartphones to navigate the use of treatments. Recently, investigators in the Netherlands and Great Britain have started to test the principles of the free market in the context of mental health service provision by examining which services are accessed, and how frequently, when patients are allowed to manage their own therapeutic budget. However, much more research is needed to ensure that such a business model would be viable, effective, affordable, and equitable.

Promoting Greater Investment in Treatment Development

Greater investment in treatment development is essential. This should involve multiple stakeholders to fund or to act in concert to promote the research agenda (see Figure 17.4). In addition to scientific peers, these stakeholders include research funders, service users and carers (family and friends),

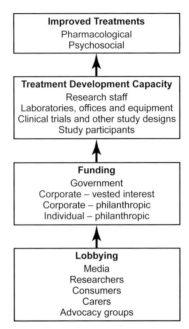

Figure 17.4 Engaging stakeholders in treatment development (after Morgan et al. 2006).

pharmaceutical and other commercial interests (e.g., biotechnology companies), and health care providers (both public and private). Despite a shared vision on what investments are needed, dialog with each of the stakeholders needs to be tailored to their specific needs and a format and language that is familiar to them needs to be employed. One key emphasis is the benefit that can result from such investments; these can be described in the vocabulary of health economics as cost benefit, cost utility, and cost effectiveness. There is current evidence that investment in mental health care can result in substantial benefits and impacts. A review of mental health research investment by the Academy of Medical Sciences in the U.K. found that for every one pound invested in mental health research, there was a return of 37 pence each year in perpetuity. Thus, after three years, the investment has been repaid and the following years actually produce a profit in terms of health care savings and reduced disability.

Engaging the media to promote information on research attainments is critical. Closer ties with science journalists are vital, so that they can better appreciate current scientific breakthroughs and grasp where we want to go next. Likewise, background and news briefings will generate greater media coverage of mental health issues and provide key science and health reporters in national news with more in-depth understanding. This is not only important for promoting investment in treatment research but also has benefits for all key stakeholders and goes hand in hand with the suggestions we make at the end of this chapter for improving the social context for people with schizophrenia. In sum, the general benefits of a media-based program include its potential to:

- Raise public awareness of issues in mental health research.
- Decrease stigma through reporting of incremental changes in understanding causation and treatment developments.
- Increase public understanding of mental ill health (the science press seems to do this better than the health press).
- Increase hope and optimism for people with mental health problems (and their families) by publicizing incremental changes.
- Increase the public profile of charities who can comment on the work being published. We know that when a charity comments on a mental health issue that it is more likely to be given a public profile report on the BBC website. This, of course, increases the visibility of the charity to potential donors. This same argument can be put to public funders who need a media profile to support future funding from national governments.

In addition, specific approaches need to be made to public investors. Research funders need to be aware of the particular areas that have been agreed as vital for improving the understanding and treatment of schizophrenia. For example, naturalistic studies that examine heterogeneous samples are not funded by the pharmaceutical industry; this raises the possibility of public investors funding these types of studies.

Finally, academic outputs can be harnessed to build a program of research that would benefit from larger-scale future investment. The leverage of larger investment will require specific approaches to foundations as they may become the change managers or charismatic leaders for such philanthropy.

Ultimately, promoting greater investment in treatment research involves a concerted effort to strengthen the nexus between lobbyists, funders, and researchers.

The Role of Social Context in Improving Treatment Outcomes

Here, we consider social context on three levels. The macro level addresses broad societal aspects that impact on people with schizophrenia. The meso level is concerned with the social networks, including family and other caregivers, within which people with schizophrenia find themselves. The micro level operates at the level of the individual: a person with schizophrenia may be designated variously as an affected person, patient, client, consumer, and so forth, with each designation implying a different perspective (Pescosolido et al. 2008).

Social Context at the Macro Level

One of the major impediments to the treatment and care of patients with schizophrenia is the stigma associated with the disorder. Stigma is a general term that describes the process of assigning a certain characteristic to a person (e.g., dangerousness), independent of the person him- or herself. Prejudice characterizes the affective component of assigning the negative characteristic (e.g., being afraid of and avoiding a person with mental illness because of his or her assumed dangerousness). Discrimination relates to the behavioral component of stigma that typically reduces opportunities of the person to gain access to resources that others in society can generally tap (Link and Phelan 2001). Examples include attempting to prevent a person with mental illness from renting a nearby apartment, obtaining a job, voting, or getting health services.

These socially constructed labels have important consequences for people with schizophrenia. In particular, labeling theory provides a useful framework for understanding their impact. Labeling theory, centered on the social construction of deviant behavior, evolved in the 1960s (Goffman 1963). The sociologist Thomas Scheff applied the theory to people with a mental illness, arguing that mental illness is a social construction and questioning its existence (Scheff 1966). In the 1980s, Link et al. (1989) presented a modified form of labeling theory that did not question the existence of mental illness. In a series of empirical studies, they described how the process of labeling people with a mental illness has a negative impact on their lives and leads to a cycle of social rejection and isolation. Today, labeling theory, as outlined by Link, is widely accepted in social psychiatric research.

Numerous studies have examined stigma in mental illness, particularly schizophrenia. In recent years, there has been an accumulation of empirical evidence of the negative consequences of labeling and perceived stigmatization. These include demoralization, low quality of life, unemployment, and reduced social networks (Graf et al. 2004). Affected individuals, once they have been labeled as having a mental illness and become aware of the related negative stereotypes, expect to be rejected, devalued, and discriminated against. Such individuals often incorporate these negative stereotypes into their own self-perceptions (called *self-stigma*), with associated problems of demoralization, avoidance, and a pervasive sense of hopelessness. This vicious cycle decreases the chance of recovery and normal life.

Successful initiatives make clear that efforts to reintegrate persons with serious mental illness into community life must be accompanied by measures on the societal level. On the basis of comprehensive research over the past decade, several strategies have been developed to fight the stigma and discrimination suffered by this group. Contact with mentally ill people reduces social distance, with those in contact often having a more positive attitude toward people with mental illness: this is a strong argument in favor of community psychiatry. Social distance from mental illness also decreases, and stigma is reduced, if mental disorders are presented as a life crisis, not as a brain disease (Lauber et al. 2004b). In addition, some research centers have developed destigmatization interventions directed at relevant target groups (e.g., students or police officers).

Social policy also needs to recognize that stigma operates at an intrapersonal as well as interpersonal level, often to the detriment of individual patients. As previously mentioned, individuals in stigmatized groups sometimes incorporate stigmatizing into their own beliefs. Self-stigmatization is a personal perspective, but also a social cognitive process, subject to therapeutic change with cognitive behavioral therapy and related methods (Link et al. 1991). For example, skillful assertiveness is an important determinant of self-esteem, self-worth, and other perspectives incompatible with stigmatization. Social policy needs to promote inclusion of these treatment resources in service systems as yet another way of combating stigmatization.

Social Context at the Meso Level

Wider Social Networks

The beneficial effects of work for a person's mental health have been known for centuries. Vocational rehabilitation has been a core element of psychiatric rehabilitation since its beginning. Employment is seen as a step toward independence and integration into society. Vocational rehabilitation is based on the assumption that work not only improves activity, financial standing, social contacts, and so forth, it also promotes gains in related areas such as self-esteem and quality of life. Enhanced self-esteem, in turn, improves adherence to rehabilitation in individuals with impaired insight (Rössler 2006).

Today, the most empirically supported vocational rehabilitation model is supported employment. In supported employment, persons with a disability are given assistance to find competitive employment based on their preferences as soon as possible, and they receive the support needed to maintain their jobs. Participation in supported employment programs is associated with greater success in finding and keeping work than other approaches to vocational rehabilitation (Burns et al. 2007). Positive relationships have been found between obtaining competitive work and nonvocational outcomes (e.g., improved self-esteem, reduced symptoms, social integration and relationships, improved cognition, quality of life).

Although findings regarding supported employment are encouraging, some critical questions remain. Many individuals in supported employment obtain unskilled part-time jobs. Thus, further research is needed into the role of supported education and career development as strategies for helping patients obtain higher paying and more interesting jobs. In addition, since most studies evaluate only short (12–18 months) follow-up periods, long-term impact remains unclear. Currently we do not know which individuals benefit from supported employment and which do not. It is important to realize that integration into the labor market depends not only on the ability of the persons affected to fulfill a work role and on the provision of sophisticated vocational training and support techniques, but also on the willingness of society to integrate its most disabled members (Rössler 2006). One indicator of such willingness is the enactment of laws prohibiting discrimination against people with a mental illness in obtaining work.

Family and Caregiver Relationship

As a consequence of deinstitutionalization, the burden of care has increasingly fallen on the relatives of the mentally ill (Schulze and Rössler 2005): 50–90% of people with a disability live with their relatives following acute psychiatric treatment. This is a task many families do not choose voluntarily. Caregiving may impose a significant strain on families (Lauber et al. 2003). Those providing informal care face considerable adverse health effects, including higher levels of stress and depression, and lower levels of subjective well-being, physical health, and self-efficacy. In addition, not all families are equally capable of giving full support to their disabled member or are willing to be a substitute for an insufficient health care system.

Still, families are an often untapped resource in the treatment of schizophrenia. They represent support systems that provide natural settings for context-dependent learning, which is important for recovery of functioning. Therefore, since the beginning of care reforms, there has been a growing interest in supporting affected families. Family intervention programs have produced promising results. Family interventions are effective in lowering relapse rates and also in improving outcomes (e.g., psychosocial functioning). Furthermore, treatment gains are fairly stable (Pilling et al. 2002). However, more data are needed to clarify the effective components of different models, which may differ in content, frequency, and length of treatment (Barbato and D'Avanzo 2000).

Social Context at the Micro Level

In some parts of the world in recent years, social skills training in psychiatric rehabilitation has become very popular and has been widely promulgated. Social skills training programs focus on areas such as medication management, symptom management, substance abuse management, basic conversational skills, interpersonal problem solving, friendship and intimacy, recreation and leisure, workplace fundamentals, community (re-)entry, and family involvement (Liberman and Kopelowicz 2002).

The results of multiple controlled studies suggest that individuals with schizophrenia can be taught a wide range of social skills (Kern et al. 2009). Social and community functioning improve when these skills are relevant for the patient's daily life, and changes in behavior are recognized and reinforced.

Unlike medication effects, benefits from skills training occur more slowly. Furthermore, long-term training has to be provided for positive effects. Overall, however, social skills training has been shown to be effective in the acquisition and maintenance of skills and their transfer to community life and improvement of psychosocial functioning (Kurtz and Mueser 2008).

Bibliography

Aarons, G. A., M. Hurlburt, and S. M. Howitz. 2011. Advancing a Conceptual Model of Evidence-Based Practice Implementation in Public Service Sectors. *Adm. Policy Ment. Health* **38**:4–23. [15]

Abbott, A. 2008. Psychiatric Genetics: The Brains of the Family. *Nature* **454**:154–157. [08]

———. 2010. Schizophrenia: The Drug Deadlock. *Nature* **468**:158–159. [13, 17]

Abel, K. M., M. P. Allin, D. R. Hemsley, and M. A. Geyer. 2003. Low Dose Ketamine Increases Prepulse Inhibition in Healthy Men. *Neuropharmacology* **44**:729–737. [07]

Abi-Dargham, A., and H. Moore. 2003. Prefrontal DA Transmission at D1 Receptors and the Pathology of Schizophrenia. *Neuroscientist* **9**:404–416. [12]

Addington, J., B. A. Cornblatt, K. S. Cadenhead, et al. 2011. At Clinical High Risk for Psychosis: Outcome for Nonconverters. *Am. J. Psychiatr.* **168**:800–805. [02]

Addington, J., H. Saeedi, and D. Addington. 2005. The Course of Cognitive Functioning in First Episode Psychosis: Changes over Time and Impact on Outcome. *Schizophr. Res.* **78**:35–43. [07]

Addington, J., S. Van Mastrigt, and D. Addington. 2004. Duration of Untreated Psychosis: Impact on 2-Year Outcome. *Psychol. Med.* **34**:277–284. [07]

Aggernaes, B., B. Y. Glenthoj, B. H. Ebdrup, et al. 2010. Sensorimotor Gating and Habituation in Antipsychotic-Naive, First-Episode Schizophrenia Patients before and after 6 Months' Treatment with Quetiapine. *Int. J. Neuropsychopharmacol.* **13**:1383–1395. [07]

Agid, O., S. Kapur, T. Arenovich, and R. B. Zipursky. 2003. Delayed Onset Hypothesis of Antipsychotic Action: A Hypothesis Tested and Rejected. *Arch. Gen. Psychiatry* **60**:1228–1235. [17]

Aguilar-Valles, A., C. Flores, and G. N. Luheshi. 2010. Prenatal Inflammation-Induced Hypoferremia Alters Dopamine Function in the Adult Offspring in Rat: Relevance for Schizophrenia. *PLoS One* **5**:e10967. [07]

Ahveninen, J., S. Kahkonen, S. Pennanen, et al. 2002. Tryptophan Depletion Effects on EEG and MEG Responses Suggest Serotonergic Modulation of Auditory Involuntary Attention in Humans. *NeuroImage* **16**:1052–1061. [07]

Aleman, A., R. Higman, E. H. F. de Haan, and R. S. Kahn. 1999. Memory Impairment in Schizophrenia: A Meta-Analysis. *Am. J. Psychiatr.* **156**:1358–1366. [04]

Alexander-Bloch, A. F., P. E. Vértes, R. Stidd, et al. 2012. The Anatomical Distance of Functional Connections Predicts Brain Network Topology in Health and Schizophrenia. *Cereb. Cortex* **23**:127–138. [06]

Allardyce, J., and J. Boydell. 2006. Review: The Wider Social Environment and Schizophrenia. *Schizophr. Bull.* **32**:592–598. [01]

Allen, P., J. Luigjes, O. D. Howes, et al. 2012. Transition to Psychosis Associated with Prefrontal and Subcortical Dysfunction in Ultra High-Risk Individuals. *Schizophr. Bull.* **38**:1268–1276. [07]

Allen, P., M. L. Seal, I. Valli, et al. 2011. Altered Prefrontal and Hippocampal Function During Verbal Encoding and Recognition in People with Prodromal Symptoms of Psychosis. *Schizophr. Bull.* **37**:746–756. [07]

Allon Therapeutics Inc. 2009. Press Release (May 11, 2013): Phase IIa Clinical Data Confirms Potential of Allon's Davunetide as a Treatment for Cognitive Impairment in Schizophrenia Patients. http://www.allontherapeutics.com/2009/12/phase-iia-clinical-data-confirms-potential-of-allon%e2%80%99s-davunetide-as-a-treatment-for-cognitive-impairment-in-schizophrenia-patients/. (accessed June 14, 2013). [04]

Allport, G. 1962. The General and the Unique in Psychological Science. *J. Pers.* **30**:405–422. [01]

Alonso, J., M. C. Angermeyer, S. Bernert, et al. 2004. Use of Mental Health Services in Europe: Results from the European Study of the Epidemiology of Mental Disorders (ESEMeD) Project. *Acta Psychiatr. Scand.* **420**:47–54. [16]

Alonso, J., M. Codony, V. Kovess, et al. 2007. Population Level of Unmet Need for Mental Healthcare in Europe. *Br. J. Psychiatry* **190**:299–306. [16]

American Psychiatric Association. 2000. Diagnostic and Statistical Manual of Mental Disorders (DSM IV). Arlington, VA: APA Publishing. [05]

———. 2013. DSM-5. Arlington, VA: APA Publishing. [02, 03]

Amminger, G. P., M. R. Schäfer, K. Papageorgiou, et al. 2010. Long-Chain Omega-3 Fatty Acids for Indicated Prevention of Psychotic Disorders: A Randomized, Placebo-Controlled Trial. *Arch. Gen. Psychiatry* **67**:146–154. [01, 07]

Anderson, C. M., D. J. Reiss, and G. E. Hogarty, eds. 1986. Schizophrenia and the Family. New York: Guilford Press. [15]

Ando, J., Y. Ono, and M. J. Wright. 2001. Genetic Structure of Spatial and Verbal Working Memory. *Behav. Genet.* **31**:615–624. [07]

Andrade, L., J. J. Caraveo-Anduaga, P. Berglund, et al. 2003. The Epidemiology of Major Depressive Episodes: Results from the International Consortium of Psychiatric Epidemiology (ICPE) Surveys. *Int. J. Meth. Psychiatr. Res.* **12**:3–21. [16]

Andreasen, N. C. 1982. Negative Symptoms in Schizophrenia. Definition and Reliability. *Arch. Gen. Psychiatry* **39**:784–788. [10, 17]

Andreasen, N. C., W. T. Carpenter, Jr., J. M. Kane, et al. 2005. Remission in Schizophrenia: Proposed Criteria and Rationale for Consensus. *Am. J. Psychiatr.* **162**:441–449. [17]

Andreasson, S., P. Allebeck, A. Engstrom, and U. Rydberg. 1987. Cannabis and Schizophrenia. A Longitudinal Study of Swedish Conscripts. *Lancet* **2**:1483–1486. [07]

Andrews, A., M. Knapp, P. McCrone, M. Parsonage, and M. Trachtenberg. 2012. Effective Interventions in Schizophrenia: The Economic Case. A Report Prepared for the Schizophrenia Commission. London: Rethink Mental Illness. http://www2.lse.ac.uk/LSEHealthAndSocialCare/pdf/LSE-economic-report-FINAL-12-Nov.pdf (accessed 25 May, 2013). [01]

Andrews, G., S. Henderson, and W. Hall. 2001. Prevalence, Comorbidity, Disability and Service Utilisation: Overview of the Australian National Mental Health Survey. *Br. J. Psychiatry* **178**:145–153. [16]

Angst, J., A. Gamma, F. Benazzi, et al. 2003. Toward a Re-Definition of Subthreshold Bipolarity: Epidemiology and Proposed Criteria for Bipolar-II, Minor Bipolar Disorders and Hypomania. *J. Affect. Disord.* **73**:133–146. [16]

Anthony, W. 1993. Recovery from Mental Illness: The Guiding Vision of the Mental Health Service System in the 1990s. *Psychosocial Rehab. J.* **16**:11–23. [15, 17]

Anthony, W., M. Cohen, M. Farkas, and C. Gagne. 2002. Psychiatric Rehabilitation (2nd edition). Boston: Boston Univ. Center for Psychiatric Rehabilitation. [15]

Archer, T. 2010. Neurodegeneration in Schizophrenia. *Expert. Rev. Neurother.* **10**:1131–1141. [07]

Arguello, P. A., and J. A. Gogos. 2006. Modeling Madness in Mice: One Piece at a Time. *Neuron* **52**:179–196. [08]
Arnold, S. E., L. Y. Han, P. J. Moberg, et al. 2001. Dysregulation of Olfactory Receptor Neuron Lineage in Schizophrenia. *Arch. Gen. Psychiatry* **58**:829–835. [10]
Arnold, S. E., E. B. Lee, P. J. Moberg, et al. 2010. Olfactory Epithelium Amyloid-Beta and Paired Helical Filament-Tau Pathology in Alzheimer Disease. *Ann. Neurol.* **67**:462–469. [10]
Arranz, M. J., and J. de Leon. 2007. Pharmacogenetics and Pharmacogenomics of Schizophrenia: A Review of Last Decade of Research. *Mol. Psychiatry* **12**:707–747. [07, 08]
Arseneault, L., M. Cannon, R. Poulton, et al. 2002. Cannabis Use in Adolescence and Risk for Adult Psychosis: Longitudinal Prospective Study. *BMJ* **325**:1212–1213. [07]
Arseneault, L., M. Cannon, J. Witton, and R. M. Murray. 2004. Causal Association between Cannabis and Psychosis: Examination of the Evidence. *Br. J. Psychiatry* **184**:110–117. [09]
AstraZeneca and Targacept. 2008. Astrazeneca and Targacept Announce Results from Trial of AZD3480 for Cognitive Dysfunction in Schizophrenia. http://www.news-medical.net/news/2008/12/10/43968.aspx. (accessed June 16, 2013). [04]
Atkins, D. 2009. Queri and Implementation Research: Emerging from Adolescence into Adulthood: Queri Series. *Implement. Sci.* **4**:12. [15]
Atkinson, R. J., P. T. Michie, and U. Schall. 2012. Duration Mismatch Negativity and P3a in First-Episode Psychosis and Individuals at Ultra-High Risk of Psychosis. *Biol. Psychiatry* **71**:98–104. [07, 09]
Aukes, M. F., B. Z. Alizadeh, M. M. Sitskoorn, et al. 2009. Genetic Overlap among Intelligence and Other Candidate Endophenotypes for Schizophrenia. *Biol. Psychiatry* **65**:527–534. [14]
Averbeck, B. B., T. Bobin, S. Evans, and S. S. Shergill. 2011. Emotion Recognition and Oxytocin in Patients with Schizophrenia. *Psychol. Med.* **42**:1–8. [07]
Ayllon, T., and N. Azrin. 1968. The Token Economy: A Motivational System for Therapy and Rehabilitation. New York: Appleton-Century-Crofts. [15]
Balaguer-Ballester, E., C. Lapish, J. Seamans, and D. Durstewitz. 2011. Attracting Dynamics of Frontal Cortex Ensembles During Memory-Guided Decision Making. *PLoS Comput. Biol.* **7**:e1002057. [12]
Baldeweg, T., D. Wong, and K. E. Stephan. 2006. Nicotinic Modulation of Human Auditory Sensory Memory: Evidence from Mismatch Negativity Potentials. *Int. J. Psychophysiol.* **59**:49–58. [07]
Bandura, A. 1969. Principles of Behavior Modification. New York: Holt, Rinehart and Winston, Inc. [15]
Barak, S., and I. Weiner. 2011. Putative Cognitive Enhancers in Preclinical Models Related to Schizophrenia: The Search for an Elusive Target. *Pharmacol. Biochem. Behav.* **99**:164–189. [07]
Barbato, A., and B. D'Avanzo. 2000. Family Interventions in Schizophrenia and Related Disorders: A Critical Review of Clinical Trials. *Acta Psychiatr. Scand.* **102**:81–97. [17]
Barch, D. M. 2010. Pharmacological Strategies for Enhancing Cognition in Schizophrenia. *Curr. Top. Behav. Neurosci.* **4**:43–96. [07]
Barch, D. M., and A. Ceaser. 2012. Cognition in Schizophrenia: Core Psychological and Neural Mechanisms. *Trends Cogn. Sci.* **16**:27–34. [01]
Barger, S. W., and A. S. Basile. 2001. Activation of Microglia by Secreted Amyloid Precursor Protein Evokes Release of Glutamate by Cystine Exchange and Attenuates Synaptic Function. *J. Neurochem.* **76**:846–854. [07]

Barlati, S., L. De Peri, G. Deste, P. Fusar-Poli, and A. Vita. 2012. Cognitive Remediation in the Early Course of Schizophrenia: A Critical Review. *Curr. Pharm. Des.* **18**:534–541. [07]

Barnett, J. H., F. McDougall, M. K. Xu, et al. 2012. Childhood Cognitive Function and Adult Psychopathology: Associations with Psychotic and Non-Psychotic Symptoms in the General Population. *Br. J. Psychiatry* **201**:124–130. [06]

Barnett, J. H., C. H. Salmond, P. B. Jones, and B. J. Sahakian. 2006. Cognitive Reserve in Neuropsychiatry. *Psychol. Med.* **36**:1053–1064. [06]

Barnsley, N., J. H. McAuley, R. Mohan, et al. 2011. The Rubber Hand Illusion Increases Histamine Reactivity in the Real Arm. *Curr. Biol.* **21**:R945–R946. [01]

Barr, M. S., F. Farzan, T. K. Rajji, et al. 2013. Can Repetitive Magnetic Stimulation Improve Cognition in Schizophrenia? Pilot Data from a Randomized Controlled Trial. *Biol. Psychiatry* **73**:510–517. [09]

Barr, R. S., D. A. Pizzagalli, M. A. Culhane, D. C. Goff, and A. E. Evins. 2008. A Single Dose of Nicotine Enhances Reward Responsiveness in Nonsmokers: Implications for Development of Dependence. *Biol. Psychiatry* **63**:1061–1065. [17]

Barrowclough, C., G. Haddock, T. Wykes, et al. 2010. Integrated Motivational Interviewing and Cognitive Behavioural Therapy for People with Psychosis and Comorbid Substance Misuse: Randomised Controlled Trial. *BMJ* **341**:c6325. [15, 17]

Barrowclough, C., and N. Tarrier. 1992. Families of Schizophrenic Patients: Cognitive Behavioural Intervention. London: Chapman & Hall. [15]

Bar-Yam, Y. 1997. Dynamics of Complex Systems. Cambridge, MA: Perseus. [01]

———. 2002. Unifying Principles in Complex Systems. *N. Engl. Complex Sys. Inst.* http://necsi.edu/projects/yaneer/complexsystems.pdf. (accessed December 31, 2012). [01]

Basar-Eroglu, C., A. Brand, H. Hildebrandt, et al. 2007. Working Memory Related Gamma Oscillations in Schizophrenia Patients. *Int. J. Psychophysiol.* **64**:39–45. [04]

Bassett, D. S., D. L. Greenfield, A. Meyer-Lindenberg, et al. 2010. Efficient Physical Embedding of Topologically Complex Information Processing Networks in Brains and Computer Circuits. *PLoS Comput. Biol.* **6**:e1000748. [08]

Baxter, L. R., Jr., J. M. Schwartz, K. S. Bergman, et al. 1992. Caudate Glucose Metabolic Rate Changes with Both Drug and Behavior Therapy for Obsessive-Compulsive Disorder. *Arch. Gen. Psychiatry* **49**:681–689. [07]

Bazan, N. G. 2005. Neuroprotectin D1 (NPD1): A DHA-Derived Mediator That Protects Brain and Retina against Cell Injury-Induced Oxidative Stress. *Brain Pathol.* **15**:159–166. [07]

Bearden, C. E., P. M. Thompson, M. Dalwani, et al. 2007. Greater Cortical Gray Matter Density in Lithium-Treated Patients with Bipolar Disorder. *Biol. Psychiatry* **62**:7–16. [07]

Bebbington, P., S. Jonas, E. Kuipers, et al. 2011. Childhood Sexual Abuse and Psychosis: Data from a Cross-Sectional National Psychiatric Survey in England. *Br. J. Psychiatry* **199**:29–37. [09]

Beck, A. T., N. A. Rector, N. Stolar, and P. Grant. 2009. Schizophrenia: Cognitive Theory, Research, and Therapy. New York: Guilford Press. [15]

Beck, A. T., A. J. Rush, B. F. Shaw, and G. Emery. 1979. Cognitive Therapy of Depression. New York: Guilford Press. [15]

Becker, D. R., and R. E. Drake. 2003. A Working Life for People with Severe Mental Illness. New York: Oxford Univ. Press. [15]

Behrens, M. M., S. S. Ali, and L. L. Dugan. 2008. Interleukin-6 Mediates the Increase in NADPH-Oxidase in the Ketamine Model of Schizophrenia. *J. Neurosci.* **28**:13,957–13,966. [11]

Behrens, M. M., and T. J. Sejnowski. 2009. Does Schizophrenia Arise from Oxidative Dysregulation of Parvalbumin-Interneurons in the Developing Cortex? *Neuropharmacology* **57**:193–200. [13]

Belforte, J. E., V. Zsiros, E. R. Sklar, et al. 2010. Postnatal NMDA Receptor Ablation in Corticolimbic Interneurons Confers Schizophrenia-Like Phenotypes. *Nat. Neurosci.* **13**:76–83. [11, 13]

Bellack, A. S., K. T. Mueser, S. Gingerich, and J. Agresta. 2004a. Social Skills Training for Schizophrenia: A Step-by-Step Guide (2nd edition). New York: Guilford Press. [15]

Bellack, A. S., N. R. Schooler, S. R. Marder, et al. 2004b. Do Clozapine and Risperidone Affect Social Competence and Problem Solving? *Am. J. Psychiatr.* **161**:364–367. [04]

Belsky, J., C. Jonassaint, M. Pluess, et al. 2009. Vulnerability Genes or Plasticity Genes? *Mol. Psychiatry* **14**:746–754. [09]

Ben-Ari, Y. 2008. Neuro-Archaeology: Pre-Symptomatic Architecture and Signature of Neurological Disorders. *Trends Neurosci.* **31**:626–636. [13]

Ben-Ari, Y., and N. C. Spitzer. 2010. Phenotypic Checkpoints Regulate Neuronal Development. *Trends Neurosci.* **33**:485–492. [13]

Bender, L. 1959. The Concept of Pseudopsychopathic Schizophrenia in Adolescents. *Am. J. Orthopsychiat.* **29**:491–512. [01]

———. 1966. The Concept of Plasticity in Childhood Schizophrenia. *Proc. Annu. Meet. Am. Psychopathol. Assoc.* **54**:354–365. [01]

Benes, F. M., and S. Berretta. 2001. Gabaergic Interneurons: Implications for Understanding Schizophrenia and Bipolar Disorder. *Neuropsychopharmacology* **25**:1–27. [12]

Bentall, R. P. 1993. Deconstructing the Concept of "Schizophrenia." *J. Ment. Health* **2**:223–228. [15]

Ben-Zeev, D., R. E. Drake, P. W. Corrigan, et al. 2012 Using Contemporary Technologies in the Assessment and Treatment of Serious Mental Illness. *Am. J. Psychiatr. Rehabil.* **15**:357–376. [15]

Berger, G. E., S. Wood, and P. D. McGorry. 2003. Incipient Neurovulnerability and Neuroprotection in Early Psychosis. *Psychopharmacol. Bull.* **37**:79–101. [07]

Berger, G. E., S. J. Wood, M. Ross, et al. 2012. Neuroprotective Effects of Low-Dose Lithium in Individuals at Ultra-High Risk for Psychosis. A Longitudinal MRI/MRS Study. *Curr. Pharm. Des.* **18**:570–575. [07]

Berger, G. E., S. J. Wood, R. M. Wellard, et al. 2008. Ethyl-Eicosapentaenoic Acid in First-Episode Psychosis. A 1H-MRS Study. *Neuropsychopharmacology* **33**:2467–2473. [07]

Bertolino, A., G. Caforio, G. Blasi, et al. 2004. Interaction of COMT (Val(108/158)Met) Genotype and Olanzapine Treatment on Prefrontal Cortical Function in Patients with Schizophrenia. *Am. J. Psychiatr.* **161**:1798–1805. [07]

Bertram, L., C. M. Lill, and R. E. Tanzi. 2010. The Genetics of Alzheimer Disease: Back to the Future. *Neuron* **68**:270–281. [13]

Betancur, C., T. Sakurai, and J. D. Buxbaum. 2009. The Emerging Role of Synaptic Cell-Adhesion Pathways in the Pathogenesis of Autism Spectrum Disorders. *Trends Neurosci.* **32**:402–412. [13]

Bickman, L. 2008. A Measurement Feedback System (MFS) Is Necessary to Improve Mental Health Outcomes. *J. Am. Acad. Child Adolesc. Psychiatry* **47**:1114–1119. [15]

Bigos, K. L., V. S. Mattay, J. H. Callicott, et al. 2010. Genetic Variation in CACNA1C Affects Brain Circuitries Related to Mental Illness. *Arch. Gen. Psychiatry* **67**:939–945. [08]
Bigos, K. L., and D. R. Weinberger. 2010. Imaging Genetics: Days of Future Past. *NeuroImage* **53**:804–809. [08]
Bijl, R. V., R. de Graaf, E. Hiripi, et al. 2003. The Prevalence of Treated and Untreated Mental Disorders in Five Countries. *Health Affairs* **22**:122–133. [16]
Bijl, R. V., and A. Ravelli. 2000. Psychiatric Morbidity, Service Use, and Need for Care in the General Population: Results of the Netherlands Mental Health Survey and Incidence Study. *Am. J. Public Health* **90**:602–607. [16]
Bijou, S. W., and R. F. Peterson. 1971. Functional Analysis in the Assessment of Children. In: Advances in Psychological Assessment, ed. P. McReynolds, vol. 2, pp. 63–78. Palo Alto: Science and Behavior Books. [15]
Bilbo, S. D., and J. M. Schwarz. 2009. Early-Life Programming of Later-Life Brain and Behavior: A Critical Role for the Immune System. *Front. Behav. Neurosci.* **3**:14. [07]
Bilder, R. M., L. Lipschutz-Broch, G. Reiter, et al. 1992. Intellectual Deficits in First-Episode Schizophrenia: Evidence for Progressive Deterioration. *Schizophr. Bull.* **18**:437–448. [04]
Bilder, R. M., G. Reiter, J. Bates, et al. 2006. Cognitive Development in Schizophrenia: Follow-Back from the First Episode. *J. Clin. Exp. Neuropsychol.* **28**:270–282. [07]
Birchwood, M., R. Mason, F. MacMillian, and J. Healy. 1993. Depression, Demoralization and Control over Psychotic Illness: A Comparison and Non-Depressed Patients with a Chronic Psychosis. *Psychol. Med.* **23**:387–395. [15]
Bittner, R. 2003. Notebook 10 (1887). In: Friedrich Nietzsche: Writings from the Late Notebooks, p. 188. Cambridge: Cambridge Univ. Press. [06]
Blackwood, D. H., A. Fordyce, M. T. Walker, et al. 2001. Schizophrenia and Affective Disorders: Cosegregation with a Translocation at Chromosome 1q42 That Directly Disrupts Brain-Expressed Genes: Clinical and P300 Findings in a Family. *Am. J. Hum. Genet.* **69**:428–433. [08]
Blake, R., and M. Shiffrar. 2007. Perception of Human Motion. *Annu. Rev. Psychol.* **58**:47–73. [07]
Bleuler, E. 1911/1950. Dementia Praecox or the Group of Schizophrenias (translated by J. Zinkin). New York: Intl. Universities Press. [02, 03, 14]
Bleuler, M. 1972. Die Schizophrenen Geistesstorungen: Im Lichte Langjähriger Kranken und Familiengeschichten. Stuttgart: Georg Thieme Verlag. [06]
———. 1974. The Offspring of Schizophrenics. *Schizophr. Bull.* **8**:93–107. [06]
Block, M. L., and J. S. Hong. 2007. Chronic Microglial Activation and Progressive Dopaminergic Neurotoxicity. *Biochem. Soc. Trans.* **35**:1127–1132. [07]
Bodatsch, M., S. Ruhrmann, M. Wagner, et al. 2011. Prediction of Psychosis by Mismatch Negativity. *Biol. Psychiatry* **69**:959–966. [07, 09]
Bond, G. R. 1992. Vocational Rehabilitation. In: Handbook of Psychiatric Rehabilitation, ed. R. P. Liberman, pp. 244–275. New York: MacMillan. [15]
———. 2004. Supported Employment: Evidence for an Evidence-Based Practice. *Psychiatr. Rehabil. J.* **27**:345–359. [15]
Bond, G. R., R. E. Drake, and D. R. Becker. 2008. An Update on Randomized Controlled Trials of Evidence-Based Supported Employment. *Psychiatr. Rehabil. J.* **31**:280–290. [15]
———. 2012. Generalizability of the Individual Placement and Support (IPS) Model of Supported Employment Outside the US. *World Psychiatry* **11**:32–39. [15]

Bond, G. R., R. E. Drake, G. J. McHugo, C. A. Rapp, and E. Whitley. 2009. Strategies for Improving Fidelity in the National Evidence-Based Practices Project. *Res. Social Work Prac.* **19**:569–581. [15]

Bond, G. R., L. Evans, M. P. Salyers, J. Williams, and H.-W. Kim. 2000. Measurement of Fidelity in Psychiatric Rehabilitation. *Ment. Health Serv. Res.* **2**:75–87. [15]

Bond, G. R., and M. Kukla. 2011. Impact of Follow-Along Support on Job Tenure in IPS Supported Employment. *J. Nerv. Ment. Dis.* **199**:150–155. [15]

Bonini, L., and P. F. Ferrari. 2011. Evolution of Mirror Systems: A Simple Mechanism for Complex Cognitive Functions. *Ann. NY Acad. Sci.* **1225**:166–175. [07]

Borgmann-Winter, K. E., N. E. Rawson, H. Y. Wang, et al. 2009. Human Olfactory Epithelial Cells Generated *in Vitro* Express Diverse Neuronal Characteristics. *Neuroscience* **158**:642–653. [10]

Borgwardt, S. J., A. Riecher-Rossler, P. Dazzan, et al. 2007. Regional Gray Matter Volume Abnormalities in the at Risk Mental State. *Biol. Psychiatry* **61**:1148–1156. [07]

Borkovec, T. D. 2004. Research in Training Clinics and Practice Research Networks: A Route to the Integration of Science and Practice. *Clin. Psychol.* **11**:211–215. [15]

Boroojerdi, B., M. Phipps, L. Kopylev, et al. 2001. Enhancing Analogic Reasoning with rTMS over the Left Prefrontal Cortex. *Neurology* **56**:526–528. [09]

Bosker, F. J., C. A. Hartman, I. M. Nolte, et al. 2011. Poor Replication of Candidate Genes for Major Depressive Disorder Using Genome-Wide Association Data. *Mol. Psychiatry* **16**:516–532. [08]

Bouffard, L. A., and L. R. Muftić. 2006. The "Rural Mystique": Social Disorganization and Violence Beyond Urban Communities. *West. Crim. Rev.* **7**:56–66. [01]

Boulanger, L. M. 2009. Immune Proteins in Brain Development and Synaptic Plasticity. *Neuron* **64**:93–109. [08]

Bourque, F., E. van der Ven, and A. Malla. 2011. A Meta-Analysis of the Risk for Psychotic Disorders among First- and Second-Generation Immigrants. *Psychol. Med.* **41**:897–910. [08]

Bowie, C. R., A. Reichenberg, T. L. Patterson, R. K. Heaton, and P. D. Harvey. 2006. Determinants of Real-World Functional Performance in Schizophrenia Subjects: Correlations with Cognition, Functional Capacity, and Symptoms. *Am. J. Psychiatr.* **163**:418–425. [14]

Bowlby, J. 1969. Attachment and Loss I: Attachment. London: Hogarth Press. [06]

———. 1973. Attachment and Loss II: Separation, Anxiety and Anger. London: Hogarth Press. [06]

———. 1980. Attachment and Loss III: Loss, Sadness, and Depression. New York: Basic Books. [06]

Boyle, D. J., and C. Hassett-Walker. 2008. Individual-Level and Socio-Structural Characteristics of Violence. *J. Interpers. Violence* **23**:1011–1026. [01]

Brady, A. M., R. D. Saul, and M. K. Wiest. 2010. Selective Deficits in Spatial Working Memory in the Neonatal Ventral Hippocampal Lesion Rat Model of Schizophrenia. *Neuropharmacology* **59**:605–611. [11]

Braff, D. L., R. Freedman, N. J. Schork, and I. I. Gottesman. 2007. Deconstructing Schizophrenia: An Overview of the Use of Endophenotypes in Order to Understand a Complex Disorder. *Schizophr. Bull.* **33**:21–32. [03]

Braff, D. L., C. Grillon, and M. A. Geyer. 1992. Gating and Habituation of the Startle Reflex in Schizophrenic Patients. *Arch. Gen. Psychiatry* **49**:206–215. [07]

Brandon, N. J., and A. Sawa. 2011. Linking Neurodevelopmental and Synaptic Theories of Mental Illness through DISC1. *Nat. Rev. Neurosci.* **12**:707–722. [10, 13]

Brans, R. G., R. S. Kahn, H. G. Schnack, et al. 2010. Brain Plasticity and Intellectual Ability Are Influenced by Shared Genes. *J. Neurosci.* **30**:5519–5524. [17]

Brase-Smith, S. B. 1995. Restraints: Retraumatization for Rape Victims? *J. Psychosoc. Nurs. Ment. Health Serv.* **33**:24–28. [15]

Breitborde, N. J., F. A. Moreno, N. Mai-Dixon, et al. 2011. Multifamily Group Psychoeducation and Cognitive Remediation for First-Episode Psychosis: A Randomized Controlled Trial. *BMC Psychiatry* **11**:9. [07]

Bremner, J. D., M. Vythilingam, E. Vermetten, et al. 2003. MRI and PET Study of Deficits in Hippocampal Structure and Function in Women with Childhood Sexual Abuse and Posttraumatic Stress Disorder. *Am. J. Psychiatr.* **160**:924–932. [01]

Brenhouse, H. C., K. C. Sonntag, and S. L. Andersen. 2008. Transient D1 Dopamine Receptor Expression on Prefrontal Cortex Projection Neurons: Relationship to Enhanced Motivational Salience of Drug Cues in Adolescence. *J. Neurosci.* **28**:2375–2382. [13]

Brennand, K. J., A. Simone, J. Jou, et al. 2011. Modelling Schizophrenia Using Human Induced Pluripotent Stem Cells. *Nature* **473**:221–225. [10, 17]

Brenner, H. D., V. Roder, B. Hodel, et al. 1994. Integrated Psychological Therapy for Schizophrenic Patients. Seattle: Hogrefe & Huber Publishers. [15]

Brewer, W. J., S. M. Francey, S. J. Wood, et al. 2005. Memory Impairments Identified in People at Ultra-High Risk for Psychosis Who Later Develop First-Episode Psychosis. *Am. J. Psychiatr.* **162**:71–78. [07]

Brewer, W. J., S. J. Wood, L. J. Phillips, et al. 2006. Generalized and Specific Cognitive Performance in Clinical High-Risk Cohorts: A Review Highlighting Potential Vulnerability Markers for Psychosis. *Schizophr. Bull.* **32**:538–555. [02]

Brockhaus-Dumke, A., I. Tendolkar, R. Pukrop, et al. 2005. Impaired Mismatch Negativity Generation in Prodromal Subjects and Patients with Schizophrenia. *Schizophr. Res.* **73**:297–310. [07]

Broen, W. E. J., and L. H. Storms. 1966. Lawful Disorganization: The Process Underlying a Schizophrenic Syndrome. *Psychol. Rev.* **73**:265–279. [01]

Bromet, E. J., R. Kotov, L. J. Fochtmann, et al. 2011. Diagnostic Shifts During the Decade Following First Admission for Psychosis. *Am. J. Psychiatr.* **168**:1186–1194. [02]

Broome, M. R., P. Matthiasson, P. Fusar-Poli, et al. 2010. Neural Correlates of Movement Generation in the "At-Risk Mental State." *Acta Psychiatr. Scand.* **122**:295–301. [07]

Brown, A. S. 2006. Prenatal Infection as a Risk Factor for Schizophrenia. *Schizophr. Bull.* **32**:200–202. [11]

Brown, A. S., and E. J. Derkits. 2010. Prenatal Infection and Schizophrenia: A Review of Epidemiologic and Translational Studies. *Am. J. Psychiatr.* **167**:261–280. [08]

Brown, A. S., and P. H. Patterson. 2011. Maternal Infection and Schizophrenia: Implications for Prevention. *Schizophr. Bull.* **37**:284–290. [07]

Brown, G. W., J. L. T. Birley, and J. K. Wing. 1972. A Replication Influence of Family Life on the Course of Schizophrenic Disorders: A Replication. *Br. J. Psychiatry* **121**:241–258. [15]

Brown, G. W., and T. O. Harris. 1978. Social Origins of Depression: A Study of Psychiatric Disorder in Women. London: Tavistock Publications. [09]

Brown, S., M. Kim, C. Mitchell, and H. Inskip. 2010. Twenty-Five Year Mortality of a Community Cohort with Schizophrenia. *Br. J. Psychiatry* **196**:116–121. [17]

Brune, M. 2005. Emotion Recognition, "Theory of Mind," and Social Behavior in Schizophrenia. *Psychiatry Res.* **133**:135–147. [04]

Bryant, R. A., K. Felmingham, A. Kemp, et al. 2008. Amygdala and Ventral Anterior Cingulate Activation Predicts Treatment Response to Cognitive Behaviour Therapy for Post-Traumatic Stress Disorder. *Psychol. Med.* **38**:555–561. [09]

Bryson, G., M. Bell, and P. Lysaker. 1997. Affect Recognition in Schizophrenia: A Function of Global Impairment or a Specific Cognitive Deficit. *Psychiatry Res.* **71**:105–113. [04]

Buchanan, R. W., R. R. Conley, D. Dickinson, et al. 2008. Galantamine for the Treatment of Cognitive Impairments in People with Schizophrenia. *Am. J. Psychiatr.* **165**:82–89. [04]

Buchanan, R. W., M. Davis, D. Goff, et al. 2005. A Summary of the FDA-NIMH-MATRICS Workshop on Clinical Trial Design for Neurocognitive Drugs for Schizophrenia. *Schizophr. Bull.* **31**:5–19. [03]

Buchanan, R. W., D. C. Javitt, S. R. Marder, et al. 2007. The Cognitive and Negative Symptoms in Schizophrenia Trial (CONSIST): The Efficacy of Glutamatergic Agents for Negative Symptoms and Cognitive Impairments. *Am. J. Psychiatr.* **164**: 1593–1602. [04]

Buchanan, R. W., R. S. Keefe, J. A. Lieberman, et al. 2011. A Randomized Clinical Trial of MK-0777 for the Treatment of Cognitive Impairments in People with Schizophrenia. *Biol. Psychiatry* **69**:442–449. [04, 11]

Buckholtz, J. W., and A. Meyer-Lindenberg. 2012. Psychopathology and the Human Connectome: Toward a Transdiagnostic Model of Risk for Mental Illness. *Neuron* **74**:990–1004. [17]

Burdick, K. E., C. B. Gopin, and A. K. Malhotra. 2011. Pharmacogenetic Approaches to Cognitive Enhancement in Schizophrenia. *Harv. Rev. Psychiatry* **19**:102–108. [07]

Burne, T., E. Scott, B. van Swinderen, et al. 2011. Big Ideas for Small Brains: What Can Psychiatry Learn from Worms, Flies, Bees and Fish? *Mol. Psychiatry* **16**:7–16. [08, 10]

Burns, T., J. Catty, T. Becker, et al. 2007. The Effectiveness of Supported Employment for People with Severe Mental Illness: A Randomised Controlled Trial. *Lancet* **370**:1146–1152. [17]

Bustillo, J. R., H. Chen, C. Gasparovic, et al. 2011. Glutamate as a Marker of Cognitive Function in Schizophrenia: A Proton Spectroscopic Imaging Study at 4 Tesla. *Biol. Psychiatry* **69**:19–27. [07]

Bustillo, J. R., L. M. Rowland, P. Mullins, et al. 2010. 1H-MRS at 4 Tesla in Minimally Treated Early Schizophrenia. *Mol. Psychiatry* **15**:629–636. [07]

Butler, P. D., and D. C. Javitt. 2005. Early-Stage Visual Processing Deficits in Schizophrenia. *Curr. Opin. Psychiatry* **18**:151–157. [13]

Butler, P. D., S. M. Silverstein, and S. C. Dakin. 2008. Visual Perception and Its Impairment in Schizophrenia. *Biol. Psychiatry* **64**:40–47. [01]

Byrd, C. A., and P. C. Brunjes. 2001. Neurogenesis in the Olfactory Bulb of Adult Zebrafish. *Neuroscience* **105**:793–801. [07]

Cabral, P., H. B. Meyer, and D. Ames. 2011. Effectiveness of Yoga Therapy as a Complementary Treatment for Major Psychiatric Disorders: A Meta-Analysis. *Prim. Care Companion CNS Disord.* **13**: [07]

Cadenhead, K. S. 2011. Startle Reactivity and Prepulse Inhibition in Prodromal and Early Psychosis: Effects of Age, Antipsychotics, Tobacco and Cannabis in a Vulnerable Population. *Psychiatry Res.* **188**:208–216. [07]

Cadenhead, K. S., B. Carasso, N. R. Swerdlow, M. A. Geyer, and D. L. Braff. 1999a. Prepulse Inhibition and Habituation of the Startle Response Are Stable Neurobiological Measures in a Normal Male Population. *Biol. Psychiatry* **45**:360–364. [07]

Cadenhead, K. S., M. A. Geyer, and D. L. Braff. 1993. Impaired Startle Prepulse Inhibition and Habituation in Patients with Schizotypal Personality Disorder. *Am. J. Psychiatr.* **150**:1862–1867. [07]

Cadenhead, K. S., W. Perry, K. Shafer, and D. L. Braff. 1999b. Cognitive Functions in Schizotypal Personality Disordered Subjects. *Schizophr. Res.* **37**:123–132. [07]

Cadenhead, K. S., N. R. Swerdlow, K. Shafer, M. Diaz, and D. L. Braff. 2000. Modulation of the Startle Response and Startle Laterality in Relatives of Schizophrenia Patients and Schizotypal Personality Disordered Subjects: Evidence of Inhibitory Deficits. *Am. J. Psychiatr.* **157**:1660–1668. [07]

Caiazzo, M., M. T. Dell'Anno, E. Dvoretskova, et al. 2011. Direct Generation of Functional Dopaminergic Neurons from Mouse and Human Fibroblasts. *Nature* **476**:224–227. [10]

Callicott, J. H., A. Bertolino, V. S. Mattay, et al. 2000. Physiological Dysfunction of the Dorsolateral Prefrontal Cortex in Schizophrenia Revisited. *Cereb. Cortex* **10**:1078–1092. [07, 11]

Callicott, J. H., M. F. Egan, V. S. Mattay, et al. 2003. Abnormal fMRI Response of the Dorsolateral Prefrontal Cortex in Cognitively Intact Siblings of Patients with Schizophrenia. *Am. J. Psychiatr.* **160**:709–719. [07]

Cannon, T. D., K. Cadenhead, B. Cornblatt, et al. 2008. Prediction of Psychosis in Youth at High Clinical Risk: A Multisite Longitudinal Study in North America. *Arch. Gen. Psychiatry* **65**:28–37. [02, 07, 09]

Cannon, T. D., and M. C. Keller. 2006. Endophenotypes in the Genetic Analyses of Mental Disorders. *Annu. Rev. Clin. Psychol.* **2**:267–290. [02]

Cannon, T. D., L. E. Zorrilla, D. Shtasel, et al. 1994. Neuropsychological Functioning in Siblings Discordant for Schizophrenia and Healthy Volunteers. *Arch. Gen. Psychiatry* **51**:651–666. [07]

Capper, E. A., and L. A. Marshall. 2001. Mammalian Phospholipases A(2): Mediators of Inflammation, Proliferation and Apoptosis. *Prog. Lipid Res.* **40**:167–197. [07]

Carandini, M., and D. J. Heeger. 2012. Normalization as a Canonical Neural Computation. *Nat. Rev. Neurosci.* **13**:51–62. [01, 02]

Cardin, J. A., M. Carlen, K. Meletis, et al. 2009. Driving Fast-Spiking Cells Induces Gamma Rhythm and Controls Sensory Responses. *Nature* **459**:663–667. [13]

Cardno, A. G., E. J. Marshall, B. Coid, et al. 1999. Heritability Estimates for Psychotic Disorders: The Maudsley Twin Psychosis Series. *Arch. Gen. Psychiatry* **56**:162–168. [06]

Carlsson, A., N. Waters, S. Holm-Waters, et al. 2001. Interactions between Monoamines, Glutamate, and GABA in Schizophrenia: New Evidence. *Annu. Rev. Pharmacol. Toxicol.* **41**:237–260. [07]

Carlsson, M., and A. Carlsson. 1990a. Interactions between Glutamatergic and Monoaminergic Systems within the Basal Ganglia: Implications for Schizophrenia and Parkinson's Disease. *Trends Neurosci.* **13**:272–276. [07]

———. 1990b. Schizophrenia: A Subcortical Neurotransmitter Imbalance Syndrome? *Schizophr. Bull.* **16**:425–430. [12]

Carpenter, W. T., Jr., and R. W. Buchanan. 1989. Domains of Psychopathology Relevant to the Study of Etiology and Treatment of Schizophrenia. In: Schizophrenia: Scientific Progress, ed. S. C. Schulz and C. T. Tamminga, pp. 13–22. New York: Oxford Univ. Press. [03]

Carpenter, W. T., Jr., D. W. Heinrichs, and A. M. I. Wagman. 1988. Deficit and Non-Deficit Forms of Schizophrenia: The Concept. *Am. J. Psychiatr.* **145**:578–583. [03, 17]

Carpenter, W. T., Jr., and J. I. Koenig. 2008. The Evolution of Drug Development in Schizophrenia: Past Issues and Future Opportunities. *Neuropsychopharmacology* **33**:2061–2079. [13]

Carpenter, W. T., Jr., J. S. Strauss, and J. J. Bartko. 1973. Flexible System for the Diagnosis of Schizophrenia: Report from the WHO International Pilot Study of Schizophrenia. *Science* **182**:1275–1278. [03]

Carrà, G., S. Johnson, P. Bebbington, et al. 2012. The Lifetime and Past-Year Prevalence of Dual Diagnosis in People with Schizophrenia across Europe: Findings from the European Schizophrenia Cohort (EuroSC). *Eur. Arch. Psychiatry Clin. Neurosci.* **262**:607–616. [05]

Carter, C. S., and D. M. Barch. 2007. Cognitive Neuroscience-Based Approaches to Measuring and Improving Treatment Effects on Cognition in Schizophrenia: The Cntrics Initiative. *Schizophr. Bull.* **33**:1131–1137. [02, 07]

Cascella, N. G., M. Takaki, S. Lin, and A. Sawa. 2007. Neurodevelopmental Involvement in Schizophrenia: The Olfactory Epithelium as an Alternative Model for Research. *J. Neurochem.* **102**:587–594. [10]

Casey, B. J., S. Duhoux, and M. Malter Cohen. 2010. Adolescence: What Do Transmission, Transition, and Translation Have to Do with It? *Neuron* **67**:749–760. [13]

Caspi, A., and T. E. Moffitt. 2006. Gene-Environment Interactions in Psychiatry: Joining Forces with Neuroscience. *Nat. Rev. Neurosci.* **7**:583–590. [08]

Caspi, A., T. E. Moffitt, M. Cannon, et al. 2005. Moderation of the Effect of Adolescent-Onset Cannabis Use on Adult Psychosis by a Functional Polymorphism in the Catechol-O-Methyltransferase Gene: Longitudinal Evidence of a Gene X Environment Interaction. *Biol. Psychiatry* **57**:1117–1127. [07]

Catty, J., M. Koletsi, S. White, et al. 2010. Therapeutic Relationships: Their Specificity in Predicting Outcomes for People with Psychosis Using Clinical and Vocational Services. *Soc. Psychiatry Psychiatr. Epidem.* **45**:1187–1193. [16]

Catty, J., S. White, M. Koletsi, et al. 2011. Therapeutic Relationships in Vocational Rehabilitation: Predicting Good Relationships for People with Psychosis. *Psychiatry Res.* **187**:68–73. [16]

Cech, T. R., and G. M. Rubin. 2004. Nurturing Interdisciplinary Research. *Nat. Struct. Mol. Biol.* **11**:1166–1169. [08]

Ceglar, A., and J. F. Roddick. 2006. Association Mining. *ACM Comput. Surveys* **38**:58–64. [05]

Cephalon Inc. 2010. Treatment of Cognitive Impairment in Men with Schizophrenia (MK5757-005) (Completed). *Natl. Library Med.* http://www.clinicaltrials.gov/ct2/show/record/NCT00848484?term=NCT00848484&rank=1 (accessed May 11, 2013). [04]

Chadwick, P. 2006. Person-Based Cognitive Therapy for Distressing Psychosis. Chichester: Wiley. [15]

Chadwick, P., M. J. Birchwood, and P. Trower. 1996. Cognitive Therapy for Delusions, Voices, and Paranoia. Chichester: Wiley. [05]

Chamberlin, J. 1997a. Confessions of a Non-Compliant Patient. *Natl. Empowerment Center Newsletter* **8/9**:13. [15]

Chamberlin, J. 1997b. A Working Definition of Empowerment. *Psychiatr. Rehabil. J.* **20**:43–46. [17]

Chambless, D. L., and T. H. Ollendick. 2001. Empirically Supported Psychological Interventions: Controversies and Evidence. *Annu. Rev. Psychol.* **52**:685–716. [15]

Chang, C.-K., R. D. Hayes, G. Perera, et al. 2011. Life Expectancy at Birth for People with Serious Mental Illness and Other Major Disorders from a Secondary Mental Health Case Register in London. *PLoS One* **6**:e19590. [05]

Chapman, L. J., and J. P. Chapman. 1973a. Disordered Thought in Schizophrenia. In: Century Psychology, ed. K. MacCorquodale et al. Englewood Cliffs, NJ: Prentice-Hall. [02]

———. 1973b. Problems in the Measurement of Cognitive Deficit. *Psychol. Bull.* **79**:380–385. [02]

———. 1978. The Measurement of Differential Deficit. *J. Psychiatr. Res.* **14**:303–311. [02]

Chapman, L. J., J. P. Chapman, T. R. Kwapil, M. Eckblad, and M. C. Zinser. 1994. Putative Psychosis-Prone Subjects 10 Years Later. *J. Abnorm. Psychol.* **103**:171–183. [02]

Charrier, C., K. Joshi, J. Coutinho-Budd, et al. 2012. Inhibition of SRGAP2 Function by Its Human-Specific Paralogs Induces Neoteny During Spine Maturation. *Cell* **149**:923–935. [10]

Chen, H. H., A. Stoker, and A. Markou. 2010. The Glutamatergic Compounds Sarcosine and N-Acetylcysteine Ameliorate Prepulse Inhibition Deficits in Metabotropic Glutamate 5 Receptor Knockout Mice. *Psychopharmacology* **209**:343–350. [07]

Chinman, M., A. S. Young, T. Schell, J. Hassell, and J. Mintz. 2004. Computer-Assisted Self-Assessment in Persons with Severe Mental Illness. *J. Clin. Psychiatry* **65**:1343–1351. [15]

Choi, S. M., H. Liu, P. Chaudhari, et al. 2011. Reprogramming of EBV-Immortalized B-Lymphocyte Cell Lines into Induced Pluripotent Stem Cells. *Blood* **118**:1801–1805. [10]

Chorpita, B. F., and E. L. Daleiden. 2009. Mapping of Evidence-Based Treatment for Children and Adolescents: Application of the Distillation and Matching Model to 615 Treatments from 22 Randomized Trials. *J. Consult. Clin. Psychol.* **77**:566–579. [15]

Cicchetti, D., and T. D. Cannon. 1999. Neurodevelopmental Processes in the Ontogenesis and Epigenesis of Psychopathology. *Dev. Psychopathol.* **11**:375–393. [02]

Cichon, S., N. Craddock, M. Daly, et al. 2009. Genomewide Association Studies: History, Rationale, and Prospects for Psychiatric Disorders. *Am. J. Psychiatr.* **166**:540–556. [08]

Clarke, H., J. Flint, A. S. Attwood, and M. R. Munafo. 2010. Association of the 5-HTTLPR Genotype and Unipolar Depression: A Meta-Analysis. *Psychol. Med.* **40**:1767–1778. [08]

Clarke, M. C., M. Harley, and M. Cannon. 2006. The Role of Obstetric Events in Schizophrenia. *Schizophr. Bull.* **32**:3–8. [09]

Cohen, J. 1992. A Power Primer. *Psychol. Bull.* **112**:155–159. [15]

Cohen, J. D., D. M. Barch, C. Carter, and D. Servan-Schreiber. 1999. Context-Processing Deficits in Schizophrenia: Converging Evidence from Three Theoretically Motivated Cognitive Tasks. *J. Abnorm. Psychol.* **108**:120–133. [02]

Cohen, J. D., and D. Servan-Schreiber. 1992. Context, Cortex and Dopamine: A Connectionist Approach to Behavior and Biology in Schizophrenia. *Psychol. Rev.* **99**:45–77. [02]

Coldwell, C. M., and W. S. Bender. 2007. The Effectiveness of Assertive Community Treatment for Homeless Populations with Severe Mental Illness: A Meta-Analysis. *Am. J. Psychiatr.* **164**:393–399. [15]

Cole, G. M., Q. L. Ma, and S. A. Frautschy. 2009. Omega-3 Fatty Acids and Dementia. *Prostaglandins Leukot. Essent. Fatty Acids* **81**:213–221. [07]

Collip, D., I. Myin-Germeys, and J. Van Os. 2008. Does the Concept of "Sensitization" Provide a Plausible Mechanism for the Putative Link between the Environment and Schizophrenia? *Schizophr. Bull.* **34**:220–225. [09]

Conklin, H. M., C. E. Curtis, M. E. Calkins, and W. G. Iacono. 2005. Working Memory Functioning in Schizophrenia Patients and Their First-Degree Relatives: Cognitive Functioning Shedding Light on Etiology. *Neuropsychologia* **43**:930–942. [01]

Cook, J. A., M. E. Copeland, J. A. Jonikas, et al. 2012. Results of a Randomized Controlled Trial of Mental Illness Self-Management Using Wellness Recovery Action Planning. *Schizophr. Bull.* **38**:881–891. [15]

Cook, J. A., S. E. Shore, J. K. Burke-Miller, et al. 2010. Participatory Action Research to Establish Self-Directed Care for Mental Health Recovery in Texas. *Psychiatr. Rehabil. J.* **34**:137–144. [15]

Cools, R., L. Clark, A. M. Owen, and T. W. Robbins. 2002. Defining the Neural Mechanisms of Probabilistic Reversal Learning Using Event-Related Functional Magnetic Resonance Imaging. *J. Neurosci.* **22**:4563–4567. [05]

Cooper, B., and B. Singh. 2000. Population Research and Mental Health Policy. Bridging the Gap. *Br. J. Psychiatry* **176**:407–411. [16]

Corey, L. A., J. M. Pellock, M. J. Kjeldsen, and K. O. Nakken. 2011. Importance of Genetic Factors in the Occurrence of Epilepsy Syndrome Type: A Twin Study. *Epilepsy Res.* **97**:103–111. [13]

Corlett, P. 2013. Sleight of Hand, Sleight of Mind: Illusions, Delusions and the Immune System. Scientific American SciCurious Blog 2013 [cited January 30, 2013 2013]. Available from http://blogs.scientificamerican.com/scicurious-brain/2013/01/30/scicurious-guest-writer-sleight-of-hand-sleight-of-mind-illusions-delusions-and-the-immune-system/. [01]

Cornblatt, B. A., and L. Erlenmeyer-Kimling. 1985. Global Attentional Deviance as a Marker of Risk for Schizophrenia: Specificity and Predictive Validity. *J. Abnorm. Psychol.* **94**:470–486. [02]

Correll, C. U., and A. K. Malhotra. 2004. Pharmacogenetics of Antipsychotic-Induced Weight Gain. *Psychopharmacology* **174**:477–489. [07]

Corrigan, P. W., S. G. McCracken, and E. P. Holmes. 2001. Motivational Interviews as Goal Assessment for Persons with Psychiatric Disability. *Commun. Ment. Health J.* **37**:113–122. [15]

Corrigan, P. W., K. T. Mueser, G. R. Bond, R. E. Drake, and P. Solomon. 2008. The Principles and Practice of Psychiatric Rehabilitation: An Empirical Approach. New York: Guilford Press. [15]

Cosway, R., M. Byrne, R. Clafferty, et al. 2000. Neuropsychological Change in Young People at High Risk for Schizophrenia: Results from the First Two Neuropsychological Assessments of the Edinburgh High Risk Study. *Psychol. Med.* **30**:1111–1121. [07]

Coyle, P., N. Tran, J. N. Fung, B. L. Summers, and A. M. Rofe. 2009. Maternal Dietary Zinc Supplementation Prevents Aberrant Behaviour in an Object Recognition Task in Mice Offspring Exposed to Lps in Early Pregnancy. *Behav. Brain Res.* **197**:210–218. [07]

Crespi, B., and C. Badcock. 2008. Psychosis and Autism as Diametrical Disorders of the Social Brain. *Behav. Brain Sci.* **31**:241–261. [01]

Crossley, N. A., A. Mechelli, P. Fusar-Poli, et al. 2009. Superior Temporal Lobe Dysfunction and Frontotemporal Dysconnectivity in Subjects at Risk of Psychosis and in First-Episode Psychosis. *Hum. Brain Mapp.* **30**:4129–4137. [09]

Csermely, P. 2008. Creative Elements: Network-Based Predictions of Active Centres in Proteins and Cellular and Social Networks. *Trends Biochem. Sci.* **33**:569–576. [01]

Cuesta, M. J., and V. Peralta. 1995. Psychopathological Dimensions in Schizophrenia. *Schizophr. Bull.* **21**:473–482. [03]

Cusin, C., and D. D. Dougherty. 2012. Somatic Therapies for Treatment-Resistant Depression: ECT, TMS, VNS, DBS. *Biol. Mood Anxiety Disord.* **2**:14. [17]

Cuthbert, B. N., and T. R. Insel. 2010. Toward New Approaches to Psychotic Disorders: The NIMH Research Domain Criteria Project. *Schizophr. Bull.* **36**:1061–1062. [17]

D'Agostino, R. B., Sr., S. Grundy, L. M. Sullivan, and P. Wilson. 2001. Validation of the Framingham Coronary Heart Disease Prediction Scores: Results of a Multiple Ethnic Groups Investigation. *JAMA* **286**:180–187. [09]

Das, P., A. H. Kemp, G. Flynn, et al. 2007. Functional Disconnections in the Direct and Indirect Amygdala Pathways for Fear Processing in Schizophrenia. *Schizophr. Res.* **90**:284–294. [04]

Davey Smith, G., and S. Ebrahim. 2001. Epidemiology: Is It Time to Call It a Day? *Int. J. Epidemiol.* **30**:1–11. [08]

Davidson, L., M. J. O'Connell, J. Tondora, M. Lawless, and A. C. Evans. 2005. Recovery in Serious Mental Illness: A New Wine or Just a New Bottle? *Prof. Psychol. Res. Pr.* **36** 480–487. [17]

Davidson, M., S. Galderisi, M. Weiser, et al. 2009. Cognitive Effects of Antipsychotic Drugs in First-Episode Schizophrenia and Schizophreniform Disorder: A Randomized, Open-Label Clinical Trial (EUFEST). *Am. J. Psychiatr.* **166**:675–682. [14]

Davies, L. M., S. W. Lewis, P. B. Jones, et al. 2007. Cost Effectiveness of First Generation Versus Second Generation Antipsychotic Drugs to Treat Psychosis: Results from a Randomised Controlled Trial in Schizophrenia Responding Poorly to Previous Therapy. *Br. J. Psychiatry* **191** 14–22. [01]

Decoster, J., J. van Os, I. Myin-Germeys, M. De Hert, and R. van Winkel. 2012. Genetic Variation Underlying Psychosis-Inducing Effects of Cannabis: Critical Review and Future Directions. *Curr. Pharm. Des.* **18**:5015–5023. [06]

Deegan, P. E. 1990. Spirit Breaking: When the Helping Professionals Hurt. *Hum. Psychol.* **18**:301–313. [15]

———. 1991. Recovery: The Lived Experience of Rehabilitation. In: The Psychological and Social Impact of Disability ed. R. P. Marinelli and A. E. Dell Orto, pp. 47–54. New York: Springer. [17]

Degenhardt, L., W. D. Hall, M. Lynskey, et al. 2009. Should Burden of Disease Estimates Include Cannabis Use as a Risk Factor for Psychosis? *PLoS Med.* **6**:e1000133. [09]

De Hert, M., J. Detraux, R. van Winkel, W. Yu, and C. U. Correll. 2012. Metabolic and Cardiovascular Adverse Effects Associated with Antipsychotic Drugs. *Nat. Rev. Endocrinol.* **8**: 114–126. [17]

De Hert, M., D. Vancampfort, C. U. Correll, et al. 2011. Guidelines for Screening and Monitoring of Cardiometabolic Risk in Schizophrenia: Systematic Evaluation. *Br. J. Psychiatry* **199**:99–105. [17]

Deisseroth, K. 2012. Optogenetics and Psychiatry: Applications, Challenges, and Opportunities. *Biol. Psychiatry* **71**:1030–1032. [17]

de la Fuente-Sandoval, C., P. León-Ortiz, M. Azcárraga, et al. 2013a. Glutamate in the Associative Striatum before and after 4 Weeks of Antipsychotic Treatment in First-Episode Psychosis: A Longitudinal Proton Magnetic Resonance Spectroscopy Study. *JAMA Psychiatry*, in press. [07]
———. 2013b. Striatal Glutamate and the Conversion to Psychosis: A Prospective 1H-MRS Imaging Study. *Int. J. Neuropsychopharmacol.* **16**:471–475. [07]
de la Fuente-Sandoval, C., P. Leon-Ortiz, R. Favila, et al. 2011. Higher Levels of Glutamate in the Associative-Striatum of Subjects with Prodromal Symptoms of Schizophrenia and Patients with First-Episode Psychosis. *Neuropsychopharmacology* **36**:1781–1791. [07]
Demjaha, A., J. H. MacCabe, and R. M. Murray. 2012. How Genes and Environmental Factors Determine the Different Neurodevelopmental Trajectories of Schizophrenia and Bipolar Disorder. *Schizophr. Bull.* **38**:209–214. [09]
De Silva, M. J., K. McKenzie, T. Harpham, and S. R. Huttly. 2005. Social Capital and Mental Illness: A Systematic Review. *J. Epidemiol. Community Health* **59**:619–627. [01]
Devrim-Ucok, M., H. Y. Keskin-Ergen, and A. Ucok. 2008. Mismatch Negativity at Acute and Post-Acute Phases of First-Episode Schizophrenia. *Eur. Arch. Psychiatry Clin. Neurosci.* **258**:179–185. [07]
Dickerson, D. D., A. M. Restieaux, and D. K. Bilkey. 2012. Clozapine Administration Ameliorates Disrupted Long-Range Synchrony in a Neurodevelopmental Animal Model of Schizophrenia. *Schizophr. Res.* **135**:112–115. [13]
Dickerson, D. D., A. R. Wolff, and D. K. Bilkey. 2010. Abnormal Long-Range Neural Synchrony in a Maternal Immune Activation Animal Model of Schizophrenia. *J. Neurosci.* **30**:12,424–12,431. [13]
Dickinson, D., V. N. Iannone, C. M. Wilk, and J. M. Gold. 2004. General and Specific Cognitive Deficits in Schizophrenia. *Biol. Psychiatry* **55**:826–833. [04]
Dickinson, D., J. D. Ragland, J. M. Gold, and R. C. Gur. 2008. General and Specific Cognitive Deficits in Schizophrenia: Goliath Defeats David? *Biol. Psychiatry* **64**:823–827. [01]
Dickinson, D., M. B. Ramsey, and J. M. Gold. 2007. Overlooking the Obvious: A Meta-Analytic Comparison of Digit Symbol Coding Tasks and Other Cognitive Measures in Schizophrenia. *Arch. Gen. Psychiatry* **64**:532–542. [02, 04]
Dickson, H., K. R. Laurens, A. E. Cullen, and S. Hodgins. 2012. Meta-Analysis of Cognitive and Motor Function in Youth Aged 16 Years and Younger Who Subsequently Develop Schizophrenia. *Psychol. Med.* **42**:743–755. [14]
Di Costanzo, A., F. Trojsi, M. Tosetti, et al. 2007. Proton Mr Spectroscopy of the Brain at 3 T: An Update. *Eur. Radiol.* **17**:1651–1662. [07]
Dixon, L., A. Lyles, J. Scott , et al. 1999. Services to Families of Adults with Schizophrenia: From Treatment Recommendations to Dissemination. *Psychiatr. Serv.* **50**:233–238. [16]
Dixon, L. B., F. Dickerson, A. S. Bellack, et al. 2010. The 2009 PORT Psychosocial Treatment Recommendations and Summary Statements. *Schizophr. Bull.* **36**:48–70. [15, 16]
Dixon, L. B., W. McFarlane, H. Lefley, et al. 2001. Evidence-Based Practices for Services to Family Members of People with Psychiatric Disabilities. *Psychiatr. Serv.* **52**:903–910. [15]
Dolmetsch, R., and D. H. Geschwind. 2011. The Human Brain in a Dish: The Promise of iPSC-Derived Neurons. *Cell* **145**:831–834. [10, 17]

Domingues, I., T. Alderman, and K. S. Cadenhead. 2011. Strategies for Effective Recruitment of Individuals at Risk for Developing Psychosis. *Early Interv. Psychiatry* **5**:233–241. [09]

Domschke, K., and U. Dannlowski. 2010. Imaging Genetics of Anxiety Disorders. *NeuroImage* **53**:822–831. [08]

Doty, R. L. 2012. Olfaction in Parkinson's Disease and Related Disorders. *Neurobiol. Dis.* **46**:527–552. [10]

Downar, J., and Z. J. Daskalakis. 2013. New Targets for rTMS in Depression: A Review of Convergent Evidence. *Brain Stimul.* **6**:231–240. [17]

Drake, R. E., and S. M. Essock. 2009. The Science-to-Service Gap in Real-World Schizophrenia Treatment: The 95% Problem. *Schizophr. Bull.* **35**:677–678. [15]

Drake, R. E., H. H. Goldman, H. S. Leff, et al. 2001. Implementing Evidence-Based Practices in Routine Mental Health Service Settings. *Psychiatr. Serv.* **52**:179–182. [15]

Drake, R. E., G. J. McHugo, R. R. Bebout, et al. 1999. A Randomized Clinical Trial of Supported Employment for Inner-City Patients with Severe Mental Illness. *Arch. Gen. Psychiatry* **56**:627–633. [15]

Drake, R. E., G. J. McHugo, D. R. Becker, W. A. Anthony, and R. E. Clark. 1996. The New Hampshire Study of Supported Employment for People with Severe Mental Illness: Vocational Outcomes. *J. Consult. Clin. Psychol.* **64**:391–399. [15]

Drake, R. E., M. R. Merrens, and D. W. Lynde. 2005. Evidence-Based Mental Health Practice: A Textbook. New York: Norton. [15]

Drake, R. E., E. O'Neal, and M. A. Wallach. 2008. A Systematic Review of Psychosocial Interventions for People with Co-Occurring Severe Mental and Substance Use Disorders. *J. Subst. Abuse Treat.* **34**:123–138. [15]

Drapalski, A. L., A. Lucksted, P. B. Perrin, et al. 2013. A Model of Internalized Stigma and Its Effects on People with Mental Illness. *Psychiatr. Serv.* **64**:264–269. [15]

Druss, B. G., W. D. Bradford, R. A. Rosenheck, M. J. Radford, and H. M. Krumholz. 2001. Quality of Medical Care and Excess Mortality in Older Patients with Mental Disorders. *Arch. Gen. Psychiatry* **58**:565–572. [17]

Drzyzga, L., E. Obuchowicz, A. Marcinowska, and Z. S. Herman. 2006. Cytokines in Schizophrenia and the Effects of Antipsychotic Drugs. *Brain Behav. Immun.* **20**:532–545. [07]

D'Souza, D. C. 2007. Cannabinoids and Psychosis. *Int. Rev. Neurobiol.* **78**:289–326. [07]

D'Souza, D. C., W. M. Abi-Saab, S. Madonick, et al. 2005. Delta-9-Tetrahydrocannabinol Effects in Schizophrenia: Implications for Cognition, Psychosis, and Addiction. *Biol. Psychiatry* **57**:594–608. [07]

Dunaif, S. L., and P. H. Hoch. 1955. Pseudopsychopathic Schizophrenia. *Proc. Annu. Meet. Am. Psychopathol. Assoc.* **1955**:169–195. [01]

Duncan, E. J., S. H. Madonick, A. Parwani, et al. 2001. Clinical and Sensorimotor Gating Effects of Ketamine in Normals. *Neuropsychopharmacology* **25**:72–83. [07]

Durlak, J. A., R. P. Weissberg, A. B. Dymnicki, R. D. Taylor, and K. B. Schellinger. 2011. The Impact of Enhancing Students' Social and Emotional Learning: A Meta-Analysis of School-Based Universal Interventions. *Child Dev.* **82**:405–432. [09]

Durstewitz, D. 2007. Dopaminergic Modulation of Prefrontal Cortex Network Dynamics. In: Monoaminergic Modulation of Cortical Excitability, ed. K.-Y. Tseng and M. Atzori. New York: Springer. [12]

Durstewitz, D., and E. Balaguer-Ballester. 2010. Statistical Approaches for Reconstructing Neuro-Cognitive Ensemble Dynamics from High-Dimensional Neural Recordings. *NeuroForum* **4**:266–276. [12]

Durstewitz, D., and J. K. Seamans. 2002. The Computational Role of Dopamine D1 Receptors in Working Memory. *Neural Netw.* **15**:561–572. [12]

———. 2008. The Dual-State Theory of Prefrontal Cortex Dopamine Function with Relevance to COMT Genotypes and Schizophrenia. *Biol. Psychiatry* **64**:739–749. [12]

Durstewitz, D., J. K. Seamans, and T. J. Sejnowski. 2000. Dopamine-Mediated Stabilization of Delay-Period Activity in a Network Model of Prefrontal Cortex. *J. Neurophysiol.* **83**:1733–1750. [12]

D'Zurilla, T. J., and M. R. Goldfried. 1971. Problem Solving and Behavior Modification. *J. Abnorm. Psychol.* **78**:107–126. [15]

Eack, S. M., G. E. Hogarty, R. Y. Cho, et al. 2010. Neuroprotective Effects of Cognitive Enhancement Therapy against Gray Matter Loss in Early Schizophrenia: Results from a 2-Year Randomized Controlled Trial. *Arch. Gen. Psychiatry* **67**:674–682. [01, 07]

Eastvold, A. D., R. K. Heaton, and K. S. Cadenhead. 2007. Neurocognitive Deficits in the (Putative) Prodrome and First Episode of Psychosis. *Schizophr. Res.* **93**:266–277. [07]

Eckblad, M., and L. J. Chapman. 1983. Magical Ideation as an Indicator of Schizotypy. *J. Consult. Clin. Psychol.* **51**:215–225. [02]

Edwards, J., H. J. Jackson, and P. E. Pattison. 2002. Emotion Recognition via Facial Expression and Affective Prosody in Schizophrenia: A Methodological Review. *Clin. Psychol. Rev.* **22**:789–832. [04]

Egan, M. F., M. Kojima, J. H. Callicott, et al. 2003. The BDNF val66met Polymorphism Affects Activity-Dependent Secretion of BDNF and Human Memory and Hippocampal Function. *Cell* **112**:257–269. [09]

Egerton, A., S. Brugger, M. Raffin, et al. 2012a. Anterior Cingulate Glutamate Levels Related to Clinical Status Following Treatment in First-Episode Schizophrenia. *Neuropsychopharmacology* **37**:2515–2521. [07]

Egerton, A., P. Fusar-Poli, and J. M. Stone. 2012b. Glutamate and Psychosis Risk. *Curr. Pharm. Des.* **18**:466–478. [07]

Ehrlichman, R. S., M. J. Gandal, C. R. Maxwell, et al. 2009. N-Methyl-D-Aspartic Acid Receptor Antagonist-Induced Frequency Oscillations in Mice Recreate Pattern of Electrophysiological Deficits in Schizophrenia. *Neuroscience* **158**:705–712. [07]

Eichenberger, A., and W. Rössler. 2000. Comparison of Self-Ratings and Therapist Ratings of Outpatients' Psychosocial Status. *J. Nerv. Ment. Dis.* **188**:297–300. [16]

Eldar, A., and M. B. Elowitz. 2010. Functional Roles for Noise in Genetic Circuits. *Nature* **467**:167–173. [13]

Ellett, L., D. Freeman, and P. A. Garety. 2008. The Psychological Effect of an Urban Environment on Individuals with Persecutory Delusions: The Camberwell Walk Study. *Schizophr. Res.* **99**:77–84. [01]

Elliot, A. J., K. M. Sheldon, and M. A. Church. 1997. Avoidance Personal Goals and Subjective Well-Being. *Pers. Soc. Psychol. Bull.* **23**:915–927. [15]

Engel, A. K., K. Friston, J. A. S. Kelso, et al. 2010. Coordination in Behavior and Cognition. In: Dynamic Coordination in the Brain: From Molecules to Mind, ed. C. von der Malsburg et al., Strüngmann Forum Report, J. Lupp, series ed., vol. 5, pp. 267–299. Cambridge, MA: MIT Press. [01]

Engel, A. K., P. Konig, A. K. Kreiter, and W. Singer. 1991. Interhemispheric Synchronization of Oscillatory Neuronal Responses in Cat Visual Cortex. *Science* **252**:1177–1179. [04]

Engel, A. K., and W. Singer. 2001. Temporal Binding and the Neural Correlates of Sensory Awareness. *Trends Cogn. Neurosci.* **5**:16–25. [04]

Erickson, K. I., M. W. Voss, R. S. Prakash, et al. 2011. Exercise Training Increases Size of Hippocampus and Improves Memory. *PNAS* **108**:3017–3022. [17]

Ericson, C. A., II. 2011. Fault Tree Analysis Primer. Charleston, NC: CreateSpace. [02]

Ericsson, K. A., and N. Charness. 1994. Expert Performance: Its Structure and Acquisition. *Am. Psychol.* **49**:725–747. [15]

Erlenmeyer-Kimling, L., and B. Cornblatt. 1987. The New York High-Risk Project: A Follow-Up Report. *Schizophr. Bull.* **13**:451–461. [02]

Erlenmeyer-Kimling, L., B. A. Cornblatt, and R. Golden. 1984. Early Indicators of Vulnerability to Schizophrenia in Children at High Genetic Risk. In: Childhood Psychopathology and Development, ed. S. B. Guze et al., pp. 247–261. New York: Raven Press. [02]

Esslinger, C., H. Walter, P. Kirsch, et al. 2009. Neural Mechanisms of a Genome-Wide Supported Psychosis Variant. *Science* **324**:605. [08, 09]

EU-GEI. 2008. Schizophrenia Aetiology: Do Gene-Environment Interactions Hold the Key? *Schizophr. Res.* **102**:21–26. [09]

Eysenck, S. B. G., H. J. Eysenck, and P. Barrett. 1985. A Revised Version of the Psychoticism Scale. *Pers. Individ. Dif.* **6**:21–29. [02]

Fadden, G. 1997. Implementation of Family Interventions in Routine Clinical Practice Following Staff Training Programs: A Major Cause for Concern. *J. Ment. Health* **6**:599–612. [15]

Falloon, I. R. H., J. L. Boyd, and C. W. McGill. 1984. Family Care of Schizophrenia: A Problem-Solving Approach to the Treatment of Mental Illness. New York: Guilford Press. [15]

Falloon, I. R. H., J. L. Boyd, C. W. McGill, et al. 1985. Family Management in the Prevention of Morbidity of Schizophrenia: Clinical Outcome of a Two Year Longitudinal Study. *Arch. Gen. Psychiatry* **42**:887–896. [15]

Fan, Y., G. Abrahamsen, J. J. McGrath, and A. Mackay-Sim. 2012. Altered Cell Cycle Dynamics in Schizophrenia. *Biol. Psychiatry* **71**:129–135. [10]

Fang, G., R. Kuang, G. Pandey, et al. 2010. Subspace Differential Coexpression Analysis: Problem Definition and a General Approach. *Pac. Symp. Biocomput.* **15**:145–156. [05]

Fang, G., G. Pandey, W. Wang, et al. 2012. Mining Low-Support Discriminative Patterns from Dense and High-Dimensional Data. *IEEE Trans. Knowl. Data Eng.* **24**:279–294. [05]

Fanous, A. H., F. A. Middleton, K. Gentile, et al. 2012. Genetic Overlap of Schizophrenia and Bipolar Disorder in a High-Density Linkage Survey in the Portuguese Island Population. *Am. J. Med. Genet. B Neuropsychiatr. Genet.* **159B**:383–391. [02]

Fanous, A. H., M. C. Neale, B. T. Webb, et al. 2008. Novel Linkage to Chromosome 20p Using Latent Classes of Psychotic Illness in 270 Irish High-Density Families. *Biol. Psychiatry* **64**:121–127. [03]

Faraone, S. V., A. I. Green, L. J. Seidman, and M. T. Tsuang. 2001. "Schizotaxia": Clinical Implications and New Directions for Research. *Schizophr. Bull.* **27**:1–18. [02]

Faraone, S. V., L. J. Seidman, W. S. Kremen, et al. 1999. Neuropsychological Functioning among the Nonpsychotic Relatives of Schizophrenic Patients: A 4-Year Follow-Up Study. *J. Abnorm. Psychol.* **108**: [07]

Faris, R. E. L., and H. W. Dunham. 1939. Mental Disorders in Urban Areas. Chicago: Univ. of Chicago Press. [01, 16]

Farkas, M. D. 2007. The Vision of Recovery Today: What It Is and What It Means for Services. *World Psychiatry* **6**:4–10. [15, 16]

Faulkner, G., A. A. Soundy, and K. Lloyd. 2003. Schizophrenia and Weight Management: A Systematic Review of Interventions to Control Weight. *Acta Psychiatr. Scand.* **108**:324–332. [17]

Fearon, P., and C. Morgan. 2006. Environmental Factors in Schizophrenia: The Role of Migrant Studies. *Schizophr. Bull.* **32**:405–408. [09]

Feifel, D., K. Macdonald, A. Nguyen, et al. 2010. Adjunctive Intranasal Oxytocin Reduces Symptoms in Schizophrenia Patients. *Biol. Psychiatry* **68**:678–680. [07]

Feifel, D., and T. Reza. 1999. Oxytocin Modulates Psychotomimetic-Induced Deficits in Sensorimotor Gating. *Psychopharmacology* **141**:93–98. [07]

Feifel, D., P. D. Shilling, and A. M. Belcher. 2012. The Effects of Oxytocin and Its Analog, Carbetocin, on Genetic Deficits in Sensorimotor Gating. *Eur. Neuropsychopharmacol.* **22**:374–378. [07]

Feinberg, I. 1982. Schizophrenia: Caused by a Fault in Programmed Synaptic Elimination During Adolescence? *J. Psychiatr. Res.* **17**:319–334. [07]

Fenno, L., O. Yizhar, and K. Deisseroth. 2011. The Development and Application of Optogenetics. *Annu. Rev. Neurosci.* **34**:389–412. [12]

Feron, F., C. Perry, M. H. Hirning, J. McGrath, and A. Mackay-Sim. 1999. Altered Adhesion, Proliferation and Death in Neural Cultures from Adults with Schizophrenia. *Schizophr. Res.* **40**:211–218. [10]

Feron, F., C. Perry, J. J. McGrath, and A. Mackay-Sim. 1998. New Techniques for Biopsy and Culture of Human Olfactory Epithelial Neurons. *Arch. Otolaryngol. Head Neck Surg.* **124**:861–866. [10]

Fish, B., and K. S. Kendler. 2005. Abnormal Infant Neurodevelopment Predicts Schizophrenia Spectrum Disorders. *J. Child Adolesc. Psychopharmacol.* **15**:348–361. [01]

Fisher, M., C. Holland, M. M. Merzenich, and S. Vinogradov. 2009. Using Neuroplasticity-Based Auditory Training to Improve Verbal Memory in Schizophrenia. *Am. J. Psychiatr.* **166**:805–811. [07]

Flagstad, P., A. Mork, B. Y. Glenthoj, et al. 2004. Disruption of Neurogenesis on Gestational Day 17 in the Rat Causes Behavioral Changes Relevant to Positive and Negative Schizophrenia Symptoms and Alters Amphetamine-Induced Dopamine Release in Nucleus Accumbens. *Neuropsychopharmacology* **29**:2052–2064. [11]

Flatscher-Bader, T., C. J. Foldi, S. Chong, et al. 2011. Increased *De Novo* Copy Number Variants in the Offspring of Older Males. *Translat. Psychiatry* **1**:e34. [08]

Fletcher, P. C., and C. D. Frith. 2009. Perceiving Is Believing: A Bayesian Approach to Explaining the Positive Symptoms of Schizophrenia. *Nat. Rev. Neurosci.* **10**:48–58. [13]

Foldi, C. J., D. W. Eyles, J. J. McGrath, and T. H. Burne. 2010. Advanced Paternal Age Is Associated with Alterations in Discrete Behavioural Domains and Cortical Neuroanatomy of C57BL/6J Mice. *Eur. J. Neurosci.* **31**:556–564. [08]

Folwell, K. 1995. Single Measures of Deprivation. *J. Epidemiol. Community Health* **49(Suppl. 2)**:S51–S56. [16]

Foster, A., J. Gable, and J. Buckley. 2012. Homelessness in Schizophrenia. *Psychiatr. Clin. North Am.* **35**:717–734. [05]

Foster, A., D. D. Miller, and P. Buckley. 2010. Pharmacogenetics and Schizophrenia. *Clin. Lab. Med.* **30**:975–993. [07]

Fowler, T., S. Zammit, M. J. Owen, and F. Rasmussen. 2012. A Population-Based Study of Shared Genetic Variation between Premorbid IQ and Psychosis among Male Twin Pairs and Sibling Pairs from Sweden. *Arch. Gen. Psychiatry* **69**:460–466. [02]

Frangou, S., M. Lewis, J. Wollard, and A. Simmons. 2007. Preliminary *in Vivo* Evidence of Increased N-Acetyl-Aspartate Following Eicosapentanoic Acid Treatment in Patients with Bipolar Disorder. *J. Psychopharmacol.* **21**:435–439. [07]

Frank, M. G., M. V. Baratta, D. B. Sprunger, L. R. Watkins, and S. F. Maier. 2007. Microglia Serve as a Neuroimmune Substrate for Stress-Induced Potentiation of CNS Pro-Inflammatory Cytokine Responses. *Brain Behav. Immun.* **21**:47–59. [07]

Freeman, D. 2008. Studying and Treating Schizophrenia Using Virtual Reality: A New Paradigm. *Schizophr. Bull.* **34**:605–610. [17]

Frese, F. J. I. 2008. Self-Help Activities. In: Clinical Handbook of Schizophrenia, ed. K. T. Mueser and D. V. Jeste, pp. 298–305. New York: Guilford Press. [15]

Freudenreich, O., D. C. Henderson, E. A. Macklin, and A. E. Evins. 2009. Modafinil for Clozapine-Treated Schizophrenia Patients: A Double-Blind, Placebo-Controlled Pilot Trial. *J. Clin. Psychiatry* **70**:1674–1680. [04]

Freyer, F., J. A. Roberts, P. Ritter, and B. M. 2012. A Canonical Model of Multistability and Scale-Invariance in Biological Systems. *PLoS Comput. Biol.* **8**:e1002634. [01]

Friedman, J. I., D. N. Adler, E. Howanitz, et al. 2002. A Double Blind Placebo Controlled Trial of Donepezil Adjunctive Treatment to Risperidone for the Cognitive Impairment of Schizophrenia. *Biol. Psychiatry* **51**:349–357. [04]

Friedman, J. I., D. N. Adler, H. D. Temporini, et al. 2001a. Guanfacine Treatment of Cognitive Impairment in Schizophrenia. *Neuropsychopharmacology* **25**:402–409. [04]

Friedman, J. I., D. Carpenter, J. Lu, et al. 2008. A Pilot Study of Adjunctive Atomoxetine Treatment to Second-Generation Antipsychotics for Cognitive Impairment in Schizophrenia. *J. Clin. Psychopharmacol.* **28**:59–63. [04]

Friedman, J. J., P. D. Harvey, T. Coleman, et al. 2001b. Six-Year Follow-up Study of Cognitive and Functional Status across the Lifespan in Schizophrenia: A Comparison with Alzheimer's Disease and Normal Aging. *Am. J. Psychiatr.* **158**:1441–1448. [04]

Fuller, R., P. Nopoulos, S. Arndt, et al. 2002. Longitudinal Assessment of Premorbid Cognitive Functioning in Patients with Schizophrenia through Examination of Standardized Scholastic Test Performance. *Am. J. Psychiatr.* **159**:1183–1189. [14, 17]

Fusar-Poli, P., M. R. Broome, J. B. Woolley, et al. 2011a. Altered Brain Function Directly Related to Structural Abnormalities in People at Ultra High Risk of Psychosis: Longitudinal VBM-fMRI Study. *J. Psychiatr. Res.* **45**:190–198. [07]

Fusar-Poli, P., O. D. Howes, P. Allen, et al. 2010. Abnormal Frontostriatal Interactions in People with Prodromal Signs of Psychosis: A Multimodal Imaging Study. *Arch. Gen. Psychiatry* **67**:683–691. [07]

———. 2011b. Abnormal Prefrontal Activation Directly Related to Pre-Synaptic Striatal Dopamine Dysfunction in People at Clinical High Risk for Psychosis. *Mol. Psychiatry* **16**:67–75. [07]

Fusar-Poli, P., J. Perez, M. Broome, et al. 2007. Neurofunctional Correlates of Vulnerability to Psychosis: A Systematic Review and Meta-Analysis. *Neurosci. Biobehav. Rev.* **31**:465–484. [07]

Fuster, J. 2003. Cortex and Mind: Unifying Cognition. New York: Oxford Univ. Press. [01]

Gaebel, W., T. Becker, B. Janssen, et al. 2012. EPA Guidance on the Quality of Mental Health Services. *Eur. Psychiatry* **27(2)**:87–113. [16]

Gale, C. R., G. D. Batty, A. M. McIntosh, et al. 2013. Is Bipolar Disorder More Common in Highly Intelligent People? A Cohort Study of a Million Men. *Molecular Psychiatry* **18**:190–194. [14]

Gale, C. R., G. D. Batty, D. P. J. Osborn, et al. 2012. Association of Mental Disorders in Early Adulthood and Later Psychiatric Hospital Admissions and Mortality in a Cohort Study of More Than 1 Million Men. *Arch. Gen. Psychiatry* **69**:823–831. [17]

Gallinat, J., G. Winterer, C. S. Herrmann, and D. Senkowski. 2004. Reduced Oscillatory Gamma-Band Responses in Unmedicated Schizophrenic Patients Indicate Impaired Frontal Network Processing. *Clin. Neurophysiol.* **115**:1863–1874. [04, 07]

Galvan, A. 2010. Adolescent Development of the Reward System. *Front. Hum. Neurosci.* **4**:6. [13]

Gandal, M. J., J. C. Edgar, K. Klook, and S. J. Siegel. 2012. Gamma Synchrony: Towards a Translational Biomarker for the Treatment-Resistant Symptoms of Schizophrenia. *Neuropharmacology* **62**:1504–1518. [07]

Ganju, V. 2003. Implementation of Evidence-Based Practices in State Mental Health Systems: Implications for Research and Effectiveness Studies. *Schizophr. Bull.* **29**:125–131. [15]

Ganter, B., R. Wille, and C. Franzke. 1999. Formal Concept Analysis: Mathematical Foundations. Heidelberg: Springer. [05]

Gao, W. J., and P. S. Goldman-Rakic. 2003. Selective Modulation of Excitatory and Inhibitory Microcircuits by Dopamine. *PNAS* **100**:2836–2841. [12]

Garcia-Bueno, B., J. R. Caso, and J. C. Leza. 2008. Stress as a Neuroinflammatory Condition in Brain: Damaging and Protective Mechanisms. *Neurosci. Biobehav. Rev.* **32**:1136–1151. [07]

Garcia-Palomares, S., J. F. Pertusa, J. Minarro, et al. 2009. Long-Term Effects of Delayed Fatherhood in Mice on Postnatal Development and Behavioral Traits of Offspring. *Biol. Reprod.* **80**:337–342. [08]

Garety, P. A., E. Kuipers, D. Fowler, D. Freeman, and P. E. Bebbington. 2001. A Cognitive Model of the Positive Symptoms of Psychosis. *Psychol. Med.* **31**:189–195. [09]

Garmezy, N. 1977. On Some Risks in Risk Research. *Psychol. Med.* **7**:1–6. [06]

Geyer, M. A., K. Krebs-Thomson, D. L. Braff, and N. R. Swerdlow. 2001. Pharmacological Studies of Prepulse Inhibition Models of Sensorimotor Gating Deficits in Schizophrenia: A Decade in Review. *Psychopharmacology* **156**:117–154. [07]

Gill, K. M., D. J. Lodge, J. M. Cook, S. Aras, and A. A. Grace. 2011. A Novel alpha5GABA(A)R-Positive Allosteric Modulator Reverses Hyperactivation of the Dopamine System in the Mam Model of Schizophrenia. *Neuropsychopharmacology* **36**:1903–1911. [11]

Gilligan, J. 1996. Violence: Reflections on a National Epidemic. New York: Putnam. [01]

Gilmore, J. H., C. Kang, D. D. Evans, et al. 2010. Prenatal and Neonatal Brain Structure and White Matter Maturation in Children at High Risk for Schizophrenia. *Am. J. Psychiatr.* **167**:1083–1091. [01]

Gingerich, S., and K. T. Mueser. 2011. Illness Management and Recovery: Personalized Skills and Strategies for Those with Mental Illness (3rd edition). Center City, MN: Hazelden. [15]

Girard, S., L. Tremblay, M. Lepage, and G. Sebire. 2010. Il-1 Receptor Antagonist Protects against Placental and Neurodevelopmental Defects Induced by Maternal Inflammation. *J. Immunol.* **184**:3997–4005. [07]

Girirajan, S., J. A. Rosenfeld, B. P. Coe, et al. 2012. Phenotypic Heterogeneity of Genomic Disorders and Rare Copy-Number Variants. *N. Engl. J. Med.* **367**:1321–1331. [13]

Giuffrida, A., F. M. Leweke, C. W. Gerth, et al. 2004. Cerebrospinal Anandamide Levels Are Elevated in Acute Schizophrenia and Are Inversely Correlated with Psychotic Symptoms. *Neuropsychopharmacology* **29**:2108–2114. [07]

Gladsjo, J. A., L. A. McAdams, B. W. Palmer, et al. 2004. A Six-Factor Model of Cognition in Schizophrenia and Related Psychotic Disorders: Relationships with Clinical Symptoms and Functional Capacity. *Schizophr. Bull.* **30**:739–754. [04]

Glasgow, R. E., E. Lichtenstein, and A. C. Marcus. 2003. Why Don't We See More Translation of Health Promotion Research to Practice? Rethinking the Efficacy-to-Effectiveness Transition. *Am. J. Public Health* **93**:1261–1267. [15]

Glisson, C., and S. K. Schoenwald. 2005. The Arc Organizational and Community Intervention Strategy for Implementing Evidence-Based Children's Mental Health Treatments. *Ment. Health Serv. Res.* **7**:243–259. [15]

Goff, D., C. Cather, O. Freudenreich, et al. 2009. A Placebo-Controlled Study of Sildenafil Effects on Cognition in Schizophrenia. *Psychopharmacology* **202**:411–417. [04]

Goff, D. C., C. Cather, J. D. Gottlieb, et al. 2008a. Once-Weekly D-Cycloserine Effects on Negative Symptoms and Cognition in Schizophrenia: An Exploratory Study. *Schizophr. Res.* **106**:320–327. [04]

Goff, D. C., M. Hill, and D. Barch. 2011. The Treatment of Cognitive Impairment in Schizophrenia. *Pharmacol. Biochem. Behav.* **99**:245–253. [04, 07]

Goff, D. C., J. S. Lamberti, A. C. Leon, et al. 2008b. A Placebo-Controlled Add-on Trial of the Ampakine, CX516, for Cognitive Deficits in Schizophrenia. *Neuropsychopharmacology* **33**:465–472. [04]

Goffman, E. 1963. Stigma: Notes on the Management of Spoiled Identity. Englewood Cliffs, NY: Prentice-Hall. [17]

Gold, J. M., D. M. Barch, C. S. Carter, et al. 2012. Clinical, Functional and Inter-Task Correlations of Measures Developed by the Cognitive Neuroscience Test Reliability and Clinical Applications for Schizophrenia Consortium. *Schizophr. Bull.* [02]

Gold, S., S. Arndt, P. Nopoulos, D. S. O'Leary, and N. C. Andreasen. 1999. Longitudinal Study of Cognitive Function in First-Episode and Recent-Onset Schizophrenia. *Am. J. Psychiatr.* **156**:1342–1348. [04]

Goldberg, T. E., R. S. Goldman, K. E. Burdick, et al. 2007. Cognitive Improvement after Treatment with Second-Generation Antipsychotic Medications in First-Episode Schizophrenia. *Arch. Gen. Psychiatry* **64**:1115–1122. [14]

Goldberg, T. E., R. S. E. Keefe, R. S. Goldman, D. G. Robinson, and P. D. Harvey. 2010. Circumstances under Which Practice Does Not Make Perfect: A Review of the Practice Effect Literature in Schizophrenia and Its Relevance to Clinical Treatment Studies. *Neuropsychopharmacology* **35**:1053–1062. [14]

Goldman, H. H., A. E. Skodol, and T. R. Lave. 1992. Revising Axis V for DSM-IV: A Review of Measures of Social Functioning. *Am. J. Psychiatr.* **149**:1148–1156. [04]

Goldman-Rakic, P. S. 1991. Prefrontal Cortical Dysfunction in Schizophrenia: The Relevance of Working Memory. In: Psychopathology and the Brain, ed. B. J. Carroll and J. E. Barrett, pp. 1–23. New York: Raven Press. [02]

Goldstein, J. M., S. L. Buka, L. J. Seidman, and M. T. Tsuang. 2010. Specificity of Familial Transmission of Schizophrenia Psychosis Spectrum and Affective Psychoses in the New England Family Study's High-Risk Design. *Arch. Gen. Psychiatry* **67**:458–467. [02]

Golzio, C., J. Willer, M. E. Talkowski, et al. 2012. KCTD13 Is a Major Driver of Mirrored Neuroanatomical Phenotypes of the 16p11.2 Copy Number Variant. *Nature* **485**:363–367. [10]

Gonzales, J. J., H. L. Ringeisen, and D. A. Chambers. 2002. The Tangled and Thorny Path of Science to Practice: Tensions in Interpreting and Applying "Evidence." *Clin. Psychol.* **9**:204–209. [15]

Gonzalez-Burgos, G., K. N. Fish, and D. A. Lewis. 2011. GABA Neuron Alterations, Cortical Circuit Dysfunction and Cognitive Deficits in Schizophrenia. *Neural Plast.* **2011**:Article ID 723184. [05, 13]

González-Pinto, A., S. Ruiz de Azúa, B. Ibáñez, et al. 2011. Can Positive Family Factors Be Protective against the Development of Psychosis? *Psychiatry Res.* **186**:28–33. [01]

Goodwin, N., N. Curry, C. Naylor, P. Ross, and D. W. 2010. Managing People with Long-Term Conditions. http://www.kingsfund.org.uk/sites/files/kf/field/field_document/managing-people-long-term-conditions-gp-inquiry-research-paper-mar11.pdf. (accessed 11 March, 2013). [06]

Gorczynski, P., and G. Faulkner. 2010. Exercise Therapy for Schizophrenia. *Schizophr. Bull.* **36**:665–666. [17]

Gorelova, N., J. K. Seamans, and C. R. Yang. 2002. Mechanisms of Dopamine Activation of Fast-Spiking Interneurons That Exert Inhibition in Rat Prefrontal Cortex. *J. Neurophysiol.* **88**:3150–3166. [12]

Gottesman, I. I., and A. Bertelsen. 1989. Confirming Unexpressed Genotypes for Schizophrenia. Risks in the Offspring of Fischer's Danish Identical and Fraternal Discordant Twins. *Arch. Gen. Psychiatry* **46**:867–872. [06]

Gottesman, I. I., and T. D. Gould. 2003. The Endophenotype Concept in Psychiatry: Etymology and Strategic Intentions. *Am. J. Psychiatr.* **160**:636–645. [02, 06, 08]

Gottesman, I. I., T. M. Laursen, A. Bertelsen, and P. B. Mortensen. 2010. Severe Mental Disorders in Offspring with 2 Psychiatrically Ill Parents. *Arch. Gen. Psychiatry* **160**:252–257. [06]

Gottesman, I. I., and J. Shields. 1967. A Polygenic Theory of Schizophrenia. *PNAS* **58**:199–205. [02, 08]

———. 1972. Schizophrenia and Genetics: A Twin Study Vantage Point. New York: Academic Press. [02]

Gourevitch, R., C. Rocher, G. L. Pen, M. O. Krebs, and T. M. Jay. 2004. Working Memory Deficits in Adult Rats after Prenatal Disruption of Neurogenesis. *Behav. Pharmacol.* **15**:287–292. [11]

Grace, A. A. 2000. Gating of Information Flow within the Limbic System and the Pathophysiology of Schizophrenia. *Brain. Res. Rev.* **31**:330–341. [02]

Grace, A. A. 2010. Dopamine System Dysregulation by the Ventral Subiculum as the Common Pathophysiological Basis for Schizophrenia Psychosis, Psychostimulant Abuse, and Stress. *Neurotox. Res.* **18**:367–376. [13]

Graf, J., C. Lauber, C. Nordt, et al. 2004. Perceived Stigmatization of Mentally Ill People and Its Consequences for the Quality of Life in a Swiss Population. *J. Nerv. Ment. Dis.* **192**:542–547. [17]

Graham, F. K. 1975. Presidential Address, 1974. The More or Less Startling Effects of Weak Prestimulation. *Psychophysiology* **12**:238–248. [07]

Granholm, E., D. Ben-Zeev, and P. C. Link. 2009. Social Disinterest Attitudes and Group Cognitive-Behavioral Social Skills Training for Functional Disability in Schizophrenia. *Schizophr. Bull.* **35**:874–883. [15]

Granholm, E., J. R. McQuaid, F. S. McClure, et al. 2007. Randomized Controlled Trial of Cognitive Behavioral Social Skills Training for Older People with Schizophrenia: 12-Month Follow-Up. *J. Clin. Psychiatry* **68**:730–737. [07]

Grant, P. M., and A. T. Beck. 2009. Defeatist Beliefs as a Mediator of Cognitive Impairment, Negative Symptoms, and Functioning in Schizophrenia. *Schizophr. Bull.* **35**:798–806. [15]

Green, M. F. 1996. What Are the Functional Consequences of Neurocognitive Deficits in Schizophrenia? *Am. J. Psychiatr.* **153**:321–330. [04, 07, 17]

———. 2006. Cognitive Impairment and Functional Outcome in Schizophrenia and Bipolar Disorder. *J. Clin. Psychiatry* **67(Suppl. 9)**:3–8. [15]

Green, M. F., R. S. Kern, D. L. Braff, and J. Mintz. 2000. Neurocognitive Deficits and Functional Outcome in Schizophrenia: Are We Measuring the "Right Stuff"? *Schizophr. Bull.* **26**:119–136. [07]

Green, M. F., R. S. Kern, and R. K. Heaton. 2004a. Longitudinal Studies of Cognition and Functional Outcome in Schizophrenia: Implications for MATRICS. *Schizophr. Res.* **72**:41–51. [02, 04]

Green, M. F., and K. H. Nuechterlein. 1999a. Cortical Oscillations and Schizophrenia: Timing Is of the Essence. *Arch. Gen. Psychiatry* **56**:1007–1008. [07]

———. 1999b. Should Schizophrenia Be Treated as a Neurocognitive Disorder? *Schizophr. Bull.* **25**:309–319. [07]

Green, M. F., K. H. Nuechterlein, J. M. Gold, et al. 2004b. Approaching a Consensus Cognitive Battery for Clinical Trials in Schizophrenia: The NIMH-MATRICS Conference to Select Cognitive Domains and Test Criteria. *Biol. Psychiatry* **56**:301–307. [04]

Green, M. F., K. H. Nuechterlein, R. S. Kern, et al. 2008. Functional Co-Primary Measures for Clinical Trials in Schizophrenia: Results from the MATRICS Psychometric and Standardization Studa. *Am. J. Psychiatr.* **165**:221–228. [04]

Green, M. F., B. Olivier, J. N. Crawley, D. L. Penn, and S. Silverstein. 2005. Social Cognition in Schizophrenia: Recommendations from the Measurement and Treatment Research to Improve Cognition in Schizophrenia New Approaches Conference. *Schizophr. Bull.* **31**:882–887. [04]

Greenberg, D. A., and R. Subaran. 2011. Blinders, Phenotype, and Fashionable Genetic Analysis: A Critical Examination of the Current State of Epilepsy Genetic Studies. *Epilepsia* **52**:1–9. [13]

Greenhalgh, T., G. Robert, F. MacFarlane, P. Bate, and O. Kyriakidou. 2004. Diffusion of Innovations in Service Organizations: Systematic Review and Recommendations. *Milbank Q.* **82**:581–629. [15]

Greenwood, K., C. F. Hung, M. Tropeano, P. McGuffin, and T. Wykes. 2011. No Association between the Catechol-O-Methyltransferase (COMT) val158met Polymorphism and Cognitive Improvement Following Cognitive Remediation Therapy (CRT) in Schizophrenia. *Neurosci. Lett.* **496**:65–69. [06]

Greenwood, T. A., D. L. Braff, G. A. Light, et al. 2007. Initial Heritability Analyses of Endophenotypic Measures for Schizophrenia: The Consortium on the Genetics of Schizophrenia. *Arch. Gen. Psychiatry* **64**:1242–1250. [07]

Gromann, P. M., D. K. Tracy, V. Giampietro, et al. 2012. Examining Frontotemporal Connectivity and rTMS in Healthy Controls: Implications for Auditory Hallucinations in Schizophrenia. *Neuropsychology* **26**:127–132. [09]

Gross, G., and G. Huber. 2010. The History of the Basic Symptom Concept. *Acta Clinica Croatica* **49**:47–59. [17]

Gruber, A. J., G. G. Calhoon, I. Shusterman, et al. 2010. More Is Less: A Disinhibited Prefrontal Cortex Impairs Cognitive Flexibility. *J. Neurosci.* **30**:17,102–17,110. [11]

Guest, P. C., E. Schwarz, D. Krishnamurthy, et al. 2011. Altered Levels of Circulating Insulin and Other Neuroendocrine Hormones Associated with the Onset of Schizophrenia. *Psychoneuroendocrinology* **36**:1092–1096. [07]

Guilmatre, A., D. C., A. L. Mosca, et al. 2009. Recurrent Rearrangements in Synaptic and Neurodevelopmental Genes and Shared Biologic Pathways in Schizophrenia, Autism, and Mental Retardation. *Arch. Gen. Psychiatry* **66**:947–956. [06]

Gur, R. E., M. E. Calkins, R. C. Gur, et al. 2007. The Consortium on the Genetics of Schizophrenia: Neurocognitive Endophenotypes. *Schizophr. Bull.* **33**:49–68. [03]

Gur, R. E., P. Cowell, B. I. Turetsky, et al. 1998. A Follow-up Magnetic Resonance Imaging Study of Schizophrenia. Relationship of Neuroanatomical Changes to Clinical and Neurobehavioral Measures. *Arch. Gen. Psychiatry* **55**:145–152. [07]

Gustavsson, A. M. Svensson, F. Jacobi, et al. 2011. Cost of disorders of the brain in Europe 2010. *Eur. Neuropsychopharm.* **21(10)**:718–779. [16]

GWAS Consortium. 2011. Genome-Wide Association Study Identifies Five New Schizophrenia Loci. *Nat. Genet.* **43**:969–976. [09]

Hagan, J., R. MacMillan, and B. Wheaton. 1996. New Kid in Town: Social Capital and the Life Course Effects of Family Migration on Children. *Am. Sociol. Rev.* **61**:368–385. [01]

Hahn, C. G., G. Gomez, D. Restrepo, et al. 2005. Aberrant Intracellular Calcium Signaling in Olfactory Neurons from Patients with Bipolar Disorder. *Am. J. Psychiatr.* **162**:616–618. [10]

Hakami, T., N. C. Jones, E. A. Tolmacheva, et al. 2009. NMDA Receptor Hypofunction Leads to Generalized and Persistent Aberrant Gamma Oscillations Independent of Hyperlocomotion and the State of Consciousness. *PLoS One* **4**:e6755. [07]

Hall, J., L. Romaniuk, A. M. McIntosh, et al. 2009. Associative Learning and the Genetics of Schizophrenia. *Trends Neurosci.* **32**:359–365. [02]

Hallak, J. E., S. M. Dursun, D. C. Bosi, et al. 2011. The Interplay of Cannabinoid and NMDA Glutamate Receptor Systems in Humans: Preliminary Evidence of Interactive Effects of Cannabidiol and Ketamine in Healthy Human Subjects. *Prog. Neuropsychopharmacol. Biol. Psychiatry* **35**:198–202. [07]

Hambrecht, M., M. Lammertink, J. Klosterkotter, E. Matuschek, and R. Pukrop. 2002. Subjective and Objective Neuropsychological Abnormalities in a Psychosis Prodrome Clinic. *Br. J. Psychiatry Suppl.* **43**:30–37. [07]

Han, J., H. Cheng, D. Xin, and X. Yan. 2007. Frequent Pattern Mining: Current Status and Future Directions. *Data Min. Knowl. Discov.* **15**:55–86. [05]

Han, S. S., L. A. Williams, and K. C. Eggan. 2011. Constructing and Deconstructing Stem Cell Models of Neurological Disease. *Neuron* **70**:626–644. [10]

Hansen, D. V., J. H. Lui, P. R. Parker, and A. R. Kriegstein. 2010. Neurogenic Radial Glia in the Outer Subventricular Zone of Human Neocortex. *Nature* **464**:554–561. [10]

Hansen, D. V., J. L. Rubenstein, and A. R. Kriegstein. 2011. Deriving Excitatory Neurons of the Neocortex from Pluripotent Stem Cells. *Neuron* **70**:645–660. [10]

Harding, C. M., G. W. Brooks, T. Ashikaga, J. S. Strauss, and A. Breier. 1987. The Vermont Longitudinal Study of Persons with Severe Mental Illness: II. Long-Term Outcome of Subjects Who Retrospectively Met DSM-III Criteria for Schizophrenia. *Am. J. Psychiatr.* **144**:727–735. [15]

Hare, R. D. 1993. Without Conscience: The Disturbing World of the Psychopaths among Us. New York: Guilford Press. [01]

Harrison, B. J., M. Yücel, J. Pujol, and C. Pantelis. 2007. Task-Induced Deactivation of Midline Cortical Regions in Schizophrenia Assessed with fMRI. *Schizophr. Res.* **91**:82–86. [04]

Harrison, P. J., and D. R. Weinberger. 2005. Schizophrenia Genes, Gene Expression, and Neuropathology: On the Matter of Their Convergence. *Mol. Psychiatry* **10**:40–68; image 45. [09]

Harvey, P. D., and R. S. E. Keefe. 2012. Technology, Society, and Mental Illness: Challenges and Opportunities for Assessment and Treatment. *Innov. Clin. Neurosci.* **9**:47–50. [17]

Harvey, P. D., J. Lombardi, M. Leibman, et al. 1996. Cognitive Impairment and Negative Symptoms in Geriatric Chronic Schizophrenic Patients: A Follow-up Study. *Schizophr. Res.* **22**:223–231. [04]

Harvey, P. D., T. L. Patterson, L. S. Potter, K. Zhong, and M. Brecher. 2006. Improvement in Social Competence with Short-Term Atypical Antipsychotic Treatment: A Randomized, Double-Blind Comparison of Quetiapine Versus Risperidone for Social Competence, Social Cognition and Neuropsychological Functioning. *Am. J. Psychiatr.* **163**:1918–1925. [04]

Hawkins, K. A., J. Addington, R. S. Keefe, et al. 2004. Neuropsychological Status of Subjects at High Risk for a First Episode of Psychosis. *Schizophr. Res.* **67**:115–122. [07]

Hayashi-Takagi, A., M. Takaki, N. Graziane, et al. 2010. Disrupted-in-Schizophrenia 1 (DISC1) Regulates Spines of the Glutamate Synapse via Rac1. *Nat. Neurosci.* **13**:327–332. [08]

Heaton, R., J. S. Paulsen, L. A. McAdams, et al. 1994. Neuropsychological Deficits in Schizophrenics: Relationship to Age, Chronicity, and Dementia. *Arch. Gen. Psychiatry* **51**:469–476. [15, 17]

Heaton, R. K., J. A. Gladsjo, B. W. Palmer, et al. 2001. Stability and Course of Neuropsychological Deficits in Schizophrenia. *Arch. Gen. Psychiatry* **58**:24–32. [04, 07]

Hedman, A. M., N. E. M. van Haren, C. G. M. van Baal, R. S. Kahn, and H. E. Hulshoff Pol. 2013. IQ Change over Time in Schizophrenia and Healthy Individuals: A Meta-Analysis, in Revision. *Schizophr. Res.* **146**:201–208. [14]

Hegarty, J. D., R. J. Baldessarini, M. Tohen, C. Waternaux, and G. Oepen. 1994. One Hundred Years of Schizophrenia: A Meta-Analysis of the Outcome Literature. *Am. J. Psychiatr.* **151**:1409–1416. [14]

Hegelstad, W. T. V., T. K. Larsen, B. Auestad, et al. 2012. Long-Term Follow-up of the Tips Early Detection in Psychosis Study: Effects on 10-Year Outcome. *Am. J. Psychiatr.* **169**:374–380. [17]

Heinrichs, R., and K. Zakanis. 1998. Neurocognitive Deficit in Schizophrenia: A Quantitative Review of the Evidence. *Neuropsychology* **12**:426–445. [04, 07]

Heinrichs, R. W. 2001. In Search of Madness. Oxford: Oxford Univ. Press. [01]

———. 2005. The Primacy of Cognition in Schizophrenia. *Am. Psychol.* **60**:229–242. [02]

Heinz, A., D. Goldman, D. W. Jones, et al. 2000. Genotype Influences *in Vivo* Dopamine Transporter Availability in Human Striatum. *Neuropsychopharmacology* **22**:133–139. [08]

Heinze, S., G. Sartory, B. W. Müller, et al. 2006. Neural Activation During Successful and Unsuccessful Verbal Learning in Schizophrenia. *Schizophr. Res.* **83**:121–130. [04]

Henchcliffe, C., and M. F. Beal. 2008. Mitochondrial Biology and Oxidative Stress in Parkinson Disease Pathogenesis. *Nat. Clin. Pract. Neurol.* **4**:600–609. [10]

Hennekens, C. H., A. R. Hennekens, D. Hollar, and D. E. Casey. 2005. Schizophrenia and Increased Risks of Cardiovascular Disease. *Am. Heart J.* **150**:1115–1121. [05]

Herbert, J. D. 2000. Defining Empirically Supported Treatments: Pitfalls and Possible Solutions. *Behav. Ther.* **23**:113–134. [15]

Hermens, D. F., P. B. Ward, M. A. Hodge, et al. 2010. Impaired MMN/P3a Complex in First-Episode Psychosis: Cognitive and Psychosocial Associations. *Prog. Neuropsychopharmacol. Biol. Psychiatry* **34**:822–829. [07]

Hertz, J., A. Krogh, and G. Palmer. 1991. Introduction to the Theory of Neural Computation. Reading, MA: Addison-Wesley. [12]

Hickie, I. B., E. M. Scott, D. F. Hermens, et al. 2013. Applying Clinical Staging to Young People Who Present for Mental Health Care. *Early Interv. Psychiatry* **7**:31–43. [05]

Hickman, M., P. Vickerman, J. Macleod, et al. 2009. If Cannabis Caused Schizophrenia: How Many Cannabis Users May Need to Be Prevented in Order to Prevent One Case of Schizophrenia? England and Wales Calculations. *Addiction* **104**:1856–1861. [09]

Higuchi, Y., T. Sumiyoshi, Y. Kawasaki, et al. 2010. Effect of Tandospirone on Mismatch Negativity and Cognitive Performance in Schizophrenia: A Case Report. *J. Clin. Psychopharmacol.* **30**:732–734. [07]

Hikida, T., H. Jaaro-Peled, S. Seshadri, et al. 2007. Dominant-Negative DISC1 Transgenic Mice Display Schizophrenia-Associated Phenotypes Detected by Measures Translatable to Humans. *PNAS* **104**:14501–14506. [11, 13]

Hill, S. K., J. L. Reilly, M. S. H. Harris, T. Khine, and J. A. Sweeney. 2008. Oculomotor and Neuropsychological Effects of Antipsychotic Treatment for Schizophrenia. Schizophrenia Bulletin. *Schizophr. Bull.* **34**:494–506. [04]

Ho, B. C., N. C. Andreasen, P. Nopoulos, et al. 2003. Progressive Structural Brain Abnormalities and Their Relationship to Clinical Outcome: A Longitudinal Magnetic Resonance Imaging Study Early in Schizophrenia. *Arch. Gen. Psychiatry* **60**:585–594. [07]

Hodge, M. A. R., D. Siciliano, P. Withey, et al. 2010. A Randomized Controlled Trial of Cognitive Remediation in Schizophrenia. *Schizophr. Bull.* **36**:419–427. [04]

Hodgins, S. 2008. Violent Behaviour among People with Schizophrenia: A Framework for Investigations of Causes, and Effective Treatment, and Prevention. *Phil. Trans. R. Soc. Lond. B* **363**:2505–2518. [01]

Hoexter, M. Q., F. L. de Souza Duran, C. C. D'Alcante, et al. 2012. Gray Matter Volumes in Obsessive-Compulsive Disorder before and after Fluoxetine or Cognitive-Behavior Therapy: A Randomized Clinical Trial. *Neuropsychopharmacology* **37**:734–745. [07]

Hoff, A. L., M. Sakuma, M. Wieneke, et al. 1999. Longitudinal Neuropsychological Follow-up Study of Patients with First-Episode Schizophrenia. *Am. J. Psychiatr.* **156**:1336–1341. [04]

Hoffman, R. E., and T. H. McGlashan. 1993. Parallel Distributed Processing and the Emergence of Schizophrenic Symptoms. *Schizophr. Bull.* **19**:119–140. [01]

Hogarty, G. E. 2002. Personal Therapy for Schizophrenia and Related Disorders: A Guide to Individualized Treatment. New York: Guilford Press. [15]

Hogarty, G. E., S. Flesher, R. F. Ulrich, et al. 2004. Cognitive Enhancement Therapy for Schizophrenia: Effects of a 2-Year Randomized Trial on Cognition and Behavior. *Arch. Gen. Psychiatry* **61**:866–876. [15]

Hogarty, G. E., S. J. Kornblith, D. Greenwald, et al. 1995. Personal Therapy: A Disorder-Relevant Psychotherapy for Schizophrenia. *Schizophr. Bull.* **21**:379–393. [03]

Holliday, E. G., D. E. McLean, D. R. Nyholt, and B. J. Mowry. 2009. Susceptibility Locus on Chromosome 1q23-25 for a Schizophrenia Subtype Resembling Deficit Schizophrenia Identified by Latent Class Analysis. *Arch. Gen. Psychiatry* **66**:1058–1067. [03]

Hollingshead, A. B., and F. C. Redlich. 1958. Social Class and Mental Illness: A Community Study. New York: Wiley. [09]

Homayoun, H., and B. Moghaddam. 2007. NMDA Receptor Hypofunction Produces Opposite Effects on Prefrontal Cortex Interneurons and Pyramidal Neurons. *J. Neurosci.* **27**:11,496–11,500. [11]

Honer, W. G., A. E. Thornton, E. Y. H. Chen, et al. 2006. Clozapine Alone Versus Clozapine and Risperidone with Refractory Schizophrenia. *N. Engl. J. Med.* **354**:472–482. [04]

Hong, L. E., A. Summerfelt, R. W. Buchanan, et al. 2010. Gamma and Delta Neural Oscillations and Association with Clinical Symptoms under Subanesthetic Ketamine. *Neuropsychopharmacology* **35**:632–640. [07]

Horn, S. D., and J. Gassaway. 2007. Practice-Based Evidence Study Design for Comparative Effectiveness Research. *Medical Care* **45**:S50–S57. [15]

Horn, S. D., J. Gassaway, L. Pentz, and R. James. 2010. Practice-Based Evidence for Clinical Practice Improvement: An Alternative Study Design for Evidence-Based Medicine. *Stud. Health Tech. Inform.* **151**:446–460. [15]

Horrobin, D. F., A. I. Glen, and C. J. Hudson. 1995. Possible Relevance of Phospholipid Abnormalities and Genetic Interactions in Psychiatric Disorders: The Relationship between Dyslexia and Schizophrenia. *Med. Hypotheses* **45**:605–613. [01]

Horvath, S., Z. Janka, and K. Mirnics. 2011. Analyzing Schizophrenia by DNA Microarrays. *Biol. Psychiatry* **69**:157–162. [10]

Howes, O. D., A. Egerton, V. Allan, et al. 2009. Mechanisms Underlying Psychosis and Antipsychotic Treatment Response in Schizophrenia: Insights from PET and SPECT Imaging. *Curr. Pharm. Des.* **15**:2550–2559. [07]

Howes, O. D., J. Kambeitz, E. Kim, et al. 2012a. The Nature of Dopamine Dysfunction in Schizophrenia and What This Means for Treatment. *Arch. Gen. Psychiatry* **69**:776–786. [13]

Howes, O. D., and S. Kapur. 2009. The Dopamine Hypothesis of Schizophrenia, Version III: The Final Common Pathway. *Schizophr. Bull.* **35**:549–562. [02, 09, 13]

Howes, O. D., P. Shotbolt, M. Bloomfield, et al. 2012b. Dopaminergic Function in the Psychosis Spectrum: An [18f]-Dopa Imaging Study in Healthy Individuals with Auditory Hallucinations. *Schizophr. Bull.* [09]

Huddy, V., C. Reeder, D. Kontis, T. Wykes, and D. Stahl. 2012. The Effect of Working Alliance on Adherence and Outcome in Cognitive Remediation Therapy. *J. Nerv. Ment. Dis.* **200**:614–619. [06]

Huot, P., and J. M. Brotchie. 2011. 5-Ht(1a) Receptor Stimulation and L-Dopa-Induced Dyskinesia in Parkinson's Disease: Bridging the Gap between Serotonergic and Glutamatergic Mechanisms. *Exp. Neurol.* **231**:195–198. [07]

Huttenlocker, P. 1979. Synaptic Density in the Human Frontal Cortex: Developmental Changes and Effects of Aging. *Brain. Res.* **163**:195–205. [07]

Huttenlocker, P., and A. Dabhokar. 1997. Regional Differences in Synaptogenesis in Human Cerebral Cortex. *J. Comp. Neurol.* **387**:167–178. [07]

Huttunen, M. O., and P. Niskanen. 1978. Prenatal Loss of Father and Psychiatric Disorders. *Arch. Gen. Psychiatry* **35**:429–431. [06]

Hyman, S. E. 2007. Can Neuroscience Be Integrated into the DSM-V? *Nat. Rev. Neurosci.* **8**:725–732. [08]

Hyman, S. E., and W. S. Fenton. 2003. What Are the Right Targets for Psychopharmacology? *Science* **299**:350–351. [04]

Insel, T. R. 2009. Translating Scientific Opportunity into Public Health Impact. *Arch. Gen. Psychiatry* **66**:128–133. [01, 15]

———. 2010. From Prevention to Preemption. *Biol. Psychiatry* **67**:188S. [01]

Insel, T. R., and B. N. Cuthbert. 2011. Endophenotypes: Bridging Genomic Complexity and Disorder Heterogeneity. *Biol. Psychiatry* **66**:988–989. [02]

Insel, T. R., B. Cuthbert, M. Garvey, et al. 2010. Research Domain Criteria (RDoC): Toward a New Classification Framework for Research on Mental Disorders. *Am. J. Psychiatr.* **167**:748–751. [01, 07]

Inskip, H. M., E. C. Harris, and C. Barraclough. 1998. Lifetime Risk of Suicide for Alcoholism, Affective Disorder and Schizophrenia. *Br. J. Psychiatry* **172**:35–37. [17]

Institute of Medicine. 2001. Crossing the Quality Chasm: A New Health System for the 21st Century. Washington, DC: Natl. Academies Press. [15]

———. 2006. Improving the Quality of Health Care for Mental and Substance-Use Conditions. Washington, DC: Natl. Academies Press. [15]

International Electrotechnical Commission. 1990. Chapter 191: Dependability and Quality of Service. *Electropedia: The World's Online Electrotechnical Vocabulary* http://www.electropedia.org/iev/iev.nsf/index?openform&part=191. (accessed March 27, 2013). [02]

International Schizophrenia Consortium. 2009. Common Polygenic Variation Contributes to Risk of Schizophrenia and Bipolar Disorder. *Nature* **460**:748–752. [05]

Isohanni, I., M. R. Järvelin, P. Jones, Jokelainen J, and I. M. 1999. Can Excellent School Performance Be a Precursor of Schizophrenia? A 28-Year Follow-up in the Northern Finland 1966 Birth Cohort. *Acta Psychiatr. Scand.* **100**:17–26. [06]

Isohanni, M., P. B. Jones, K. Moilanen, et al. 2001. Early Developmental Milestones in Adult Schizophrenia and Other Psychoses. A 31-Year Follow-up of the Northern Finland 1966 Birth Cohort. *Schizophr. Res.* **52**:1–19. [06]

Ison, J. R., and H. S. Hoffman. 1983. Reflex Modification in the Domain of Startle: II. The Anomalous History of a Robust and Ubiquitous Phenomenon. *Psychol. Bull.* **94**:3–17. [07]

Iyer, S. N., T. L. Rothmann, J. E. Vogler, and W. D. Spaulding. 2005. Evaluating Outcomes of Rehabilitation for Severe Mental Illness. *Rehab. Psychol.* **50**:43–55. [15]

Jackson, H. J., H. L. Whiteside, G. W. Bates, et al. 1991. Diagnosing Personality Disorders in Psychiatric Inpatients. *Acta Psychiatr. Scand.* **83**:206–213. [01]

Jacobs, B. L., H. Praag, and F. H. Gage. 2000. Adult Brain Neurogenesis and Psychiatry: A Novel Theory of Depression. *Mol. Psychiatry* **5**:262–269. [07]

Jacobsen, L. K., J. N. Giedd, F. X. Castellanos, et al. 1998. Progressive Reduction of Temporal Lobe Structures in Childhood-Onset Schizophrenia. *Am. J. Psychiatr.* **155**:678–685. [07]

Jaeger, M., and W. Rössler. 2010. Attitudes Towards Long-Acting Depot Antipsychotics: A Survey of Patients, Relatives and Psychiatrists. *Psychiatry Res.* **175**:58–62. [16]

Jaffee, S. R., T. E. Moffitt, A. Caspi, and A. Taylor. 2003. Life with (or without) Father: The Benefits of Living with Two Biological Parents Depend on the Father's Antisocial Behavior. *Child Dev.* **74**:109–126. [09]

Jahshan, C., K. S. Cadenhead, A. J. Rissling, et al. 2012. Automatic Sensory Information Processing Abnormalities across the Illness Course of Schizophrenia. *Psychol. Med.* **42**:85–97. [07, 09]

Jahshan, C., R. K. Heaton, S. Golshan, and K. S. Cadenhead. 2010. Course of Neurocognitive Deficits in the Prodrome and First Episode of Schizophrenia. *Neuropsychology* **24**:109–120. [07]

Jansma, J. M., N. F. Ramsey, N. J. van der Wee, and R. S. Kahn. 2004. Working Memory Capacity in Schizophrenia: A Parametric fMRI Study. *Schizophr. Res.* **68**:159–171. [04]

Jaspers, K. 1946/1997. General Psychopathology (translated by J. Hoenig and Marian W. Hamilton). Baltimore: Johns Hopkins Univ. Press. [05]

———. 1963. General Psychopathology (translated by J. Hoenig and M. W. Hamilton). Manchester: Manchester Univ. Press. [03]

Javitt, D. C., R. W. Buchanan, R. S. E. Keefe, et al. 2012. Effect of the Neuroprotective Peptide Davunetide (AL-108) on Cognition and Functional Capacity in Schizophrenia. *Schizophr. Res.* **136**:25–31. [04]

Javitt, D. C., A. Shelley, and W. Ritter. 2000. Associated Deficits in Mismatch Negativity Generation and Tone Matching in Schizophrenia. *Clin. Neurophysiol.* **111**:1733–1737. [07]

Javitt, D. C., and S. R. Zukin. 1991. Recent Advances in the Phencyclidine Model of Schizophrenia. *Am. J. Psychiatr.* **148**:1301–1308. [07]

Jenkins, R., G. Lewis, P. Bebbington, et al. 1997. The National Psychiatric Morbidity Surveys of Great Britain Initial Findings from the Household Survey. *Psychol. Med.* **27**:777–789. [16]

Jennings, A. F. 1994. On Being Invisible in the Mental Health System. *J. Ment. Health Admin.* **21**:374–387. [15]

Johnson, A. B. 1990. Out of Bedlam: The Truth About Deinstitutionalization. New York: Basic Books. [15]

Johnson, S. C., A. J. Saykin, L. A. Flashman, T. W. McAllister, and M. B. Sparling. 2001. Brain Activation on fMRI and Verbal Memory Ability: Functional Neuroanatomic Correlates of CVLT Performance. *J. Int. Neuropsychol. Soc.* **7**:55–62. [04]

Johnston, P. J., H. Devir, and F. Karayanidis. 2006. Facial Emotion Processing in Schizophrenia: No Evidence for a Deficit Specific to Negative Emotions in a Differential Deficit Design. *Psychiatry Res.* **143**:51–61. [04]

Jones, J. A. H., S. R. Sponheim, and A. W. MacDonald, III. 2010a. The Dot Pattern Expectancy Task: Reliability and Replication of Deficits in Schizophrenia. *Psychol. Assess.* **22**:131–141. [02]

Jones, L., P. A. Holmans, M. L. Hamshere, et al. 2010b. Genetic Evidence Implicates the Immune System and Cholesterol Metabolism in the Aetiology of Alzheimer's Disease. *PLoS One* **5**:e13950. [09]

Jones, P., B. Rodgers, R. Murray, and M. Marmot. 1994. Child Development Risk Factors for Adult Schizophrenia in the British 1946 Birth Cohort. *Lancet* **344**:1398–1402. [06]

Jones, P. B., and S. R. Marder. 2008. Psychosocial and Pharmacological Treatments for Schizophrenia. In: Cambridge Textbook of Effective Treatments in Psychiatry, ed. P. J. Tyrer and K. R. Silk, pp. 469–480. New York: Cambridge Univ. Press. [06]

Jones, P. B., and J. Van Os. 1998. Predicting Schizophrenia in Teenagers: Pessimistic Results from the British 1946 Birth Cohort. *Schizophr. Res.* **29**:11. [09]

Kahkonen, S., V. Makinen, I. P. Jaaskelainen, et al. 2005. Serotonergic Modulation of Mismatch Negativity. *Psychiatry Res.* **138**:61–74. [07]

Kaitin, K. I., and J. A. DiMasi. 2011. Pharmaceutical Innovation in the 21st Century: New Drug Approvals in the First Decade, 2000–2009. *Clin. Pharmacol. Therap.* **89**:183–188. [17]

Kaitin, K. I., and C. P. Milne. 2011. A Dearth of New Meds: Drugs to Treat Neuropsychiatric Disorders Have Become Too Risky for Big Pharma. *Sci. Am.* **305**:16–17. [17]

Kaji, K., K. Norrby, A. Paca, et al. 2009. Virus-Free Induction of Pluripotency and Subsequent Excision of Reprogramming Factors. *Nature* **458**:771–775. [10]

Kalmijn, S., E. J. Feskens, L. J. Launer, and D. Kromhout. 1997. Polyunsaturated Fatty Acids, Antioxidants, and Cognitive Function in Very Old Men. *Am. J. Epidemiol.* **145**:33–41. [07]

Kandel, E. R., J. H. Schwartz, and T. M. Jessell. 2000. Principles of Neural Science. Smell and Taste: The Chemical Senses, L. B. Buck, series ed: McGraw-Hill. [10]

Kane, J. M., M. Cohen, J. Zhao, L. M. Alphs, and J. Panagides. 2010a. Efficacy and Safety of Asenapine in a Placebo- and Haloperidol-Controlled Trial in Patients with Acute Exacerbation of Schizophrenia. *J. Clin. Psychopharmacol.* **30**:106–115. [04]

Kane, J. M., D. C. D'Souza, A. A. Patkar, et al. 2010b. Armodafinil as Adjunctive Therapy in Adults with Cognitive Deficits Associated with Schizophrenia: A 4-Week, Double-Blind, Placebo-Controlled Study. *J. Clin. Psychiatry* **71**:1475–1481. [06]

Kano, S., C. Colantuoni, F. Han, et al. 2012. Genome-Wide Profiling of Multiple Histone Methylations in Olfactory Cells: Further Implications for Cellular Susceptibility to Oxidative Stress in Schizophrenia. *Mol. Psychiatry* [10]

Kapur, S., T. Arenovich, O. Agid, et al. 2005. Evidence for Onset of Antipsychotic Effects within the First 24 Hours of Treatment. *Am. J. Psychiatr.* **162**:939–946. [17]

Kapur, S., R. Zipursky, C. Jones, G. Remington, and S. Houle. 2000. Relationship between Dopamine D(2) Occupancy, Clinical Response, and Side Effects: A Double-Blind PET Study of First-Episode Schizophrenia. *Am. J. Psychiatr.* **157**:514–520. [07]

Karaka, S., B. Bekci, and O. U. Erzengin. 2003. Early Gamma Response in Human Neuroelectric Activity Is Correlated with Neuropsychological Test Scores. *Neurosci. Lett.* **340**:37–40. [04]

Karksson, J. L. 1970. Genetic Association of Giftedness and Creativity with Schizophrenia. *Hereditas* **66**:177–182. [06]

Kates, W. R. 2010. Investigating the Cognitive Antecedents of Schizophrenia. *Am. J. Psychiatr.* **167**:122–124. [09]

Kato, M., and A. Serretti. 2010. Review and Meta-Analysis of Antidepressant Pharmacogenetic Findings in Major Depressive Disorder. *Mol. Psychiatry* **15**:473–500. [08]

Keane, B. P., S. M. Silverstein, Y. Wang, and T. V. Papathomas. 2013. Reduced Depth Inversion Illusions in Schizophrenia Are State-Specific and Occur for Multiple Object Types and Viewing Conditions. *J. Abnorm. Psychol.* **122**:506–512. [01]

Keefe, R. S., R. M. Bilder, S. M. Davis, et al. 2007. CATIE Investigators: Neurocognitive Working Group. Neurocognitive Effects of Antipsychotic Medications in Patients with Chronic Schizophrenia in the CATIE Trial. *Arch. Gen. Psychiatry* **64**:633–647. [14]

Keefe, R. S., R. W. Buchanan, S. R. Marder, et al. 2013. Clinical Trials of Potential Cognitive-Enhancing Drugs in Schizophrenia: What Have We Learned So Far? *Schizophr. Bull.* **39**:417–435. [04, 06, 07]

Keefe, R. S., C. E. Eesley, and M. P. Poe. 2005. Defining a Cognitive Function Decrement in Schizophrenia. *Biol. Psychiatry* **57**:688–691. [14]

Keefe, R. S., K. H. Fox, P. D. Harvey, et al. 2011a. Characteristics of the MATRICS Consensus Cognitive Battery in a 29-Site Antipsychotic Schizophrenia Clinical Trial. *Schizophr. Res.* **125**:161–168. [14]

Keefe, R. S., A. K. Malhotra, H. Y. Meltzer, et al. 2008. Efficacy and Safety of Donepezil in Patients with Schizophrenia or Schizoaffective Disorder: Significant Placebo/Practice Effects in a 12-Week, Randomized, Double-Blind, Placebo-Controlled Trial. *Neuropsychopharmacology* **33**:1217–1228. [14]

Keefe, R. S., D. O. Perkins, H. Gu, et al. 2006. A Longitudinal Study of Neurocognitive Function in Individuals at Risk for Psychosis. *Schizophr. Res.* **88**:26–35. [07]

Keefe, R. S., S. Vinogradov, A. Medalia, et al. 2012. Feasibility and Pilot Efficacy Results from the Multisite Cognitive Remediation in the Schizophrenia Trials Network (CRSTN) Randomized Controlled Trial. *J. Clin. Psychiatry* **73**:1016–1022. [06]

Keefe, R. S. E., and W. S. Fenton. 2007. How Should DSM-V Criteria for Schizophrenia Include Cognitive Impairment? *Schizophr. Bull.* **33**:912–920. [14]

Keefe, R. S. E., T. E. Goldberg, P. D. Harvey, et al. 2004. The Brief Assessment of Cognition in Schizophrenia: Reliability, Sensitivity, and Comparison with a Standard Neurocognitive Battery. *Schizophr. Res.* **68**:283–297. [04]

Keefe, R. S. E., S. Vinogradov, A. Medalia, et al. 2011b. Report from the Working Group Conference on Multi-Site Trial Design for Cognitive Remediation in Schizophrenia. *Schizophr. Bull.* **37**:1057–1065. [17]

Kegeles, L. S., X. Mao, A. D. Stanford, et al. 2012. Elevated Prefrontal Cortex Gamma-Aminobutyric Acid and Glutamate-Glutamine Levels in Schizophrenia Measured *in Vivo* with Proton Magnetic Resonance Spectroscopy. *Arch. Gen. Psychiatry* **69**:449–459. [07]

Kelleher, I., M. C. Clarke, C. Rawdon, J. Murphy, and M. Cannon. 2012a. Neurocognition in the Extended Psychosis Phenotype: Performance of a Community Sample of Adolescents with Psychotic Symptoms on the MATRICS Neurocognitive Battery. *Schizophr. Bull.* [09]

Kelleher, I., D. Connor, M. C. Clarke, et al. 2012b. Prevalence of Psychotic Symptoms in Childhood and Adolescence: A Systematic Review and Meta-Analysis of Population-Based Studies. *Psychol. Med.* **42**:1857–1863. [06]

Kelly, D. L., R. W. Buchanan, D. L. Boggs, and R. P. McMahon. 2009. A Randomized Double-Blind Trial of Atomoxetine for Cognitive Impairments in 32 People with Schizophrenia. *J. Clin. Psychiatry* **70**:518–525. [04]

Kemp, A. S., N. R. Schooler, A. H. Kalali, et al. 2010. What Is Causing the Reduced Drug-Placebo Difference in Recent Schizophrenia Clinical Trials and What Can Be Done About It? *Schizophr. Bull.* **36**:504–509. [01]

Kempton, M. J., J. R. Geddes, U. Ettinger, S. C. Williams, and P. M. Grasby. 2008. Meta-Analysis, Database, and Meta-Regression of 98 Structural Imaging Studies in Bipolar Disorder. *Arch. Gen. Psychiatry* **65**:1017–1032. [07]

Kendler, K. S., M. McGuire, A. M. Gruenberg, et al. 1993. The Roscommon Family Study III: Schizophrenia-Related Personality Disorder in Relatives. *Arch. Gen. Psychiatry* **50**:781–788. [05]

Keri, S., and G. Benedek. 2009. Oxytocin Enhances the Perception of Biological Motion in Humans. *Cogn. Affect. Behav. Neurosci.* **9**:237–241. [07]

Kern, R. S., S. M. Glynn, W. P. Horan, and S. R. Marder. 2009. Psychosocial Treatments to Promote Functional Recovery in Schizophrenia. *Schizophr. Bull.* **35**:347–361. [17]

Kern, R. S., K. H. Nuechterlein, M. F. Green, et al. 2008. The MATRICS Consensus Cognitive Battery. Part 2: Co-Norming and Standardization. *Am. J. Psychiatr.* **165**:214–220. [06]

Keshavan, M. S., S. Anderson, and J. W. Pettegrew. 1994. Is Schizophrenia Due to Excessive Synaptic Pruning in the Prefrontal Cortex? The Feinberg Hypothesis Revisited. *J. Psychiatr. Res.* **28**:239–265. [07]

Keshavan, M. S., R. Tandon, N. N. Boutros, and H. A. Nasrallah. 2008. Schizophrenia, "Just the Facts": What We Know in 2008. Part 3: Neurobiology. *Schizophr. Res.* **106**:89–107. [05]

Kessler, R. C., S. Avenevoli, E. J. Costello, et al. 2012. Prevalence, Persistence, and Sociodemographic Correlates of DSM-IV Disorders in the National Comorbidity Survey Replication Adolescent Supplement. *Arch. Gen. Psychiatry* **69**:372–380. [06]

Kessler, R. C., P. Berglund, O. Demler, et al. 2003. The Epidemiology of Major Depressive Disorder: Results from the National Comorbidity Survey Replication (NCS-R). *JAMA* **289**:3095–3105. [16]

Kessler, R. C., O. Demler, R. G. Frank, et al. 2005. Prevalence and Treatment of Mental Disorders, 1990 to 2003. *N. Engl. J. Med.* **352**:2515–2523. [16]

Kessler, R. C., K. A. McLaughlin, J. G. Green, et al. 2010. Childhood Adversities and Adult Psychopathology in the WHO World Mental Health Surveys. *Br. J. Psychiatry* **197**:378–385. [09]

Kety, S. S., W. Rosenthal, P. H. Wender, and F. Schulsinger. 1971. Mental Illness in the Biological and Adoptive Families of Adopted Schizophrenics. *Am. J. Psychiatr.* **128**:302–306. [02]

Keuken, M. C., A. Hardie, B. T. Dorn, et al. 2011. The Role of the Left Inferior Frontal Gyrus in Social Perception: An rTMS Study. *Brain. Res.* **1383**:196–205. [07]

Khandaker, G. M., J. H. Barnett, I. R. White, and P. B. Jones. 2011. A Quantitative Meta-Analysis of Population-Based Studies of Premorbid Intelligence and Schizophrenia. *Schizophr. Res.* **132**:220–227. [06, 14]

Khandaker, G. M., J. Zimbron, C. Dalman, G. Lewis, and P. B. Jones. 2012a. Childhood Infection and Adult Schizophrenia: A Meta-Analysis of Population-Based Studies. *Schizophr. Res.* **139**:161–168. [06]

Khandaker, G. M., J. Zimbron, G. Lewis, and P. B. Jones. 2012b. Prenatal Maternal Infection, Neurodevelopment and Adult Schizophrenia: A Systematic Review of Population-Based Studies. *Psychol. Med.* **16**:1–19. [06]

Kilbourne, A. M., N. E. Morden, K. Austin, et al. 2009. Excess Heart-Disease-Related Mortality in a National Study of Patients with Mental Disorders: Identifying Modifiable Risk Factors. *Gen. Hosp. Psychiatry* **31**:555–563. [05]

Kim, K., A. Doi, B. Wen, et al. 2010. Epigenetic Memory in Induced Pluripotent Stem Cells. *Nature* **467**:285–290. [10]

Kingdon, D. G., and D. Turkington. 2004. Cognitive Therapy of Schizophrenia. New York: Guilford Press. [15]

Kinon, B. J., L. Zhang, B. A. Millen, et al. 2011. A Multicenter, Inpatient, Phase 2, Double-Blind, Placebo-Controlled Dose-Ranging Study of LY2140023 Monohydrate in Patients with DSM-IV Schizophrenia. *J. Clin. Psychopharmacol.* **31**:349–355. [04, 07]

Kirkbride, J. B., J. Boydell, G. B. Ploubidis, et al. 2008. Testing the Association between the Incidence of Schizophrenia and Social Capital in an Urban Area. *Psychol. Med.* **38**:1083–1094. [01]

Kirkbride, J. B., A. Errazuriz, T. J. Croudace, et al. 2012. Incidence of Schizophrenia and Other Psychoses in England, 1950-2009: A Systematic Review and Meta-Analyses. *PLoS One* **7**:e31660. [09]

Kirkbride, J. B., P. Fearon, C. Morgan, et al. 2006. Heterogeneity in Incidence Rates of Schizophrenia and Other Psychotic Syndromes: Findings from the 3-Center Aesop Study. *Arch. Gen. Psychiatry* **63**:250–258. [09]

Kirkpatrick, B. 2009. Schizophrenia as a Systemic Disease. *Schizophr. Bull.* **35**:381–382. [10]

Kirkpatrick, B., R. W. Buchanan, D. E. Ross, and W. T. Carpenter, Jr. 2001. A Separate Disease within the Syndrome of Schizophrenia. *Arch. Gen. Psychiatry* **58**:165–171. [03]

Kirkpatrick, B., W. S. Fenton, W. T. Carpenter, Jr., and S. R. Marder. 2006. The NIMH-MATRICS Consensus Statement on Negative Symptoms. *Schizophr. Bull.* **32**:214–219. [03]

Kirov, G., D. Gumus, W. Chen, et al. 2008. Comparative Genome Hybridization Suggests a Role for NRXN1 and APBA2 in Schizophrenia. *Hum. Mol. Genet.* **17**:458–465. [05]

Kirov, G., A. J. Pocklington, P. Holmans, et al. 2012. De Novo CNV Analysis Implicates Specific Abnormalities of Postsynaptic Signalling Complexes in the Pathogenesis of Schizophrenia. *Mol. Psychiatry* **17**:142–153. [09]

Kirsch, I., B. J. Deacon, T. B. Huedo-Medina, et al. 2008. Initial Severity and Antidepressant Benefits: A Meta-Analysis of Data Submitted to the Food and Drug Administration. *PLoS Med.* **5**:e45, 260–268. [17]

Knight, R. A., and S. M. Silverstein. 2001. A Process-Oriented Approach for Averting Confounds Resulting from General Performance Deficiencies in Schizophrenia. *J. Abnorm. Psychol.* **110**:15–30. [01, 02, 09]

Knightbridge, S. M., R. King, and T. J. Rolfe. 2006. Using Participatory Action Research in a Community-Based Initiative Addressing Complex Mental Health Needs. *Aust. NZ J. Psychiatry* **40**:325–332. [15]

Koethe, D., A. Giuffrida, D. Schreiber, et al. 2009a. Anandamide Elevation in Cerebrospinal Fluid in Initial Prodromal States of Psychosis. *Br. J. Psychiatry* **194**:371–372. [07]

Koethe, D., C. Hoyer, and F. M. Leweke. 2009b. The Endocannabinoid System as a Target for Modelling Psychosis. *Psychopharmacology* **206**:551–561. [07]

Kohler, C. G., W. Bilker, M. Hagendoorn, R. E. Gur, and R. C. Gur. 2000. Emotion Recognition Deficit in Schizophrenia: Association with Symptomatology and Cognition. *Biol. Psychiatry* **48**:127–136. [04]

Kohler, C. G., T. H. Turner, W. B. Bilker, et al. 2003. Facial Emotion Recognition in Schizophrenia: Intensity Effects and Error Pattern. *Am. J. Psychiatr.* **160**:1768–1774. [04]

Kohler, C. G., J. B. Walker, E. A. Martin, K. M. Healey, and P. J. Moberg. 2009. Facial Emotion Perception in Schizophrenia: A Meta-Analytic Review. *Schizophr. Bull.* **36**:1009–1019. [04]

Kong, A., M. L. Frigge, G. Masson, et al. 2012. Rate of De Novo Mutations and the Importance of Father's Age to Disease Risk. *Nature* **488**:471–475. [08]

Kontis, D., V. Huddy, C. Reeder, S. Landau, and T. Wykes. 2012. Effects of Age and Cognitive Reserve on Cognitive Remediation Therapy Outcome in Patients with Schizophrenia. *Am. J. Geriatr. Psychiatry* **21**:218–230. [06]

Kopelowicz, A., C. J. Wallace, and R. Zarate. 1998. Teaching Psychiatric Inpatients to Re-Enter the Community: A Brief Method of Improving the Continuity of Care. *Psychiatr. Serv.* **49**:1313–1316. [15]

Koren, D., L. J. Seidman, M. Goldsmith, and P. D. Harvey. 2006. Real World Cognitive- and Metacognitive-Dysfunction in Schizophrenia: A New Approach for Measuring (and Remediating) More "Right Stuff." *Schizophr. Bull.* **32**:310–326. [04]

Kornack, D. R., and P. Rakic. 1999. Continuation of Neurogenesis in the Hippocampus of the Adult Macaque Monkey. *PNAS* **96**:5768–5773. [07]

Korostenskaja, M., K. Dapsys, A. Siurkute, et al. 2005. Effects of Olanzapine on Auditory P300 and Mismatch Negativity (MMN) in Schizophrenia Spectrum Disorders. *Prog. Neuropsychopharmacol. Biol. Psychiatry* **29**:543–548. [07]

Korostenskaja, M., V. V. Nikulin, D. Kicic, A. V. Nikulina, and S. Kahkonen. 2007. Effects of NMDA Receptor Antagonist Memantine on Mismatch Negativity. *Brain Res. Bull.* **72**:275–283. [07]

Koseki, T., A. Mouri, T. Mamiya, et al. 2012. Exposure to Enriched Environments During Adolescence Prevents Abnormal Behaviours Associated with Histone Deacetylation in Phencyclidine-Treated Mice. *Int. J. Neuropsychopharmacol.* **15**:1489–1501. [09]

Kosmidis, M. H., V. P. Bozikas, M. Giannakou, et al. 2007. Impaired Emotion Perception in Schizophrenia: A Differential Deficit. *Psychiatry Res.* **149**:279–284. [04]

Kotov, R., L. T. Guey, E. J. Bromet, and J. E. Schwartz. 2010. Smoking in Schizophrenia: Diagnostic Specificity, Symptom Correlates, and Illness Severity. *Schizophr. Bull.* **36**:173–181. [17]

Kotov, R., C. J. Ruggero, R. F. Krueger, et al. 2011. New Dimensions in the Quantitative Classification of Mental Illness. *Arch. Gen. Psychiatry* **68**:1003–1011. [02]

Koutsouleris, N., G. J. Schmitt, C. Gaser, et al. 2009. Neuroanatomical Correlates of Different Vulnerability States for Psychosis and Their Clinical Outcomes. *Br. J. Psychiatry* **195**:218–226. [07]

Kovelman, J. A., and A. B. Scheibel. 1984. A Neurohistological Correlate of Schizophrenia. *Biol. Psychiatry* **19**:1601–1917. [11]

Krabbendam, L., B. Arts, J. van Os, and A. Aleman. 2005. Cognitive Functioning in Patients with Schizophrenia and Bipolar Disorder: A Quantitative Review. *Schizophr. Res.* **80**:137–149. [14]

Krabbendam, L., and J. van Os. 2005. Schizophrenia and Urbanicity: A Major Environmental Influence—Conditional on Genetic Risk. *Schizophr. Bull.* **31**:795–799. [08, 09]

Kraepelin, E. 1896. Lehrbuch Für Studierende und Ärtze. Leipzig: Barth. [14]

———. 1919/1971. Dementia Praecox and Paraphrenia (translated by R.M. Barclay). Huntington, NY: Robert E. Kreiger. [02, 03, 06, 15]

Kryzhanovskaya, L. A., S. Schulz, C. McDougle, et al. 2009. Results from a Double-Blind Placebo-Controlled Trial of Olanzapine in Adolescents with Schizophrenia. *J. Am. Acad. Child Adol. Psychiatr.* **48**:60–70. [04]

Kuepper, R., J. van Os, R. Lieb, H. U. Wittchen, and C. Henquet. 2011. Do Cannabis and Urbanicity Co-Participate in Causing Psychosis? Evidence from a 10-Year Follow-up Cohort Study. *Psychol. Med.* **41**:2121–2129. [09]

Kuipers, L., J. Leff, and D. Lam. 2002. Family Work for Schizophrenia: A Practical Guide (2nd edition). London: Gaskell. [15]

Kumari, V., D. Fannon, E. R. Peters, et al. 2011. Neural Changes Following Cognitive Behaviour Therapy for Psychosis: A Longitudinal Study. *Brain* **134**:2396–2407. [07]

Kumari, V., D. Fannon, A. L. Sumich, and T. Sharma. 2007. Startle Gating in Antipsychotic-Naive First Episode Schizophrenia Patients: One Ear Is Better Than Two. *Psychiatry Res.* **151**:21–28. [07]

Kumari, V., G. H. Gudjonsson, S. Raghuvanshi, et al. 2013. Reduced Thalamic Volume in Men with Antisocial Personality Disorder or Schizophrenia and a History of Serious Violence and Childhood Abuse. *Eur. Psychiatry* **28**:225–234. [01]

Kumari, V., E. R. Peters, D. Fannon, et al. 2009. Dorsolateral Prefrontal Cortex Activity Predicts Responsiveness to Cognitive-Behavioral Therapy in Schizophrenia. *Biol. Psychiatry* **66**:594–602. [07]

Kumra, A., J. Frazier, L. K. Jacobsen, et al. 1996. Childhood-Onset Schizophrenia. A Double-Blind Clozapine-Haloperidol Comparison. *Arch. Gen. Psychiatry* **53**:1090–1097. [04]

Kumra, S., H. Kranzler, G. Gerbino-Rosen, et al. 2008. Clozapine and "High-Dose" Olanzapine in Refractory Early-Onset Schizophrenia: A 12-Week Randomized and Double-Blind Comparison. *Biol. Psychiatry* **63**:524–529. [04]

Kurtz, M. M., and K. T. Mueser. 2008. A Meta-Analysis of Controlled Research on Social Skills Training for Schizophrenia. *J. Consult. Clin. Psychol.* **76**:491–504. [15, 17]

Kurtz, M. M., and C. L. Richardson. 2012. Social Cognitive Training for Schizophrenia: A Meta-Analytic Investigation of Controlled Research. *Schizophr. Bull.* **38**:1092–1104. [15]

Laan, W., D. E. Grobbee, J. P. Selten, et al. 2010. Adjuvant Aspirin Therapy Reduces Symptoms of Schizophrenia Spectrum Disorders: Results from a Randomized, Double-Blind, Placebo-Controlled Trial. *J. Clin. Psychiatry* **71**:520–527. [07]

Labrie, V., S. Pai, and A. Petronis. 2012. Epigenetics of Major Psychosis: Progress, Problems and Perspectives. *Trends Genet.* **28**:427–435. [09]

Ladewig, J., J. Mertens, J. Kesavan, et al. 2012. Small Molecules Enable Highly Efficient Neuronal Conversion of Human Fibroblasts. *Nat. Methods* **9**:575–578. [10]

Lahey, B. B., B. Applegate, J. K. Hakes, et al. 2012. Is There a General Factor of Prevalent Psychopathology During Adulthood? *J. Abnorm. Psychol.* **121**:971–977. [02]

Lahti, A. C., and M. A. Reid. 2011. Is There Evidence for Neurotoxicity in the Prodromal and Early Stages of Schizophrenia? *Neuropsychopharmacology* **36**:1779–1780. [07]

Lambert, M. J. 2005. Emerging Methods for Providing Clinicians with Timely Feedback on Treatment Effectiveness: An Introduction. *J. Clin. Psychol.* **61**:141–144. [15]

Lancet Global Mental Health Group, D. Chisholm, A. J. Flisher, et al. 2007. Scale up Services for Mental Disorders: A Call for Action. *Lancet* **370(9594)**:1241–1252. [16]

Landgraf, S., I. Amado, A. Berthoz, M. Krebs, and E. Van Der Meer. 2012. Cognitive Identify in Schizophrenia: Vision, Space, and Body Perception from Prodrome to Syndrome. *Curr. Psychiatry Rev.* **8**:119–139. [01, 05]

Langfeldt, G. 1939. The Schizophreniform States: A Katamnestic Study Based on Individual Re-Examinations: With Special Reference to Diagnostic and Prognostic Clues, and with a View to Presenting a Standard Material for Comparison with the Remissions Effected by Shock Treatment. Copenhagen: Munksgaard. [03]

Lante, F., J. Meunier, J. Guiramand, et al. 2007. Neurodevelopmental Damage after Prenatal Infection: Role of Oxidative Stress in the Fetal Brain. *Free Radic. Biol. Med.* **42**:1231–1245. [07]

Lau, A., and M. Tymianski. 2010. Glutamate Receptors, Neurotoxicity and Neurodegeneration. *Pflugers Arch.* **460**:525–542. [07]

Lauber, C., M. Anthony, V. Ajdacic-Gross, and W. Rössler. 2004a. What About Psychiatrists' Attitude to Mentally Ill People? *Eur. Psychiatry* **19**:423–427. [16]

Lauber, C., A. Eichenberger, P. Luginbühl, C. Keller, and W. Rössler. 2003. Determinants of Burden in Caregivers of Patients with Exacerbating Schizophrenia. *Eur. Psychiatry* **18**:285–289. [16, 17]

Lauber, C., C. Nordt, L. Falcato, and W. Rössler. 2001. Lay Recommendations on How to Treat Mental Disorders. *Soc. Psychiatry Psychiatr. Epidem.* **36**:553–556. [16]

———. 2002. Public Attitude to Compulsory Admission of Mentally Ill People. *Acta Psychiatr. Scand.* **105**:385–389. [16]

———. 2004b. Factors Influencing Social Distance toward People with Mental Illness. *Commun. Ment. Health J.* **40**:265–274. [17]

Lauber, C., C. Nordt, N. Sartorius, L. Falcato, and W. Rössler. 2000. Public Acceptance of Restrictions on Mentally Ill People. *Acta Psychiatr. Scand.* **102**:26–32. [16]

Laursen, T. M. 2011. Life Expectancy among Persons with Schizophrenia or Bipolar Affective Disorder. *Schizophr. Res.* **131**:101–104. [17]

Laursen, T. M., E. Agerbo, and C. B. Pedersen. 2009. Bipolar Disorder, Schizoaffective Disorder, and Schizophrenia Overlap: A New Comorbidity Index. *J. Clin. Psychiatry* **70**:1432–1438. [02]

Lavin, A., L. Nogueira, C. C. Lapish, et al. 2005. Mesocortical Dopamine Neurons Operate in Distinct Temporal Domains Using Multimodal Signaling. *J. Neurosci.* **25**:5013–5023. [12]

Lavoie, S., M. M. Murray, P. Deppen, et al. 2008. Glutathione Precursor, N-Acetyl-Cysteine, Improves Mismatch Negativity in Schizophrenia Patients. *Neuropsychopharmacology* **33**:2187–2199. [07]

Lawn, S. 2012. In It Together: Physical Health and Well-Being for People with Mental Illness. *Aust. NZ J. Psychiatry* **46**:14–17. [17]

Lawrence, D., C. D. Holman, A. Jablensky, S. Fuller, and A. Stoney. 2001. Increasing Rates of Suicide in Western Australian Psychiatric Patients: A Record Linkage Study. *Acta Psychiatr. Scand.* **104**:443–451. [17]

Lawrence, D., C. D. Holman, A. V. Jablensky, and M. Hobbs. 2003. Death Rate from Ischaemic Heart Disease in Western Australian Psychiatric Patients 1980-1998. *Br. J. Psychiatry* **182**:31–36. [17]

Lawrie, S. M., C. Buechel, H. C. Whalley, et al. 2002. Reduced Fronto-Temporal Functional Connectivity in Schizophrenia Associated with Auditory Hallucinations. *Biol. Psychiatry* **51**:1008–1011. [04]

Lederbogen, F., P. Kirsch, L. Haddad, et al. 2011. City Living and Urban Upbringing Affect Neural Social Stress Processing in Humans. *Nature* **474**:498–501. [08, 09, 12]

Lee, H., D. Dvorak, H. Y. Kao, et al. 2012. Early Cognitive Experience Prevents Adult Deficits in a Neurodevelopmental Schizophrenia Model. *Neuron* **75**:714–724. [09, 13]

Leeson, V. C., P. Sharma, M. Harrison, et al. 2011. IQ Trajectory, Cognitive Reserve and Clinical Outcome Following a First-Episode of Psychosis: A Three Year Longitudinal Study. *Schizophr. Bull.* **37**:768–777. [04]

Leff, J., L. Kuipers, R. Berkowitz, and D. Sturgeon. 1985. A Controlled Trial of Social Intervention in the Families of Schizophrenic Patients: Two-Year Follow-Up. *Br. J. Psychiatry* **146**:594–600. [15]

Lehman, A. F. 1998. The Role of Mental Health Service Research in Promoting Effective Treatment for Adults with Schizophrenia. *J. Ment. Health Policy Econ.* **1**:199–204. [16]

Lehman, A. F., and D. M. Steinwachs. 1998. Patterns of Usual Care for Schizophrenia: Initial Results from the Schizophrenia Patient Outcomes Research Team (PORT) Client Survey. *Schizophr. Bull.* **24**:11–20. [01, 15]

Lehrer, J. 2009. Accept Defeat: The Neuroscience of Screwing Up. *Wired Magazine* Jan 2010, http://www.wired.com/magazine/2009/12/fail_accept_defeat/all/1. [08]

Lehtinen, V., J. Aaltonen, T. Koffert, V. Räkköläinen, and E. Syvälahti. 2000. Two-Year Outcome in First-Episode Psychosis Treated According to an Integrated Model. Is Immediate Neuroleptisation Always Needed? *Eur. Psychiatry* **15**:312–320. [01]

Leicht, G., S. Karch, E. Karamatskos, et al. 2011. Alterations of the Early Auditory Evoked Gamma-Band Response in First-Degree Relatives of Patients with Schizophrenia: Hints to a New Intermediate Phenotype. *J. Psychiatr. Res.* **45**:699–705. [07]

Lelliot, P., A. Beevor, G. Hogman, et al. 2001. Carers' and Users' Expectations of Services: User Version (CUES-U): A New Instrument to Measure the Experience of Users of Mental Health Services. *Br. J. Psychiatry* **179**:67–72. [16]

Leonard, C. M., J. M. Kuldau, L. Maron, et al. 2008. Identical Neural Risk Factors Predict Cognitive Deficit in Dyslexia and Schizophrenia. *Neuropsychology* **22**:147–158. [01]

Leucht, S., S. Heres, W. Kissling, and J. M. Davis. 2011. Evidence-Based Pharmacotherapy of Schizophrenia. *Int. J. Neuropsychopharmacol.* **14**:269–284. [05]

Leucht, S., and R. Lasser. 2006. The Concepts of Remission and Recovery in Schizophrenia. *Pharmacopsychiatry* **39**:161–170. [17]

Levinson, D. F. 2006. The Genetics of Depression: A Review. *Biol. Psychiatry* **60**:84–92. [08]

Levkovitz, Y., S. Mendlovich, S. Riwkes, et al. 2010. A Double-Blind, Randomized Study of Minocycline for the Treatment of Negative and Cognitive Symptoms in Early-Phase Schizophrenia. *J. Clin. Psychiatry* **71**:138–149. [04, 07]

Leweke, F., D. Koethe, C. W. Gerth, et al. 2007a. Double Blind, Controlled Clinical Trial of Cannabidiol Monotherapy Versus Amisulpiride in the Treatment of Acutely Psychotic Schizophrenia Patients. *Schizophr. Bull.* **33**:310. [07]

Leweke, F. M., A. Giuffrida, D. Koethe, et al. 2007b. Anandamide Levels in Cerebrospinal Fluid of First-Episode Schizophrenic Patients: Impact of Cannabis Use. *Schizophr. Res.* **94**:29–36. [07]

Lewinsohn, P. M. 1974. A Behavioral Approach to the Treatment of Depression. In: The Psychology of Depression, ed. R. M. Friedman and M. M. Katz, pp. 157–185. New York: Wiley. [15]

Lewis, D. A., R. Y. Cho, C. S. Carter, et al. 2008. Subunit-Selective Modulation of GABA Type a Receptor Neurotransmission and Cognition in Schizophrenia. *Am. J. Psychiatr.* **165**:1585–1593. [07]

Lewis, D. A., and G. Gonzalez-Burgos. 2008. Neuroplasticity of Neocortical Circuits in Schizophrenia. *Neuropsychopharmacology* **33**:141–165. [07]

Lewis, D. A., T. Hashimoto, and D. W. Volk. 2005. Cortical Inhibitory Neurons and Schizophrenia. *Nat. Rev. Neurosci.* **6**:312–324. [13]

Lewis, D. A., and B. Moghaddam. 2006. Cognitive Dysfunction in Schizophrenia: Convergence of Gamma-Aminobutyric Acid and Glutamate Alterations. *Arch. Neurol.* **63**:1372–1376. [07]

Lewis, S. W., and J. Lieberman. 2008. CATIE and Cutlass: Can We Handle the Truth? *Br. J. Psychiatry* **192**:161–163. [01]

Li, W., H. Zhou, R. Abujarour, et al. 2009. Generation of Human-Induced Pluripotent Stem Cells in the Absence of Exogenous Sox2. *Stem Cells* **27**:2992–3000. [10]

Liberman, R. P. 1979. Social and Political Challenges to the Development of Behavioral Programs in Organizations. In: Trends in Behavior Therapy ed. P. O. Sjoden et al. New York: Academic Press. [17]

Liberman, R. P., W. J. DeRisi, and K. T. Mueser. 1989. Social Skills Training for Psychiatric Patients. Needham Heights, MA: Allyn & Bacon. [15]

Bibliography

Liberman, R. P., and A. Kopelowicz. 2002. Teaching Persons with Severe Mental Disabilities to Be Their Own Case Managers. *Psychiatr. Serv.* **53**:1377–1379. [17]

Liberman, R. P., and K. T. Mueser. 1989. Psychosocial Treatment of Schizophrenia. In: Comprehensive Textbook of Psychiatry (5th edition), ed. H. I. Kaplan and B. J. Sadock, pp. 792–806. Baltimore: Williams & Wilkins. [15]

Liberman, R. P., K. T. Mueser, C. J. Wallace, et al. 1986. Training Skills in the Psychiatrically Disabled: Learning Coping and Competence. *Schizophr. Bull.* **12**:631–647. [15]

Lichtenstein, P., B. H. Yip, C. Björk, et al. 2009. Common Genetic Determinants of Schizophrenia and Bipolar Disorder in Swedish Families: A Population-Based Study. *Lancet* **373**:234–239. [02, 05]

Lieberman, J. A., K. Papadakis, J. Csernansky, et al. 2009. A Randomized, Placebo-Controlled Study of Memantine as Adjunctive Treatment in Patients with Schizophrenia. *Neuropsychopharmacology* **34**:1322–1329. [04]

Lieberman, J. A., T. S. Stroup, J. P. McEvoy, et al. 2005. Effectiveness of Antipsychotic Drugs in Patients with Chronic Schizophrenia. *N. Engl. J. Med.* **353**:1209–1223. [01]

Light, G. A., and D. L. Braff. 2005. Stability of Mismatch Negativity Deficits and Their Relationship to Functional Impairments in Chronic Schizophrenia. *Am. J. Psychiatr.* **162**:1741–1743. [07]

Light, G. A., J. L. Hsu, M. H. Hsieh, et al. 2006. Gamma Band Oscillations Reveal Neural Network Cortical Coherence Dysfunction in Schizophrenia Patients. *Biol. Psychiatry* **60**:1231–1240. [04]

Lin, C. Y., A. Sawa, and H. Jaaro-Peled. 2012. Better Understanding of Mechanisms of Schizophrenia and Bipolar Disorder: From Human Gene Expression Profiles to Mouse Models. *Neurobiol. Dis.* **45**:48–56. [10]

Lincoln, T. M., N. Peter, M. Schäfer, and M. S. 2009. Impact of Stress on Paranoia: An Experimental Investigation of Moderators and Mediators. *Psychol. Med.* **39**:1129–1139. [01]

Lincoln, T. M., K. Wilhelma, and Y. Nestoriuca. 2007. Effectiveness of Psychoeducation for Relapse, Symptoms, Knowledge, Adherence and Functioning in Psychotic Disorders: A Meta-Analysis. *Schizophr. Res.* **96**:232–245. [15]

Lindenmayer, J.-P. 2000. Treatment Refractory Schizophrenia. *Psychiatric Q.* **71**:373–384. [15]

Lindenmayer, J. P., S. R. McGurk, K. T. Mueser, et al. 2008. A Randomized Controlled Trial of Cognitive Remediation among Inpatients with Persistent Mental Illness. *Psychiatr. Serv.* **59**:241–247. [15]

Linehan, M. M. 1993. Cognitive-Behavioral Treatment of Borderline Personality Disorder. New York: Guilford Press. [15]

Link, B. G., F. T. Cullen, E. Struening, P. E. Shrout, and B. P. Dohrenwend. 1989. A Modified Labeling Theory Approach to Mental Disorders: An Empirical Assessment. *Am. Sociol. Rev.* **54**:400–423. [17]

Link, B. G., J. Mirotznik, and F. T. Cullen. 1991. The Effectiveness of Stigma Coping Orientations: Can Negative Consequences of Mental Illness Labeling Be Avoided? *J. Health Soc. Behav.* **32**:302–320. [17]

Link, B. G., and J. C. Phelan. 2001. Conceptualizing Stigma. *Annu. Rev. Sociol.* **27**:363–385. [17]

Linke, J., A. V. King, M. Rietschel, et al. 2012. Increased Medial Orbitofrontal and Amygdala Activation: Evidence for a Systems-Level Endophenotype of Bipolar I Disorder. *Am. J. Psychiatr.* **169**:316–325. [05]

Lipina, T. V., M. Niwa, H. Jaaro-Peled, et al. 2010. Enhanced Dopamine Function in DISC1-L100p Mutant Mice: Implications for Schizophrenia. *Genes Brain Behav.* **9**:777–789. [13]

Lipska, B. K., G. E. Jaskiw, S. Chrapusta, F. Karoum, and D. R. Weinberger. 1992. Ibotenic Acid Lesion of the Ventral Hippocampus Differentially Affects Dopamine and Its Metabolites in the Nucleus Accumbens and Prefrontal Cortex in the Rat. *Brain Res.* **585**:1–6. [11]

Lisman, J. E. 2012. Excitation, Inhibition, Local Oscillations, or Large-Scale Loops: What Causes the Symptoms of Schizophrenia? *Curr. Opin. Neurobiol.* **22**:537–544. [13]

Lisman, J. E., J. T. Coyle, R. W. Green, et al. 2008. Circuit-Based Framework for Understanding Neurotransmitter and Risk Gene Interactions in Schizophrenia. *Trends Neurosci.* **31**:234–242. [07, 13]

Lodge, D. J., M. M. Behrens, and A. A. Grace. 2009. A Loss of Parvalbumin-Containing Interneurons Is Associated with Diminished Oscillatory Activity in an Animal Model of Schizophrenia. *J. Neurosci.* **29**:2344–2354. [11, 13]

Loh, Y. H., O. Hartung, H. Li, et al. 2010. Reprogramming of T Cells from Human Peripheral Blood. *Cell Stem Cell* **7**:15–19. [10]

Lohr, J. M., S. O. Lilienfeld, D. F. Tolin, and J. D. Herbert. 1999. Eye Movement Desensitization and Reprocessing: An Analysis of Specific Versus Non-Specific Treatment Factors. *J. Abnorm. Psychol.* **13**:185–207. [15]

Lonergan, P. E., D. S. Martin, D. F. Horrobin, and M. A. Lynch. 2002. Neuroprotective Effect of Eicosapentaenoic Acid in Hippocampus of Rats Exposed to Gamma-Irradiation. *J. Biol. Chem.* **277**:20,804–20,811. [07]

Lozada, A. F., X. Wang, N. V. Gounko, et al. 2012. Glutamatergic Synapse Formation Is Promoted by Alpha7-Containing Nicotinic Acetylcholine Receptors. *J. Neurosci.* **32**:7651–7661. [13]

Luby, E. D., B. D. Cohen, G. Rosenbaum, J. S. Gottlieb, and R. Kelly. 1959. Study of a New Schizophrenomimetic Drug - Sernyl. *Arch. Neurol. Psychiat.* **81**:363–369. [11]

Lucas, S., M. A. Hodge, A. Shores, J. Brennan, and A. Harris. 2009. Factors Associated with Functional Psychosocial Status after Diagnosis of First Episode Psychosis. *Early Interv. Psychiatry* **3**:35–43. [04]

Luck, S. J., and J. M. Gold. 2008. The Construct of Attention in Schizophrenia. *Biol. Psychiatry* **64**:34–39. [02]

Lucksted, A., A. Drapalski, C. Calmes, et al. 2011. Ending Self-Stigma: Pilot Evaluation of a New Intervention to Reduce Internalized Stigma among People with Mental Illnesses. *Psychiatr. Rehabil. J.* **35**:51–54. [15]

Lucksted, A., W. McFarlane, D. Downing, and L. Dixon. 2012. Recent Developments in Family Psychoeducation as an Evidence-Based Practice. *J. Marital Fam. Ther.* **38**:101–121. [15]

Ludewig, K., M. A. Geyer, and F. X. Vollenweider. 2003. Deficits in Prepulse Inhibition and Habituation in Never-Medicated, First-Episode Schizophrenia. *Biol. Psychiatry* **54**:121–128. [07]

Lugnegård, T., M. Unenge Hallerbäck, F. Hjärthag, and C. Gillberg. 2013. Social Cognition Impairments in Asperger Syndrome and Schizophrenia. *Schizophr. Res.* **143**:277–284. [01]

Lutkenhoff, E. S., T. G. van Erp, M. A. Thomas, et al. 2010. Proton MRS in Twin Pairs Discordant for Schizophrenia. *Mol. Psychiatry* **15**:308–318. [07]

Ly, M., J. C. Motzkin, C. L. Philippi, et al. 2012. Cortical Thinning in Psychopathy. *Am. J. Psychiatr.* **169**:743–749. [01]

Lynch, A. M., D. J. Loane, A. M. Minogue, et al. 2007. Eicosapentaenoic Acid Confers Neuroprotection in the Amyloid-Beta Challenged Aged Hippocampus. *Neurobiol. Aging* **28**:845–855. [07]

Lynch, D., K. R. Laws, and P. J. McKenna. 2010. Cognitive Behavioural Therapy for Major Psychiatric Disorder: Does It Really Work? A Meta-Analytical Review of Well-Controlled Trials. *Psychol. Med.* **40**:9–24. [01]

Lyon, H. M., S. Kaney, and R. P. Bentall. 1994. The Defensive Function of Persecutory Delusions. Evidence from Attribution Tasks. *Br. J. Psychiatry* **164**:637–646. [05]

Lyoo, I. K., S. R. Dager, J. E. Kim, et al. 2010. Lithium-Induced Gray Matter Volume Increase as a Neural Correlate of Treatment Response in Bipolar Disorder: A Longitudinal Brain Imaging Study. *Neuropsychopharmacology* **35**:1743–1750. [07]

Lysaker, P. H., K. D. Buck, and V. A. LaRocco. 2007. Clinical and Psychosocial Significance of Trauma History in the Treatment of Schizophrenia. *J. Psychosoc. Nurs. Ment. Health Serv.* **45**:44–51. [01]

Lysaker, P. H., and J. T. Lysaker. 2010. Schizophrenia and Alterations in Self-Experience: A Comparison of 6 Perspectives. *Schizophr. Bull.* **36**:331–340. [01]

Ma, J., and L. S. Leung. 2007. The Supramammillo-Septal-Hippocampal Pathway Mediates Sensorimotor Gating Impairment and Hyperlocomotion Induced by MK-801 and Ketamine in Rats. *Psychopharmacology* **191**:961–974. [07]

MacCabe, J. H., M. P. Lambe, S. Cnattinghius, et al. 2008. Scholastic Achievement at Age 16 and Risk of Schizophrenia and Other Psychoses: A National Cohort Study. *Psychol. Med.* **38**:1133–1140. [14]

———. 2010. Excellent School Performance at Age 16 and Risk of Adult Bipolar Disorder: National Cohort Study. *Br. J. Psychiatry* **196**:109–115. [14, 17]

MacCabe, J. H., S. Wicks, S. Löfving, et al. 2013. Decline in Cognitive Performance between Ages 13 and 18 Years and the Risk for Psychosis in Adulthood: A Swedish Longitudinal Cohort Study in Males. *JAMA Psychiatry* **70**:261–270. [15]

MacDonald, A. W., III. 2009. Is More Cognitive Experimetnal Psychopathology of Schizophrenia Really Necessary? Challenges and Opportunities. In: Neuropsychiatric Biomarkers, Endophenotypes, and Genes: Promises, Advances and Challenges, ed. M. Ristner, vol. 1, pp. 141–154. New York: Springer. [02]

MacDonald, A. W., III, and C. S. Carter. 2002. Cognitive Experimental Approaches to Investigating Impaired Cognition in Schizophrenia: A Paradigm Shift. *J. Clin. Exp. Neuropsychol.* **24**:873–882. [01, 02]

MacDonald, A. W., III, C. S. Carter, J. G. Kerns, et al. 2005a. Specificity of Prefrontal Dysfunction and Context Processing Deficits to Schizophrenia in Never-Medicated Patients with First-Episode Psychosis. *Am. J. Psychiatr.* **162**:475–484. [02, 07]

MacDonald, A. W., III, V. M. Goghari, B. M. Hicks, et al. 2005b. A Convergent-Divergent Approach to Context Processing, General Intellectual Functioning and the Genetic Liability to Schizophrenia. *Neuropsychology* **19**:814–821. [02]

Mackeprang, T., K. T. Kristiansen, and B. Y. Glenthoj. 2002. Effects of Antipsychotics on Prepulse Inhibition of the Startle Response in Drug-Naive Schizophrenic Patients. *Biol. Psychiatry* **52**:863–873. [07]

Malaspina, D., S. Harlap, S. Fennig, et al. 2001. Advancing Paternal Age and the Risk of Schizophrenia. *Arch. Gen. Psychiatry* **58**:361–367. [08]

Malberg, J. E., A. J. Eisch, E. J. Nestler, and R. S. Duman. 2000. Chronic Antidepressant Treatment Increases Neurogenesis in Adult Rat Hippocampus. *J. Neurosci.* **20**:9104–9110. [07]

Malhotra, A. K., T. Lencz, C. U. Correll, and J. M. Kane. 2007. Genomics and the Future of Pharmacotherapy in Psychiatry. *Int. Rev. Psychiatry* **19**:523–530. [07]

Malhotra, A. K., J. P. Zhang, and T. Lencz. 2012. Pharmacogenetics in Psychiatry: Translating Research into Clinical Practice. *Mol. Psychiatry* **17**:760–769. [07, 17]

Malhotra, D., and J. Sebat. 2011. CNVs: Harbingers of a Rare Variant Revolution in Psychiatric Genetics. *Cell* **148**:1223–1241. [05]

Mallinckrodt, C. H., R. N. Tamura, and Y. Tanaka. 2011. Recent Developments in Improving Signal Detection and Reducing Placebo Response in Psychiatric Clinical Trials. *J. Psychiatr. Res.* **45**:1202–1207. [17]

Manji, H. K., G. J. Moore, and G. Chen. 2000. Clinical and Preclinical Evidence for the Neurotrophic Effects of Mood Stabilizers: Implications for the Pathophysiology and Treatment of Manic-Depressive Illness. *Biol. Psychiatry* **48**:740–754. [07]

Marchetto, M. C., C. Carromeu, A. Acab, et al. 2010. A Model for Neural Development and Treatment of Rett Syndrome Using Human Induced Pluripotent Stem Cells. *Cell* **143**:527–539. [10]

Marco, E. J., O. M. Wolkowitz, S. Vinogradov, et al. 2002. Double-Blind Antiglucocorticoid Treatment in Schizophrenia and Schizoaffective Disorder: A Pilot Study. *World J. Biol. Psychiatry* **3**:156–161. [07]

Marder, S. R. 2006. A Review of Agitation in Mental Illness: Treatment Guidelines and Current Therapies. *J. Clin. Psychiatry* **67**:13–21. [04]

Marder, S. R., and W. Fenton. 2004. Measurement and Treatment Research to Improve Cognition in Schizophrenia: NIMH MATRICS Initiative to Support the Development of Agents for Improving Cognition in Schizophrenia. *Schizophr. Res.* **72**:5–9. [07]

Markon, K. E. 2010. Modeling Psychopathology Structure: A Symptom-Level Analysis of Axis I and II Disorders. *Psychol. Med.* **40**:273–288. [02]

Marshall, C. T., Z. Guo, C. Lu, et al. 2005a. Human Adult Olfactory Neuroepithelial Derived Progenitors Retain Telomerase Activity and Lack Apoptotic Activity. *Brain. Res.* **1045**:45–56. [10]

Marshall, M., S. W. Lewis, A. Lockwood, et al. 2005b. Association between Duration of Untreated Psychosis and Outcome in Cohorts of First-Episode Patients: A Systematic Review. *Arch. Gen. Psychiatry* **62**:975–983. [06]

Marsman, A., M. P. van den Heuvel, D. W. Klomp, et al. 2011. Glutamate in Schizophrenia: A Focused Review and Meta-Analysis of 1H-MRS Studies. *Schizophr. Bull.* [07]

Martin, B. C., and L. S. Miller. 1998. Expenditures for Treating Schizophrenia: A Population-Based Study of Georgia Medicaid Recipients. *Schizophr. Bull.* **24**:479–488. [01]

Marx, C. E., R. S. E. Keefe, R. W. Buchanan, et al. 2009. Proof-of-Concept Trial with the Neurosteroid Pregnenolone Targeting Cognitive and Negative Symptoms in Schizophrenia. *Neuropsychopharmacology* **34**:1885–1903. [04]

Massatti, R. R., H. A. Seweeney, P. C. Panzano, and D. Roth. 2008. The De-Adoption of Innovative Mental Health Practices (IMHP): Why Organizations Choose Not to Sustain an IMHP. *Adm. Policy Ment. Health* **35**:50–65. [15]

Masterton, G., and A. J. Mander. 1990. Psychiatric Emergencies. Scotland and the World Cup Finals. *Br. J. Psychiatry* **156**:475–478. [16]

Matheson, S. L., A. M. Shepherd, R. M. Pinchbeck, K. R. Laurens, and V. J. Carr. 2013. Childhood Adversity in Schizophrenia: A Systematic Meta-Analysis. *Psychol. Med.* **43**:225–238. [01]

Matigian, N., G. Abrahamsen, R. Sutharsan, et al. 2010. Disease-Specific, Neurosphere-Derived Cells as Models for Brain Disorders. *Dis. Model Mech.* **3**:785–798. [10]

Mattay, V. S., T. E. Goldberg, F. Fera, et al. 2003. Catechol O-Methyltransferase Val158-Met Genotype and Individual Variation in the Brain Response to Amphetamine. *PNAS* **100**:6186–6191. [09]

Mazereeuw, G., K. L. Lanctot, S. A. Chau, W. Swardfager, and N. Herrmann. 2012. Effects of Omega-3 Fatty Acids on Cognitive Performance: A Meta-Analysis. *Neurobiol. Aging* **33**:1482 e1417–1429. [07]

McCabe, R., J. Bullenkamp, L. Hansson, et al. 2012. The Therapeutic Relationship and Adherence to Antipsychotic Medication in Schizophrenia. *PLoS One* **7**:e36080. [16]

McCarley, R. W., D. F. Salisbury, Y. Hirayasu, et al. 2002. Association between Smaller Left Posterior Superior Temporal Gyrus Volume on Magnetic Resonance Imaging and Smaller Left Temporal P300 Amplitude in First-Episode Schizophrenia. *Arch. Gen. Psychiatry* **59**:321–331. [07]

McClay, J. L., D. E. Adkins, K. Aberg, et al. 2011. Genome-Wide Pharmacogenomic Study of Neurocognition as an Indicator of Antipsychotic Treatment Response in Schizophrenia. *Neuropsychopharmacology* **36**:616–626. [07]

McCormick, L. M., M. C. Brumm, J. N. Beadle, et al. 2012. Mirror Neuron Function, Psychosis, and Empathy in Schizophrenia. *Psychiatry Res.* **201**:233–239. [07]

McCreadie, R. G., K. Phillips, J. A. Harvey, et al. 1991. The Nithsdale Schizophrenia Surveys. VIII: Do Relatives Want Family Intervention—and Does It Help? *Br. J. Psychiatry* **158**:110–113. [15]

McCurdy, R. D., F. Feron, C. Perry, et al. 2006. Cell Cycle Alterations in Biopsied Olfactory Neuroepithelium in Schizophrenia and Bipolar I Disorder Using Cell Culture and Gene Expression Analyses. *Schizophr. Res.* **82**:163–173. [10]

McDannald, M. A., J. P. Whitt, G. G. Calhoon, et al. 2011. Impaired Reality Testing in an Animal Model of Schizophrenia. *Biol. Psychiatry* **70**:1122–1126. [11]

McFarlane, W. R. 2002. Multifamily Groups in the Treatment of Severe Psychiatric Disorders. New York: Guilford Press. [15]

McFarlane, W. R., S. McNary, L. Dixon, H. Hornby, and E. Cimett. 2001. Predictors of Dissemination of Family Psychoeducation in Community Mental Health Centers in Maine and Illinois. *Psychiatr. Serv.* **52**:935–942. [15]

McGahon, B. M., D. S. Martin, D. F. Horrobin, and M. A. Lynch. 1999. Age-Related Changes in Synaptic Function: Analysis of the Effect of Dietary Supplementation with Omega-3 Fatty Acids. *Neuroscience* **94**:305–314. [07]

McGlashan, T. H. 2006. Is Active Psychosis Neurotoxic? *Schizophr. Bull.* **32**:609–613. [07]

McGlashan, T. H., and R. E. Hoffman. 2000. Schizophrenia as a Disorder of Developmentally Reduced Synaptic Connectivity. *Arch. Gen. Psychiatry* **57**:637–648. [07]

McGorry, P. D. 2007. Issues for DSM-V. Clinical Staging: A Heuristic Pathway to Valid Nosology and Safer, More Effective Treatment in Psychiatry. *Am. J. Psychiatr.* **164**:859–860. [09]

McGorry, P. D., I. B. Hickie, A. R. Yung, C. Pantelis, and H. J. Jackson. 2006. Clinical Staging of Psychiatric Disorders: A Heuristic Framework for Choosing Earlier, Safer and More Effective Interventions. *Aust. NZ J. Psychiatry* **40**:616–622. [09]

McGorry, P. D., B. Nelson, S. Goldstone, and A. R. Yung. 2010. Clinical Staging: A Heuristic and Practical Strategy for New Research and Better Health and Social Outcomes for Psychotic and Related Mood Disorders. *Can. J. Psychiatry* **55**:486–497. [03]

McGowan, P. O., A. Sasaki, A. C. D'Alessio, et al. 2009. Epigenetic Regulation of the Glucocorticoid Receptor in Human Brain Associates with Childhood Abuse. *Nat. Neurosci.* **12**:342–348. [08]

McGrath, J. 2007. The Surprisingly Rich Contours of Schizophrenia Epidemiology. *Arch. Gen. Psychiatry* **64**:14–16. [09]

McGrath, J., and D. Castle. 1995. Does Influenza Cause Schizophrenia? A Five Year Review. *Aust. NZ J. Psychiatry* **29**:23–31. [08]

McGrath, J., S. Saha, D. Chant, and J. Welham. 2008. Schizophrenia: A Concise Overview of Incidence, Prevalence, and Mortality. *Epidemiol. Rev.* **30**:67–76. [01]

McGrath, J., S. Saha, J. Welham, et al. 2004. A Systematic Review of the Incidence of Schizophrenia: The Distribution of Rates and the Influence of Sex, Urbanicity, Migrant Status and Methodology. *BMC Medical* **2**:13. [09]

McGrath, J. J., D. W. Eyles, C. B. Pedersen, et al. 2010. Neonatal Vitamin D Status and Risk of Schizophrenia: A Population-Based Case-Control Study. *Arch. Gen. Psychiatry* **67**:889–894. [06]

McGrath, J. J., and L. J. Richards. 2009. Why Schizophrenia Epidemiology Needs Neurobiology—and Vice Versa. *Schizophr. Bull.* **35**:577–581. [01, 08]

McGrath, J. J., and E. S. Susser. 2009. New Directions in the Epidemiology of Schizophrenia. *Med. J. Aust.* **190**:S7–9. [13]

McGurk, S. R., and K. T. Mueser. 2004. Cognitive Functioning, Symptoms, and Work in Supported Employment: A Review and Heuristic Model. *Schizophr. Res.* **70**:147–174. [15]

McGurk, S. R., K. T. Mueser, T. DeRosa, and R. Wolfe. 2009. Work, Recovery, and Comorbidity in Schizophrenia: A Randomized Controlled Trial of Cognitive Remediation. *Schizophr. Bull.* **35**:319–335. [15]

McGurk, S. R., K. T. Mueser, and A. Pascaris. 2005. Cognitive Training and Supported Employment for Persons with Severe Mental Illness: One Year Results from a Randomized Controlled Trial. *Schizophr. Bull.* **31**:898–909. [15]

McGurk, S. R., E. W. Twamley, D. I. Sitzer, G. J. McHugo, and K. T. Mueser. 2007. A Meta-Analysis of Cognitive Remediation in Schizophrenia. *Am. J. Psychiatr.* **164**:1791–1802. [04, 15]

McHugo, G. J., R. E. Drake, R. Whitley, et al. 2007. Fidelity Outcomes in the National Implementing Evidence-Based Practices Project. *Psychiatr. Serv.* **58**:1279–1284. [15]

McLaughlin, K. A., J. G. Green, M. J. Gruber, et al. 2010. Childhood Adversities and Adult Psychiatric Disorders in the National Comorbidity Survey Replication II: Associations with Persistence of DSM-IV Disorders. *Arch. Gen. Psychiatry* **67**:124–132. [09]

McLean, A. 1995. Empowerment and the Psychiatric Consumer/Ex-Patient Movement in the United States: Contradictions, Crisis and Change. *Soc. Sci. Med.* **40**:1053–1071. [15, 17]

Mechelli, A., A. Riecher-Rossler, E. M. Meisenzahl, et al. 2011. Neuroanatomical Abnormalities That Predate the Onset of Psychosis: A Multicenter Study. *Arch. Gen. Psychiatry* **68**:489–495. [07]

Meehl, P. E. 1962. Schizotaxia, Schizotypy, Schizophrenia. *Am. Psychol.* **17**:827–838. [02]

———. 1990. Toward an Integrated Theory of Schizotaxia, Schizotypy and Schizophrenia. *J. Pers. Disord.* **4**:1–99. [02]

Meincke, U., G. A. Light, M. A. Geyer, D. L. Braff, and E. Gouzoulis-Mayfrank. 2004. Sensitization and Habituation of the Acoustic Startle Reflex in Patients with Schizophrenia. *Psychiatry Res.* **126**:51–61. [07]

Meise, U., G. Kemmler, M. Kurz, and W. Rössler. 1996. Die Standortqualität als Grundlage Psychiatrischer Versorgungsplanung. *Gesundheitswesen* **58**:29–37. [16]

Melle, I., T. K. Larsen, U. Haahr, et al. 2004. Reducing the Duration of Untreated First-Episode Psychosis: Effects on Clinical Presentation. *Arch. Gen. Psychiatry* **61**:143–150. [07]

Meltzer, H. Y., L. Arvanitis, D. Bauer, and W. Rein. 2004. Placebo-Controlled Evaluation of Four Novel Compounds for the Treatment of Schizophrenia and Schizoaffective Disorder. *Am. J. Psychiatr.* **161**:975–984. [07]

Memory Pharmaceuticals Corp. 2008. Press Release (May 11, 2013): Memory Pharmaceuticals Achieves Enrollment Goal for Phase 2 Study of MEM 3454 in Cognitive Impairment Associated with Schizophrenia. http://www.thefreelibrary.com/Memory+Pharmaceuticals+Achieves+Enrollment+Goal+for+Phase+2+Study+of ...-a0189242108. (accessed June 14, 2013). [04]

Mendel, P., L. S. Meredith, M. Schoenbaum, C. D. Sherbourne, and K. B. Wells. 2008. Interventions in Organizational and Community Context: A Framework for Building Evidence on Dissemination and Implementation in Health Services Research. *Adm. Policy Ment. Health* **35**:21–37. [15]

Mendrek, A., K. A. Kiehl, A. M. Smith, et al. 2005. Dysfunction of a Distributed Neural Circuitry in Schizophrenia Patients During a Working-Memory Performance. *Psychol. Med.* **35**:187–196. [07]

Merck & Co. 2011. MK0249 for the Treatment of Cognitive Impairment in Patients with Schizophrenia (0249-016). *Natl. Library Med.* http://www.clinicaltrials.gov/ct2/show/NCT00506077?term=NCT00506077&rank=1. (accessed May 11, 2013). [04]

Messias, E., B. Kirkpatrick, E. Bromet, et al. 2004. Summer Birth and Deficit Schizophrenia. A Pooled Analysis from 6 Countries. *Arch. Gen. Psychiatry* **61**:985–989. [03]

Meyer, J. M., V. G. Davis, D. C. Goff, et al. 2008. Change in Metabolic Syndrome Parameters with Antipsychotic Treatment in the CATIE Schizophrenia Trial: Prospective Data from Phase 1. *Schizophr. Res.* **101**:273–286. [17]

Meyer, U. 2011. Developmental Neuroinflammation and Schizophrenia. *Prog. Neuropsychopharmacol. Biol. Psychiatry* **42**:20–34. [07]

Meyer, U., and J. Feldon. 2012. To Poly(I:C) or Not to Poly(I:C): Advancing Preclinical Schizophrenia Research through the Use of Prenatal Immune Activation Models. *Neuropharmacology* **62(3)**:1308–1321. [11]

Meyer, U., J. Feldon, and B. K. Yee. 2009. A Review of the Fetal Brain Cytokine Imbalance Hypothesis of Schizophrenia. *Schizophr. Bull.* **35(5)**:959–972 [08]

Meyer, U., S. Schwendener, J. Feldon, and B. K. Yee. 2006. Prenatal and Postnatal Maternal Contributions in the Infection Model of Schizophrenia. *Exp. Brain Res.* **173**:243–257. [11]

Meyer, U., I. Weiner, G. M. McAlonan, and J. Feldon. 2011. The Neuropathological Contribution of Prenatal Inflammation to Schizophrenia. *Expert. Rev. Neurother.* **11**:29–32. [07]

Meyer-Lindenberg, A. 2010a. From Maps to Mechanisms through Neuroimaging of Schizophrenia. *Nature* **468**:194–202. [09, 17]

———. 2010b. Imaging Genetics of Schizophrenia. *Dialog. Clin. Neurosci.* **12**:449–456. [08]

———. 2011. Neuroimaging and the Question of Neurodegeneration in Schizophrenia. *Prog. Neurobiol.* **95**:514–516. [09]

Meyer-Lindenberg, A., J.-B. Poline, P. D. Kohn, et al. 2001. Evidence for Abnormal Cortical Functional Connectivity During Working Memory in Schizophrenia. *Am. J. Psychiatr.* **158**:1809–1817. [04, 12, 13]

Meyer-Lindenberg, A., and H. Tost. 2012. Neural Mechanisms of Social Risk for Psychiatric Disorders. *Nat. Neurosci.* **15**:663–668. [08, 09, 17]

Meyer-Lindenberg, A., and D. R. Weinberger. 2006. Intermediate Phenotypes and Genetic Mechanisms of Psychiatric Disorders. *Nat. Rev. Neurosci.* **7**:818–827. [09, 12]

Meyer-Lindenberg, A. S., R. K. Olsen, P. D. Kohn, et al. 2005. Regionally Specific Disturbance of Dorsolateral Prefrontal-Hippocampal Functional Connectivity in Schizophrenia. *Arch. Gen. Psychiatry* **62**:379–386. [08, 09]

Mier, D., P. Kirsch, and A. Meyer-Lindenberg. 2010. Neural Substrates of Pleiotropic Action of Genetic Variation in COMT: A Meta-Analysis. *Mol. Psychiatry* **15**:918–927. [08]

Millar, J. K., J. C. Wilson-Annan, S. Anderson, et al. 2000. Disruption of Two Novel Genes by a Translocation Co-Segregating with Schizophrenia. *Hum. Mol. Genet.* **9**:1415–1423. [11]

Miller, G. 2010. Is Pharma Running out of Brainy Ideas? *Science* **329**:502–504. [17]

Miller, T. J., T. H. McGlashan, J. L. Rosen, et al. 2003. Prodromal Assessment with the Structured Interview for Prodromal Syndromes and the Scale of Prodromal Symptoms: Predictive Validity, Interrater Reliability, and Training to Reliability. *Schizophr. Bull.* **29**:703–715. [07]

Miller, W. R., and S. Rollnick. 2002. Motivational Interviewing: Preparing People for Change (2nd edition). New York: Guilford Press. [17]

Mishara, A. L., and T. E. Goldberg. 2004. A Meta-Analysis and Critical Review of the Effects of Conventional Neuroleptic Treatment on Cognition in Schizophrenia: Opening a Closed Book. *Biol. Psychiatry* **55**:1013–1022. [07, 14]

Mitchell, K. J. 2007. The Genetics of Brain Wiring: From Molecule to Mind. *PLoS Biol.* **5**:e113. [13]

———. 2011a. The Genetics of Neurodevelopmental Disease. *Curr. Opin. Neurobiol.* **21**:197–203. [13]

———. 2011b. The Miswired Brain: Making Connections from Neurodevelopment to Psychopathology. *BMC Biol.* **9**:23. [13]

Mitchell, K. J., Z. J. Huang, B. Moghaddam, and A. Sawa. 2011. Following the Genes: A Framework for Animal Modeling of Psychiatric Disorders. *BMC Biol.* **9**:76. [13]

Mitchell, K. J., and D. J. Porteous. 2011. Rethinking the Genetic Architecture of Schizophrenia. *Psychol. Med.* **41**:19–32. [01, 13]

Moghaddam, B., B. Adams, A. Verma, and D. Daly. 1997. Activation of Glutamatergic Neurotransmission by Ketamine: A Novel Step in the Pathway from NMDA Receptor Blockade to Dopaminergic and Cognitive Disruptions Associated with the Prefrontal Cortex. *J. Neurosci.* **17**:2921–2927. [11]

Moghaddam, B., and D. Javitt. 2012. From Revolution to Evolution: The Glutamate Hypothesis of Schizophrenia and Its Implication for Treatment. *Neuropsychopharmacology* **37**:4–15. [07]

Mohamed, S., R. A. Rosenheck, M. S. Swartz, et al. 2008. Relationship of Cognition and Psychopathology to Functional Impairment in Schizophrenia. *Am. J. Psychiatr.* **165**:978–987. [14]

Mojtabai, R., R. A. Nicholson, and B. N. Carpenter. 1998. Role of Psychosocial Treatments in Management of Schizophrenia: A Meta-Analytic Review of Controlled Outcome Studies. *Schizophr. Bull.* **24**:569–587. [15, 16]

Mondelli, V., P. Dazzan, N. Hepgul, et al. 2010a. Abnormal Cortisol Levels During the Day and Cortisol Awakening Response in First-Episode Psychosis: The Role of Stress and of Antipsychotic Treatment. *Schizophr. Res.* **116**:234–242. [09]

Mondelli, V., C. M. Pariante, S. Navari, et al. 2010b. Higher Cortisol Levels Are Associated with Smaller Left Hippocampal Volume in First-Episode Psychosis. *Schizophr. Res.* **119**:75–78. [09]

Montoya, I. D., and F. Vocci. 2007. Medications Development for the Treatment of Nicotine Dependence in Individuals with Schizophrenia. *J. Dual Diagn.* **3**:113–150. [17]

Moore, G. J., B. M. Cortese, D. A. Glitz, et al. 2009. A Longitudinal Study of the Effects of Lithium Treatment on Prefrontal and Subgenual Prefrontal Gray Matter Volume in Treatment-Responsive Bipolar Disorder Patients. *J. Clin. Psychiatry* **70**:699–705. [07]

Moore, H. 2010. The Role of Rodent Models in the Discovery of New Treatments for Schizophrenia: Updating Our Strategy. *Schizophr. Bull.* **36**:1066–1072. [13]

Moore, H., J. D. Jentsch, M. Ghajarnia, M. A. Geyer, and A. A. Grace. 2006. A Neurobehavioral Systems Analysis of Adult Rats Exposed to Methylazoxymethanol Acetate on E17: Implications for the Neuropathology of Schizophrenia. *Biol. Psychiatry* **60**:253–264. [11]

Moore, T. H., S. Zammit, A. Lingford-Hughes, et al. 2007. Cannabis Use and Risk of Psychotic or Affective Mental Health Outcomes: A Systematic Review. *Lancet* **370**:319–328. [07, 09]

Moran, M. 2006. U.S. Insurance Crisis Gets Pessimistic Prognosis. *Psychiatric News* **41**:16. [04]

Morgan, C., J. Kirkbride, G. Hutchinson, et al. 2008. Cumulative Social Disadvantage, Ethnicity and First-Episode Psychosis: A Case-Control Study. *Psychol. Med.* **38**:1701–1715. [09]

Morgan, C., and A. Kleinman. 2010. Social Science Perspectives: A Failure of the Sociological Imagination. In: Principles of Social Psychiatry, ed. C. Morgan and D. Bhugra. London: Wiley-Blackwell. [09]

Morgan, V. A., J. Badcock, and A. Jablensky. 2006. Submission 277: Prepared on Behalf of the Australasian Society for Psychiatric Research, May 2005. In: The Senate Select Committee on Mental Health. A National Approach to Mental Health from Crisis to Community. Canberra: Commonwealth of Australia. http://www.aph.gov.au/Parliamentary_Business/Committees/Senate_Committees?url=mentalhealth_ctte/submissions/sub277.pdf. (accessed 13 June, 2013). [17]

Morgan, V. A., G. M. Valuri, M. L. Croft, et al. 2011. Cohort Profile: Pathways of Risk from Conception to Disease: The Western Australian Schizophrenia High-Risk E-Cohort. *Int. J. Epidemiol.* **40**:1477–1485. [17]

Morgan, V. A., A. Waterreus, A. Jablensky, et al. 2012. People Living with Psychotic Illness in 2010: The Second Australian National Survey of Psychosis. *Aust. NZ J. Psychiatry* **46**:735–752. [17]

Moritz, S., N. Van Quaquebeke, and T. M. Lincoln. 2012. Jumping to Conclusions Is Associated with Paranoia but Not General Suspiciousness: A Comparison of Two Versions of the Probabilistic Reasoning Paradigm. *Schizophr. Res. Treast.* **2012**:Article ID 384039. [05]

Morrison, A. P., P. French, S. L. Stewart, et al. 2012. Early Detection and Intervention Evaluation for People at Risk of Psychosis: Multisite Randomised Controlled Trial. *BMJ* **344**:e2233. [01, 06]

Morrison, A. P., P. French, L. Walford, et al. 2004. Cognitive Therapy for the Prevention of Psychosis in People at Ultra-High Risk: Randomised Controlled Trial. *Br. J. Psychiatry* **185**:291–297. [07]

Mors, O., P. B. Mortensen, and H. Ewald. 1999. A Population-Based Register Study of the Association between Schizophrenia and Rheumatoid Arthritis. *Schizophr. Res.* **40**:67–74. [01]

Mortensen, P. B., M. G. Pedersen, and C. B. Pedersen. 2010. Psychiatric Family History and Schizophrenia Risk in Denmark: Which Mental Disorders Are Relevant? *Psychol. Med.* **40**:201–210. [02, 05]

Mortensen, P. B., C. B. Pedersen, T. Westergaard, et al. 1999. Effects of Family History and Place and Season of Birth on the Risk of Schizophrenia. *N. Engl. J. Med.* **340**:603–608. [09]

Motzkin, J. C., J. P. Newman, K. A. Kiehl, and M. Koenigs. 2011. Reduced Prefrontal Connectivity in Psychopathy. *J. Neurosci.* **31**:17,348–17,357. [01]

Mowry, B. J., and J. Gratten. 2013. The Emerging Spectrum of Allelic Variation in Schizophrenia: Current Evidence and Strategies for the Identification and Functional Characterization of Common and Rare Variants. *Mol. Psychiatry* **18**:38–52. [07]

Mueser, K. T., S. Aalto, D. R. Becker, et al. 2005. The Effectiveness of Skills Training for Improving Outcomes in Supported Employment. *Psychiatr. Serv.* **56**:1254–1260. [15]

Mueser, K. T., A. S. Bellack, M. S. Douglas, and J. H. Wade. 1991. Prediction of Social Skill Acquisition in Schizophrenic and Major Affective Disorder Patients from Memory and Symptomatology. *Psychiatry Res.* **37**:281–296. [15]

Mueser, K. T., R. E. Clark, M. Haines, et al. 2004. The Hartford Study of Supported Employment for Severe Mental Illness. *J. Consult. Clin. Psychol.* **72**:479–490. [15]

Mueser, K. T., P. W. Corrigan, D. Hilton, et al. 2002. Illness Management and Recovery for Severe Mental Illness: A Review of the Research. *Psychiatr. Serv.* **53**:1272–1284. [15]

Mueser, K. T., F. Deavers, D. L. Penn, and J. Cassisi. 2013a. Psychosocial Treatments for Schizophrenia. *Annu. Rev. Clin. Psychol.* **9**:465–497. [15]

Mueser, K. T., S. M. Silverstein, and M. Farkas. 2013b. Should the Training of Clinical Psychologists Require Competence in the Treatment and Rehabilitation of Individuals with a Serious Mental Illness? *Psychiatr. Rehabil. J.* **36**:54–59. [15]

Mueser, K. T., W. C. Torrey, D. Lynde, P. Singer, and R. E. Drake. 2003. Implementing Evidence-Based Practices for People with Severe Mental Illness. *Behav. Mod.* **27**:387–411. [15]

Mueser, K. T., P. R. Yarnold, S. D. Rosenberg, et al. 2000. Substance Use Disorder in Hospitalized Severely Mentally Ill Psychiatric Patients: Prevalence, Correlates, and Subgroups. *Schizophr. Bull.* **26**:179–192. [17]

Muller, N., D. Krause, S. Dehning, et al. 2010. Celecoxib Treatment in an Early Stage of Schizophrenia: Results of a Randomized, Double-Blind, Placebo-Controlled Trial of Celecoxib Augmentation of Amisulpride Treatment. *Schizophr. Res.* **121**:118–124. [07]

Munafo, M. R., S. M. Brown, and A. R. Hariri. 2008. Serotonin Transporter (5-HTTLPR) Genotype and Amygdala Activation: A Meta-Analysis. *Biol. Psychiatry* **63**:852–857. [08]

Munafo, M. R., N. B. Freimer, W. Ng, et al. 2009. 5-HTTLPR Genotype and Anxiety-Related Personality Traits: A Meta-Analysis and New Data. *Am. J. Med. Genet. B Neuropsychiatr. Genet.* **150B**:271–281. [08]

Munro, C. A., M. E. McCaul, D. F. Wong, et al. 2006. Sex Differences in Striatal Dopamine Release in Healthy Adults. *Biol. Psychiatry* **59**:966–974. [07]

Murphy, K. C. 2002. Schizophrenia and Velo-Cardio-Facial Syndrome. *Lancet* **359**:426–430. [05]

Murray, R. M. 2003. The Epidemiology of Schizophrenia. Cambridge, U.K., New York: Cambridge Univ. Press. [09]

Murray, R. M., and S. W. Lewis. 1987. Is Schizophrenia a Neurodevelopmental Disorder? *BMJ (Clin. Res. Ed.)* **295**:681–682. [03, 08, 09]

Murray, R. M., P. Sham, J. Van Os, et al. 2004. A Developmental Model for Similarities and Dissimilarities between Schizophrenia and Bipolar Disorder. *Schizophr. Res.* **71**:405–416. [09]

Myin-Germeys, I., and J. van Os. 2008. Adult Adversity: Do Early Environment and Genotype Create Lasting Vulnerability for Adult Social Adversity in Psychosis? In: Society and Psychosis, ed. C. Morgan et al., pp. 127–142. Cambridge: Cambridge Univ. Press. [09]

Myin-Germeys, I., J. van Os, J. E. Schwartz, A. A. Stone, and P. A. Delespaul. 2001. Emotional Reactivity to Daily Life Stress in Psychosis. *Arch. Gen. Psychiatry* **58**:1137–1144. [09]

Naatanen, R., A. W. Gaillard, and S. Mantysalo. 1978. Early Selective-Attention Effect on Evoked Potential Reinterpreted. *Acta Psychol.* **42**:313–329. [07]

National Collaborating Centre for Mental Health. 2009. Schizophrenia: Core Interventions in the Treatment and Management of Schizophrenia in Adults in Primary and Secondary Care (updated edition). Natl. Clinical Guideline Nr. 82. London: British Psychological Society and The Royal College of Psychiatrists. [15]

Need, A. C., R. S. Keefe, D. Ge, et al. 2009. Pharmacogenetics of Antipsychotic Response in the CATIE Trial: A Candidate Gene Analysis. *Eur. J. Hum. Genet.* **17**:946–957. [07]

Nelson, B., A. Thompson, and A. R. Yung. 2012. Basic Self-Disturbance Predicts Psychosis Onset in the Ultra High Risk for Psychosis "Prodromal" Population. *Schizophr. Bull.* **38**:1277–1287. [01]

———. 2013. Not All First-Episode Psychosis Is the Same: Preliminary Evidence of Greater Basic Self-Disturbance in Schizophrenia Spectrum Cases. *Early Interv. Psychiatry* **7**:200–204. [01]

Nelson, G., T. Aubry, and A. Lafrance. 2007. A Review of the Literature on the Effectiveness of Housing and Support, Assertive Community Treatment, and Intensive Case Management Interventions for Persons with Mental Illness Who Have Been Homeless. *Am. J. Orthopsychiat.* **77**:350–361. [15]

Newman, K., C. Fox, D. Harding, J. Mehta, and W. Roth. 2004. Rampage: The Social Roots of School Shootings. New York: Basic Books. [01]

Nordt, C., W. Rössler, and C. Lauber. 2006. Attitudes of Mental Health Professionals toward People with Schizophrenia and Major Depression. *Schizophr. Bull.* **32**:709–714. [16]

Nuechterlein, K. H., D. M. Barch, J. M. Gold, et al. 2004. Identification of Separable Cognitive Factors in Schizophrenia. *Schizophr. Res.* **72**:29–39. [04]

Nuechterlein, K. H., and M. E. Dawson. 1984. A Heuristic Vulnerability/Stress Model of Schizophrenic Episodes. *Schizophr. Bull.* **10**:300–312. [15]

Nuechterlein, K. H., M. F. Green, R. S. Kern, et al. 2008. The MATRICS Consensus Cognitive Battery, Part 1: Test Selection, Reliability, and Validity. *Am. J. Psychiatr.* **165**:203–213. [06]

Nuechterlein, K. H., K. L. Subotnik, M. Gitlin, et al. 1999. Neurocognitive and Environmental Contributors to Work Recovery after Initial Onset of Schizophrenia: Answers from Path Analyses. In: International Congress on Schizophrenia Research. Santa Fe, NM. [04]

Nutt, D., and G. Goodwin. 2011. ECNP Summit on the Future of CNS Drug Research in Europe 2011. *Eur. Neuropsychopharmacol.* **21**:495–499. [17]

O'Doherty, J., M. L. Kringelbach, E. T. Rolls, J. Hornak, and C. Andrews. 2001. Abstract Reward and Punishment Representation in the Human Orbitofrontal Cortex. *Nat. Neurosci.* **4**:95–102. [05]

O'Donnell, P. 2011. Adolescent Onset of Cortical Disinhibition in Schizophrenia: Insights from Animal Models. *Schizophr. Bull.* **37**:484–492. [13]

———. 2012. Cortical Disinhibition in the Neonatal Ventral Hippocampal Lesion Model of Schizophrenia: New Vistas on Possible Therapeutic Approaches. *Pharmacol. Ther.* **133**:19–25. [11]

O'Donnell, P., J. H. Cabungcal, P. T. Piantadosi, et al. 2011. **Oxidative Stress During Development in Prefrontal Cortical Interneurons in Developmental Animal Models of Schizophrenia.** *Schizophr. Bull.* **37**:111. [11]

O'Donovan, M. C., G. Kirov, and M. J. Owen. 2008. Phenotypic Variations on the Theme of CNVs. *Nat. Genet.* **40**:1392–1393. [08]

OECD. 2012. Sick on the Job? Myths and Realities About Mental Health and Work. Paris: OECD Publishing. [16]

Oh, G., and A. Petronis. 2008. Environmental Studies of Schizophrenia through the Prism of Epigenetics. *Schizophr. Bull.* **34**:1122–1129. [09]

Oken, R. J., and M. Schulzer. 1999. At Issue: Schizophrenia and Rheumatoid Arthritis: The Negative Association Revisited. *Schizophr. Bull.* **25**:625–638. [01]

Oliver, P. L. 2011. Challenges of Analysing Gene-Environment Interactions in Mouse Models of Schizophrenia. *ScientificWorldJournal* **11**:1411–1420. [13]

Olney, J. W., and N. B. Farber. 1995a. Glutamate Receptor Dysfunction and Schizophrenia. *Arch. Gen. Psychiatry* **52**:998–1007. [07]

———. 1995b. NMDA Antagonists as Neurotherapeutic Drugs, Psychotogens, Neurotoxins, and Research Tools for Studying Schizophrenia. *Neuropsychopharmacology* **13**:335–345. [07]

Ongur, D., J. E. Jensen, A. P. Prescot, et al. 2008. Abnormal Glutamatergic Neurotransmission and Neuronal-Glial Interactions in Acute Mania. *Biol. Psychiatry* **64**:718–726. [07]

Onstad, S., I. Skre, J. Edvardsen, and S. Torgersen. 1991. Mental Disorders in First-Degree Relatives of Schizophrenics. *Acta Psychiatr. Scand.* **83**:463–467. [02]

O'Reilly, R., and Y. Munakata. 2000. Understanding the Mind by Simulating the Brain: Computational Explorations in Cognitive Neuroscience. Cambridge MA: MIT Press. [12]

Oshima, I., Y. Mino, and Y. Inomata. 2003. Institutionalisation and Schizophrenia in Japan: Social Environments and Negative Symptoms: Nationwide Survey of In-Patients. *Br. J. Psychiatry* **183**:50–56. [01]

———. 2005. Effects of Environmental Deprivation on Negative Symptoms of Schizophrenia: A Nationwide Survey in Japan's Psychiatric Hospitals. *Psychiatry Res.* **136**:163–171. [01]

Oswald, L. M., D. F. Wong, M. McCaul, et al. 2005. Relationships among Ventral Striatal Dopamine Release, Cortisol Secretion, and Subjective Responses to Amphetamine. *Neuropsychopharmacology* **30**:821–832. [07]

Ota, M., M. Ishikawa, N. Sato, et al. 2012. Glutamatergic Changes in the Cerebral White Matter Associated with Schizophrenic Exacerbation. *Acta Psychiatr. Scand.* **126**:72–78. [07]

Othman, M., C. Lu, K. Klueber, W. Winstead, and F. Roisen. 2005. Clonal Analysis of Adult Human Olfactory Neurosphere Forming Cells. *Biotech. Histochem.* **80**:189–200. [10]

Owen, M. J., N. M. Williams, and M. C. O'Donovan. 2004. The Molecular Genetics of Schizophrenia: New Findings Promise New Insights. *Mol. Psychiatry* **9**:14–27. [08]

Ozyurt, B., M. Sarsilmaz, N. Akpolat, et al. 2007. The Protective Effects of Omega-3 Fatty Acids against MK-801-Induced Neurotoxicity in Prefrontal Cortex of Rat. *Neurochem. Int.* **50**:196–202. [07]

Pacher, P., S. Batkai, and G. Kunos. 2006. The Endocannabinoid System as an Emerging Target of Pharmacotherapy. *Pharmacol. Rev.* **58**:389–462. [07]

Pajonk, F. G., T. Wobrock, O. Gruber, et al. 2010. Hippocampal Plasticity in Response to Exercise in Schizophrenia. *Arch. Gen. Psychiatry* **67**:133–143. [17]

Pang, Y., S. Rodts-Palenik, Z. Cai, W. A. Bennett, and P. G. Rhodes. 2005. Suppression of Glial Activation Is Involved in the Protection of Il-10 on Maternal *E. coli* Induced Neonatal White Matter Injury. *Brain Res. Dev. Brain Res.* **157**:141–149. [07]

Pang, Z. P., N. Yang, T. Vierbuchen, et al. 2011. Induction of Human Neuronal Cells by Defined Transcription Factors. *Nature* **476**:220–223. [10]

Pantelis, C., D. Velakoulis, P. D. McGorry, et al. 2003. Neuroanatomical Abnormalities before and after Onset of Psychosis: A Cross-Sectional and Longitudinal MRI Comparison. *Lancet* **361**:281–288. [07]

Pantelis, C., M. Yucel, S. J. Wood, et al. 2005. Structural Brain Imaging Evidence for Multiple Pathological Processes at Different Stages of Brain Development in Schizophrenia. *Schizophr. Bull.* **31**:672–696. [07, 11]

Papassotiropoulos, A., K. Henke, E. Stefanova, et al. 2011. A Genome-Wide Survey of Human Short-Term Memory. *Mol. Psychiatry* **16**:184–192. [08]

Pariante, C. M., K. Vassilopoulou, D. Velakoulis, et al. 2004. Pituitary Volume in Psychosis. *Br. J. Psychiatry* **185**:5–10. [09]

Parnas, J., P. Vianin, D. Saebye, et al. 2001. Visual Binding Abilities in the Initial and Advanced Stages of Schizophrenia. *Acta Psychiatr. Scand.* **103**:171–180. [01, 07]

Pasca, S. P., T. Portmann, I. Voineagu, et al. 2011. Using iPSC-Derived Neurons to Uncover Cellular Phenotypes Associated with Timothy Syndrome. *Nat. Med.* **17**:1657–1662. [10]

Patil, S. T., L. Zhang, F. Martenyi, et al. 2007. Activation of mGlu2/3 Receptors as a New Approach to Treat Schizophrenia: A Randomized Phase 2 Clinical Trial. *Nat. Med.* **13**:1102–1107. [07, 11]

Patterson, P. H. 2009. Immune Involvement in Schizophrenia and Autism: Etiology, Pathology and Animal Models. *Behav. Brain Res.* **204**:313–321. [08]

Patterson, T. L., S. Goldman, C. L. McKibbin, T. Hughs, and D. V. Jeste. 2001. UCSD Performance-Based Skills Assessment: Development of a New Measure of Everyday Functioning for Severely Mentally Ill Adults. *Schizophr. Bull.* **27**:235–245. [04]

Paul, G. L. 1986. Principles and Methods to Support Cost-Effective Quality Operations: Assessment in Residential Treatment Settings, Part 1. Champaign: Research Press. [15]

Paul, G. L., and R. J. Lentz. 1977. Psychosocial Treatment of Chronic Mental Patients: Milieu Versus Social-Learning Programs. Cambridge, MA: Harvard Univ. Press. [15, 17]

Paulus, F. M., S. Krach, J. Bedenbender, et al. 2013. Partial Support for ZNF804A Genotype-Dependent Alterations in Prefrontal Connectivity. *Hum. Brain Mapp.* **34**:304–313. [09]

Pedersen, C. A., C. M. Gibson, S. W. Rau, et al. 2011. Intranasal Oxytocin Reduces Psychotic Symptoms and Improves Theory of Mind and Social Perception in Schizophrenia. *Schizophr. Res.* **132**:50–53. [07]

Pedersen, C. B., and P. B. Mortensen. 2001. Evidence of a Dose-Response Relationship between Urbanicity During Upbringing and Schizophrenia Risk. *Arch. Gen. Psychiatry* **58**:1039–1046. [08, 09]

Peleg-Raibstein, D., and J. Feldon. 2008. Effects of Withdrawal from an Escalating Dose of Amphetamine on Conditioned Fear and Dopamine Response in the Medial Prefrontal Cortex. *Behav. Brain Res.* **186**:12–22. [13]

Penadés, R., N. Pujol, R. Catalán, et al. 2013. Brain Effects of Cognitive Remediation Therapy in Schizophrenia: A Structural and Functional Neuroimaging Study. *Biol. Psychiatry* **73**:1015–1023. [01]

Penn, D. L., P. W. Corrigan, R. P. Bentall, J. M. Racenstein, and L. Newman. 1997. Social Cognition in Schizophrenia. *Psychol. Bull.* **121**:114–132. [04]

Penschuck, S., P. Flagstad, M. Didriksen, M. Leist, and A. T. Michael-Titus. 2006. Decrease in Parvalbumin-Expressing Neurons in the Hippocampus and Increased Phencyclidine-Induced Locomotor Activity in the Rat Methylazoxymethanol (MAM) Model of Schizophrenia. *Eur. J. Neurosci.* **23**:279–284. [11]

Penzes, P., M. E. Cahill, K. A. Jones, J. E. VanLeeuwen, and K. M. Woolfrey. 2011. Dendritic Spine Pathology in Neuropsychiatric Disorders. *Nat. Neurosci.* **14**:285–293. [12]

Peralta, V., and M. J. Cuesta. 2001. How Many and Which Are the Psychopathological Dimensions of Schizophrenia? Issues Influencing Their Ascertainment. *Schizophr. Res.* **49**:269–285. [02, 03]

Perkins, R. 2001. What Constitutes Success? The Relative Priority of Service Users' and Clinicians' Views of Mental Health Services. *Br. J. Psychiatry* **179**:9–10. [16]

Perry, A., S. Bentin, I. Shalev, et al. 2010. Intranasal Oxytocin Modulates EEG Mu/Alpha and Beta Rhythms During Perception of Biological Motion. *Psychoneuroendocrinology* **35**:1446–1453. [07]

Perry, V. H. 2007. Stress Primes Microglia to the Presence of Systemic Inflammation: Implications for Environmental Influences on the Brain. *Brain Behav. Immun.* **21**:45–46. [07]

Pescosolido, B. A., J. K. Martin, A. Lang, and S. Olafsdottir. 2008. Rethinking Theoretical Approaches to Stigma: A Framework Integrating Normative Influences on Stigma (FINIS). *Soc. Sci. Med.* **67**:431–440. [17]

Petersen, L., M. Nordentoft, P. Jeppesen, et al. 2005. Improving 1-Year Outcome in First-Episode Psychosis: Opus Trial. *Br. J. Psychiatry Suppl.* **48**:s98–103. [07]

Petronis, A. 2010. Epigenetics as a Unifying Principle in the Aetiology of Complex Traits and Diseases. *Nature* **465**:721–727. [09]

Pezawas, L., and A. Meyer-Lindenberg. 2010. Imaging Genetics: Progressing by Leaps and Bounds. *NeuroImage* **53**:801–803. [08]

Pfammatter, M., U. M. Junghan, and H. D. Brenner. 2006. Efficacy of Psychological Therapy in Schizophrenia: Conclusions from Meta-Analyses. *Schizophr. Bull.* **32(Suppl. 1)**:S64–S68. [15]

Pfisterer, U., A. Kirkeby, O. Torper, et al. 2011. Direct Conversion of Human Fibroblasts to Dopaminergic Neurons. *PNAS* **108**:10343–10348. [10]

Pharoah, F., J. Mari, J. Rathbone, and W. Wong. 2010. Family Intervention for Schizophrenia. *Cochrane Database Syst. Rev.* **12**:CD000088. [15]
Phillips, W. A., and S. M. Silverstein. 2003. Convergence of Biological and Psychological Perspectives on Cognitive Coordination in Schizophrenia. *Behav. Brain Sci.* **26**:65–137. [01, 02, 07]
———. 2013. The Coherent Organization of Mental Life Depends on Mechanisms for Context-Sensitive Gain-Control That Are Impaired in Schizophrenia. *Front. Psychol.* **4**:307. [01]
Picton, T. W., C. Alain, L. Otten, W. Ritter, and A. Achim. 2000. Mismatch Negativity: Different Water in the Same River. *Audiol. Neurootol.* **5**:111–139. [07]
Pidsley, R., and J. Mill. 2011. Epigenetic Studies of Psychosis: Current Findings, Methodological Approaches, and Implications for Postmortem Research. *Biol. Psychiatry* **69**:146–156. [09]
Pilling, S., P. Bebbington, E. Kuipers, et al. 2002. Psychological Treatments in Schizophrenia: I. Meta-Analysis of Family Intervention and Cognitive Behaviour Therapy. *Psychol. Med.* **32**:763–782. [17]
Pilowsky, L. S., D. C. Costa, P. J. Ell, et al. 1993. Antipsychotic Medication, D2 Dopamine Receptor Blockade and Clinical Response: A 123i IBZM SPET (Single Photon Emission Tomography) Study. *Psychol. Med.* **23**:791–797. [07]
Pineda, J. A., D. Brang, E. Hecht, et al. 2008. Positive Behavioral and Electrophysiological Changes Following Neurofeedback Training in Children with Autism. *Res. Autism Spectr. Disord.* **2**:557–581. [07]
Piomelli, D., C. Pilon, B. Giros, et al. 1991. Dopamine Activation of the Arachidonic Acid Cascade as a Basis for D1/D2 Receptor Synergism. *Nature* **353**:164–167. [07]
Pitschel-Walz, G., S. Leucht, J. Bäuml, W. Kissling, and R. R. Engel. 2001. The Effect of Family Interventions on Relapse and Rehospitalization in Schizophrenia: A Meta-Analysis. *Schizophr. Bull.* **27**:73–92. [15]
Planck, M. 1949. The Meaning and Limits of Exact Science. *Science* **110**:325. [05]
Poduri, A., and D. Lowenstein. 2011. Epilepsy Genetics: Past, Present, and Future. *Curr. Opin. Genet. Dev.* **21**:325–332. [13]
Pollmacher, T., M. Haack, A. Schuld, T. Kraus, and D. Hinze-Selch. 2000. Effects of Antipsychotic Drugs on Cytokine Networks. *J. Psychiatr. Res.* **34**:369–382. [07]
Pompili, M., X. Amador, P. Girardi, et al. 2007. Suicide Risk in Schizophrenia: Learning from the Past to Change the Future. *Ann Gen Psychiatry* **6**:1–22. [17]
Pool, D., W. Bloom, D. Mielke, J. J. Roniger, and D. Callant. 1976. A Controlled Evaluation of Lozitane in Seventy-Five Adolescent Schizophrenia Patients. *Curr. Ther. Res. Clin. Exp.* **19**:99–104. [04]
Porteous, D. J., J. K. Millar, N. J. Brandon, and A. Sawa. 2011. DISC1 at 10: Connecting Psychiatric Genetics and Neuroscience. *Trends Molec. Med.* **17**:699–706. [08]
Posthuma, D., E. J. Mulder, D. I. Boomsma, and E. J. de Geus. 2002. Genetic Analysis of IQ, Processing Speed and Stimulus-Response Incongruency Effects. *Biol. Psychol.* **61**:157–182. [07]
Powell, S. B., and M. A. Geyer. 2002. Developmental Markers of Psychiatric Disorders as Identified by Sensorimotor Gating. *Neurotox. Res.* **4**:489–502. [07]
Power, P. J., R. J. Bell, R. Mills, et al. 2003. Suicide Prevention in First Episode Psychosis: The Development of a Randomised Controlled Trial of Cognitive Therapy for Acutely Suicidal Patients with Early Psychosis. *Aust. NZ J. Psychiatry* **37**:414–420. [07]

President's New Freedom Commission on Mental Health. 2003. Achieving the Promise: Transforming Mental Health Care in America: Final Report. http://www.promotingexcellence.org/downloads/mass/presidents_summary.pdf. (accessed June 11, 2013). [15, 17]

Prikryl, R., M. Mikl, H. Prikrylova Kucerova, et al. 2012. Does Repetitive Transcranial Magnetic Stimulation Have a Positive Effect on Working Memory and Neuronal Activation in Treatment of Negative Symptoms of Schizophrenia? *Neuro. Endocrinol. Lett.* **33**:90–97. [09]

Proctor, E. K., J. Landsverk, G. A. Aarons, et al. 2009. Implementation Research in Mental Health Services: An Emerging Science with Conceptual, Methodological, and Training Challenges. *Adm. Policy Ment. Health* **36**:24–34. [15]

Pruessner, J. C., F. Champagne, M. J. Meaney, and A. Dagher. 2004. Dopamine Release in Response to a Psychological Stress in Humans and Its Relationship to Early Life Maternal Care: A Positron Emission Tomography Study Using [11C]Raclopride. *J. Neurosci.* **24**:2825–2831. [08]

Psychiatric GWAS Consortium Steering Committee. 2009. A Framework for Interpreting Genome-Wide Association Studies of Psychiatric Disorders. *Mol. Psychiatry* **14**:10–17. [08]

Pukrop, R., F. Schultze-Lutter, S. Ruhrmann, et al. 2006. Neurocognitive Functioning in Subjects at Risk for a First Episode of Psychosis Compared with First- and Multiple-Episode Schizophrenia. *J. Clin. Exp. Neuropsychol.* **28**:1388–1407. [07]

Purcell, S. M., N. R. Wray, J. L. Stone, et al. 2009. Common Polygenic Variation Contributes to Risk of Schizophrenia and Bipolar Disorder. *Nature* **460**:748–752. [07, 09]

Qiang, L., R. Fujita, T. Yamashita, et al. 2011. Directed Conversion of Alzheimer's Disease Patient Skin Fibroblasts into Functional Neurons. *Cell* **146**:359–371. [10]

Quednow, B. B., I. Frommann, J. Berning, et al. 2008. Impaired Sensorimotor Gating of the Acoustic Startle Response in the Prodrome of Schizophrenia. *Biol. Psychiatry* **64**:766–773. [07]

Racenstein, J. M., M. Harrow, R. Reed, et al. 2002. The Relationship between Positive Symptoms and Instrumental Work Functioning in Schizophrenia: A 10-Year Follow-up Study. *Schizophr. Res.* **56**:95–103. [15]

Raj, A., and A. van Oudenaarden. 2008. Nature, Nurture, or Chance: Stochastic Gene Expression and Its Consequences. *Cell* **135**:216–226. [13]

Rajesh, D., S. J. Dickerson, J. Yu, et al. 2011. Human Lymphoblastoid B-Cell Lines Reprogrammed to EBV-Free Induced Pluripotent Stem Cells. *Blood* **118**:1797–1800. [10]

Ralph, R. O. 2000. Recovery. *Psychiatric Rehab. Skills* **4**:480–517. [15]

Ramsden, S., F. M. Richardson, G. Josse, et al. 2011. Verbal and Non-Verbal Intelligence Changes in the Teenage Brain. *Nature* **479**:113–116. [17]

Ransohoff, R. M., and V. H. Perry. 2009. Microglial Physiology: Unique Stimuli, Specialized Responses. *Annu. Rev. Immunol.* **27**:119–145. [07]

Rapaport, M. H., K. K. Delrahim, C. J. Bresee, et al. 2005. Celecoxib Augmentation of Continuously Ill Patients with Schizophrenia. *Biol. Psychiatry* **57**:1594–1596. [07]

Rasetti, R., F. Sambataro, Q. Chen, et al. 2011. Altered Cortical Network Dynamics: A Potential Intermediate Phenotype for Schizophrenia and Association with ZNF804A. *Arch. Gen. Psychiatry* **68**:1207–1217. [09]

Rasser, P. E., U. Schall, J. Todd, et al. 2011. Gray Matter Deficits, Mismatch Negativity, and Outcomes in Schizophrenia. *Schizophr. Bull.* **37**:131–140. [07]

Rawson, N. E., G. Gomez, B. Cowart, et al. 1997. Selectivity and Response Characteristics of Human Olfactory Neurons. *J. Neurophysiol.* **77**:1606–1613. [10]

RDoC. 2011. NIMH Research Domain Criteria. http://www.nimh.nih.gov/research-funding/rdoc/nimh-research-domain-criteria-rdoc.shtml. (accessed June 4, 2013). [03, 05]

Realmuto, G., W. Erickson, A. Yellin, J. H. Hopwood, and L. Greenberg. 1984. Clinical Comparison of Thiothixene and Thiorizadine in Schizophrenia Adolescents. *Am. J. Psychiatr.* **141**:440–442. [04]

Reddy, F., W. D. Spaulding, M. A. Jansen, A. A. Menditto, and S. Pickett. 2010. Psychologists' Roles and Opportunities in Rehabilitation and Recovery for Serious Mental Illness: A Survey of Council of University Directors of Clinical Psychology (CUDCP) Clinical Psychology Training and Doctoral Education. *Train. Educ. Prof. Psychol.* **4**:254–263. [15]

Reeder, C., E. Newton, S. Frangou, and T. Wykes. 2004. Which Cognitive Skills Should We Target to Effect Social Functioning Change? A Study of a Cognitive Remediation Therapy Programme. *Schizophr. Bull.* **30**:87–100. [01]

Regier, D. A., C. T. Kaelber, D. S. Rae, et al. 1998. Limitations of Diagnostic Criteria and Assessment Instrument for Mental Disorders. Implications for Research and Policy. *Arch. Gen. Psychiatry* **55**:109–115. [16]

Regier, D. A., E. A. Kuhl, and D. J. Kupfer. 2013. The DSM-5: Classification and Criteria Changes. *World Psychiatry* **12**:92–99. [16]

Regier, D. A., W. E. Narrow, D. S. Rae, et al. 1993. The *De Facto* US Mental and Addictive Disorders Service System. Epidemiologic Catchment Area Prospective 1-Year Prevalence Rates of Disorders and Services. *Arch. Gen. Psychiatry* **50**:85–94. [16]

Reichenberg, A., A. Caspi, H. Harrington, et al. 2010. Static and Dynamic Cognitive Deficits in Childhood Preceding Adult Schizophrenia: A 30-Year Study. *Am. J. Psychiatr.* **167**:160–169. [09, 14, 17]

Reichenberg, A., M. Weiser, J. Rabinowitz, et al. 2002. A Population-Based Cohort Study of Premorbid Intellectual, Language, and Behavioral Functioning in Patients with Schizophrenia, Schizoaffective Disorder, and Nonpsychotic Bipolar Disorder. *Am. J. Psychiatr.* **159**:2027–2035. [14, 17]

Reid, M. A., L. E. Stoeckel, D. M. White, et al. 2010. Assessments of Function and Biochemistry of the Anterior Cingulate Cortex in Schizophrenia. *Biol. Psychiatry* **68**:625–633. [07]

ReliaSoft. 2012. Fault Tree Diagrams and System Analysis: Reliawiki. http://reliawiki.com/index.php/Fault_Tree_Diagrams_and_System_Analysis. (accessed May 25, 2012). [02]

Remijnse, P. L., M. M. Nielen, H. B. Uylings, and D. J. Veltman. 2005. Neural Correlates of a Reversal Learning Task with an Affectively Neutral Baseline: An Event-Related fMRI Study. *NeuroImage* **26**:609–618. [05]

Renes, R. A., L. Vermeulen, R. S. Kahn, H. Aarts, and N. E. van Haren. 2013. Abnormalities in the Establishment of Feeling of Self-Agency in Schizophrenia. *Schizophr. Res.* **143**:50–54. [01]

Reser, J. E. 2007. Schizophrenia and Phenotypic Plasticity: Schizophrenia May Represent a Predictive, Adaptive Response to Severe Environmental Adversity That Allows Both Bioenergetic Thrift and a Defensive Behavioral Strategy. *Med. Hypotheses* **69**:383–394. [01]

Resnick, S. G., R. A. Rosenheck, L. Dixon, and A. F. Lehman. 2005. Correlates for Family Contact with the Mental Health System: Allocation of a Scarce Resource. *Ment. Health Serv. Res.* **7**:113–121. [15]

Resnick, S. G., R. A. Rosenheck, and A. F. Lehmann. 2004. An Exploratory Analysis of Correlates of Recovery. *Psychiatr. Serv.* **55**:540–547. [16]

Restrepo, D., Y. Okada, J. H. Teeter, et al. 1993. Human Olfactory Neurons Respond to Odor Stimuli with an Increase in Cytoplasmic Ca2+. *Biophys. J.* **64**:1961–1966. [10]

Revheim, N., P. D. Butler, I. Schechter, et al. 2006. Reading Impairment and Visual Processing Deficits in Schizophrenia. *Schizophr. Res.* **87**:238–245. [01]

Richardson, A. J. 1994. Dyslexia, Handedness and Syndromes of Psychosis-Proneness. *Int. J. Psychophysiol.* **18**:251–263. [01]

Richfield, E. K., J. B. Penney, and A. B. Young. 1989. Anatomical and Affinity State Comparisons between Dopamine D1 and D2 Receptors in the Rat Central Nervous System. *Neuroscience* **30**:767–777. [12]

Richtand, N. M., and R. K. McNamara. 2008. Serotonin and Dopamine Interactions in Psychosis Prevention. *Prog. Brain Res.* **172**:141–153. [07]

Ridler, K., J. M. Veijola, P. Tanskanen, et al. 2006. Fronto-Cerebellar Systems Are Associated with Infant Motor and Adult Executive Functions in Healthy Adults but Not in Schizophrenia. *PNAS* **103**:15651–15656. [06]

Rietschel, M., M. Mattheisen, J. Frank, et al. 2010. Genome-Wide Association-, Replication-, and Neuroimaging Study Implicates Homer1 in the Etiology of Major Depression. *Biol. Psychiatry* **68**:578–585. [08]

Riley, E. M., D. McGovern, D. Mockler, et al. 2000. Neuropsychological Functioning in First-Episode Psychosis: Evidence of Specific Deficits. *Schizophr. Res.* **43**:47–55. [04]

Ring, K. L., L. M. Tong, M. E. Balestra, et al. 2012. Direct Reprogramming of Mouse and Human Fibroblasts into Multipotent Neural Stem Cells with a Single Factor. *Cell Stem Cell* **11**:100–109. [10]

Ripke, S., A. R. Sanders, K. S. Kendler, et al. 2011. Genome-Wide Association Study Identifies Five New Schizophrenia Loci. *Nat. Genet.* **43**:969–976. [08]

Risch, N., and K. Merikangas. 1996. The Future of Genetic Studies of Complex Human Diseases. *Science* **273**:1516–1517. [08]

Ritsner, M. S., A. Gibel, T. Shleifer, and I. Boguslavsky. 2010. Pregnenolone and Dehydroepiandrosterone as an Adjunctive Treatment in Schizophrenia and Schizoaffective Disorder. *J. Clin. Psychiatry* **71**:1351–1362. [04]

Robertson, S. A., A. S. Care, and R. J. Skinner. 2007. Interleukin 10 Regulates Inflammatory Cytokine Synthesis to Protect against Lipopolysaccharide-Induced Abortion and Fetal Growth Restriction in Mice. *Biol. Reprod.* **76**:738–748. [07]

Robinson, D. G., M. G. Woerner, J. M. Alvir, et al. 1999. Predictors of Relapse Following Response from a First Episode of Schizophrenia or Schizoaffective Disorder. *Arch. Gen. Psychiatry* **56**:241–247. [06]

Robinson, D. G., M. G. Woerner, M. McMeniman, A. Mendelowitz, and R. M. Bilder. 2004. Symptomatic and Functional Recovery from a First Episode of Schizophrenia or Schizoaffective Disorder. *Am. J. Psychiatr.* **161**:473–479. [06]

Rocha, K. K., A. M. Ribeiro, K. C. Rocha, et al. 2012. Improvement in Physiological and Psychological Parameters after 6 Months of Yoga Practice. *Conscious Cogn.* **21**:843–850. [07]

Roder, C. H., J. M. Hoogendam, and F. M. van der Veen. 2010. fMRI, Antipsychotics and Schizophrenia. Influence of Different Antipsychotics on BOLD-Signal. *Curr. Pharm. Des.* **16**:2012–2025. [07]

Roder, V., D. R. Mueller, and S. J. Schmidt. 2011a. Effectiveness of Integrated Psychological Therapy (IPT) for Schizophrenia Patients: A Research Update. *Schizophr. Bull.* **37(Suppl. 2)**:S71–S79. [15]

Roder, V., D. R. Müller, H. D. Brenner, and W. D. Spaulding. 2011b. Integrated Psychological Therapy (IPT) for the Treatment of Neurocognition, Social Cognition, and Social Competency in Schizophrenia Patients. Cambridge, MA: Hogreffe. [15]

Roe, D. 2001. Progressing from "Patienthood" to "Personhood" across the Multi-Dimensional Outcomes in Schizophrenia and Related Disorders. *J. Nerv. Ment. Dis.* **189**:691–699. [06]

Roe, D., and M. Chopra. 2003. Beyond Coping with Mental Illness: Toward Personal Growth. *Am. J. Orthopsychiat.* **73**:334–344. [15]

Rogers, E. 2003. Diffusion of Innovations (5th edition). New York: Free Press. [15]

Rogers, E. S., J. Chamberlin, M. L. Ellison, and T. Crean. 1997. A Consumer-Constructed Scale to Measure Empowerment among Users of Mental Health Services. *Psychiatr. Serv.* **48**:1042–1047. [16]

Roisen, F. J., K. M. Klueber, C. L. Lu, et al. 2001. Adult Human Olfactory Stem Cells. *Brain. Res.* **890**:11–22. [10]

Rorick-Kehn, L. M., B. G. Johnson, J. L. Burkey, et al. 2007. Pharmacological and Pharmacokinetic Properties of a Structurally Novel, Potent, and Selective Metabotropic Glutamate 2/3 Receptor Agonist: *In Vitro* Characterization of Agonist (-)-(1r,4S,5S,6S)-4-Amino-2-Sulfonylbicyclo[3.1.0]-Hexane-4,6-Dicarboxylic Acid (LY404039). *J. Pharmacol. Exp. Ther.* **321**:308–317. [07]

Rose, G. 1985. Sick Individuals and Sick Populations. *Int. J. Epidemiol.* **14**:32–38. [09]

Rosen, A., K. T. Mueser, and M. Teeson. 2007. Assertive Community Treatment: Issues from Scientific and Clinical Literature with Implications for Practice. *J. Rehabil. Res. Dev.* **44**:813–826. [15]

Rosenberg, D. R., and D. A. Lewis. 1995. Postnatal Maturation of the Dopaminergic Innervation of Monkey Prefrontal and Motor Cortices: A Tyrosine Hydroxylase Immunohistochemical Analysis. *J. Comp. Neurol.* **358**:383–400. [13]

Roser, P., F. X. Vollenweider, and W. Kawohl. 2010. Potential Antipsychotic Properties of Central Cannabinoid (CB1) Receptor Antagonists. *World J. Biol. Psychiatry* **11**:208–219. [07]

Rössler, W. 2000. Schizophrenie: Psychosoziale Einflussfaktoren. In: Psychiatrie Der Gegenwart V, ed. N. Sartorius, pp. 181–192. Heidelberg: Springer. [16]

———. 2006. Psychiatric Rehabilitation Today: An Overview. *World Psychiatry* **5**:151–157. [16, 17]

———. 2012. Stress, Burnout, and Job Dissatisfaction in Mental Health Workers. *Eur. Arch. Psychiatry Clin. Neurosci.* **262**:S65–S69. [16]

Rössler, W., M. P. Hengartner, V. Ajdacic-Gross, H. Haker, and J. Angst. 2013. Lifetime and 12-Month Prevalence Rates of Sub-Clinical Psychosis Symptoms in a Community Cohort of 50-Year-Old Individuals. *Eur. Psychiatry* **28**:202–307. [16]

Rössler, W., M. P. Hengartner, V. Ajdacic-Gross, et al. 2011. Sub-Clinical Psychosis Symptoms in Young Adults Are Risk Factors for Subsequent Common Mental Disorders. *Schizophr. Res.* **131**:18–23. [16]

Rössler, W., M. P. Hengartner, J. Angst, and V. Ajdacic-Grosset. 2012. Linking Substance Use with Symptoms of Subclinical Psychosis in a Community Cohort over 30 Years. *Addiction* **107**:1174–1184. [16]

Rössler, W., A. Riecher, W. Löffler, and B. Fätkenheuer. 1991. Community Care in Child Psychiatry: An Empirical Approach Using the Concept of Travel Time. *Soc. Psychiatry Psychiatr. Epidem.* **26**:28–33. [16]

Rössler, W., A. Riecher-Rössler, J. Angst, et al. 2007. Psychotic Experiences in the General Population: A Twenty-Year Prospective Community Study. *Schizophr. Res.* **92**:1–14. [01, 16]

Rounis, E., K. E. Stephan, L. Lee, et al. 2006. Acute Changes in Frontoparietal Activity after Repetitive Transcranial Magnetic Stimulation over the Dorsolateral Prefrontal Cortex in a Cued Reaction Time Task. *J. Neurosci.* **26**:9629–9638. [09]

Rowland, L. M., K. Kontson, J. West, et al. 2012. *In Vivo* Measurements of Glutamate, GABA, and NAAG in Schizophrenia. *Schizophr. Bull.* doi: 10.1093/schbul/sbs1092. [07]

Rujescu, D., A. Ingason, S. Cichon, et al. 2008. Disruption of the Neurexin 1 Gene Is Associated with Schizophrenia. *Hum. Mol. Genet.* **18**:988–996. [05]

Rund, B. R. 1998. A Review of Longitudinal Studies of Cognitive Function in Schizophrenia Patients. *Schizophr. Bull.* **24**: [07]

Rupp, A., and S. J. Keith. 1993. The Costs of Schizophrenia: Assessing the Burden. *Psychiatr. Clin. North Am.* **16**:413–423. [01]

Rüsch, N., P. W. Corrigan, and A. Wassel. 2009. Self-Stigma, Group Identification, Perceived Legitimacy of Discrimination and Mental Health Service Use. *Br. J. Psychiatry* **195**:551–552. [16]

Russell-Smith, S. N., M. T. Maybery, and D. M. Bayliss. 2010. Are the Autism and Positive Schizotypy Spectra Diametrically Opposed in Local Versus Global Processing? *J. Autism Dev. Disord.* **40**:968–977. [01]

Rutten, B. P., and J. Mill. 2009. Epigenetic Mediation of Environmental Influences in Major Psychotic Disorders. *Schizophr. Bull.* **35**:1045–1056. [07, 09]

Rutter, M. 1985. Resilience in the Face of Adversity. Protective Factors and Resistance to Psychiatric Disorder. *Br. J. Psychiatry* **147**:598–611. [06]

Sabb, F. W., T. G. van Erp, M. E. Hardt, et al. 2010. Language Network Dysfunction as a Predictor of Outcome in Youth at Clinical High Risk for Psychosis. *Schizophr. Res.* **116**:173–183. [07]

Sachs, G., D. Steger-Wuchse, I. Kryspin-Exner, R. C. Gur, and H. Katschnig. 2004. Facial Recognition Deficits and Cognition in Schizophrenia. *Schizophr. Res.* **68**:27–35. [04]

Sackett, D. L., W. S. Richardson, W. Rosenberg, and R. B. Haynes. 1997. Evidence-Based Medicine. New York: Churchill Livingstone. [15]

Saldivia, S., B. Vicente, R. Kohn, P. Rioseco, and S. Torres. 2004. Use of Mental Health Services in Chile. *Psychiatr. Serv.* **55**:71–76. [16]

Salisbury, D. F., N. Kuroki, K. Kasai, M. E. Shenton, and R. W. McCarley. 2007. Progressive and Interrelated Functional and Structural Evidence of Post-Onset Brain Reduction in Schizophrenia. *Arch. Gen. Psychiatry* **64**:521–529. [07]

Salisbury, D. F., M. E. Shenton, C. B. Griggs, A. Bonner-Jackson, and R. W. McCarley. 2002. Mismatch Negativity in Chronic Schizophrenia and First-Episode Schizophrenia. *Arch. Gen. Psychiatry* **59**:686–694. [07]

Salize, H. J., M. R., J. Bullenkamp, et al. 2009. Cost of Treatment in Six European Countries. *Schizophr. Res.* **111**:70–77. [01, 16]

Salize, H. J., R. McCabe, J. Bullenkamp, et al. 2009. Cost of Treatment in Six European countries. *Schiz. Res.* **111**:70–77. [16]

Salize, H. J., and W. Rössler. 1996. The Cost of Comprehensive Care of People with Schizophrenia Living in the Community. A Cost Evaluation from a German Catchment Area. *Br. J. Psychiatry* **169**:42–48. [16]

Salkever, D. S., M. C. Karakus, E. P. Slade, et al. 2007. Measures and Predictors of Community-Based Employment and Earnings of Persons with Schizophrenia in a Multisite Study. *Psychiatr. Serv.* **58**:315–324. [05]

Sampson, R. J., and W. B. Groves. 1989. Community Structure and Crime: Testing Social Disorganization Theory. *Am. J. Sociol.* **94**:774–802. [01]

Sandefur, R., and E. O. Laumann. 1998. A Paradigm for Social Capital. *Rational. Soc.* **10**:481–501. [01]

Sass, L. A., and J. Parnas. 2009. Schizophrenia, Consciousness, and the Self. *Schizophr. Bull.* **29**:427–444. [01]

Sato, M. 2006. Renaming Schizophrenia: A Japanese Perspective. *World Psychiatry* **5**:53–55. [05]

Sattler, R., Y. Ayukawa, L. Coddington, et al. 2011. Human Nasal Olfactory Epithelium as a Dynamic Marker for CNS Therapy Development. *Exp. Neurol.* **232**:203–211. [10]

Sawa, A., and N. G. Cascella. 2009. Peripheral Olfactory System for Clinical and Basic Psychiatry: A Promising Entry Point to the Mystery of Brain Mechanism and Biomarker Identification in Schizophrenia. *Am. J. Psychiatr.* **166**:137–139. [10]

Sawamura, N., T. Ando, Y. Maruyama, et al. 2008. Nuclear DISC1 Regulates Cre-Mediated Gene Transcription and Sleep Homeostasis in the Fruit Fly. *Mol. Psychiatry* **13**:1138–1148, 1069. [10]

Saxena, S., E. Gorbis, J. O'Neill, et al. 2009. Rapid Effects of Brief Intensive Cognitive-Behavioral Therapy on Brain Glucose Metabolism in Obsessive-Compulsive Disorder. *Mol. Psychiatry* **14**:197–205. [07]

Schall, U., S. V. Catts, F. Karayanidis, and P. B. Ward. 1999. Auditory Event-Related Potential Indices of Fronto-Temporal Information Processing in Schizophrenia Syndromes: Valid Outcome Prediction of Clozapine Therapy in a Three-Year Follow-Up. *Int. J. Neuropsychopharmacol.* **2**:83–93. [07]

Scharinger, C., U. Rabl, H. H. Sitte, and L. Pezawas. 2010. Imaging Genetics of Mood Disorders. *NeuroImage* **53**:810–821. [08]

Scheff, T. 1966. Being Mentally Ill: A Sociology Theory. Chicago: Aldine. [17]

Schenkel, L. S., and S. Silverstein. 2004. Dimensions of Premorbid Functioning in Schizophrenia: A Review of Neuromotor, Cognitive, Social, and Behavioral Domains. *Genet. Soc. Gen. Psychol. Monogr.* **130**:241–270. [01]

Schiffman, J., J. A. Maeda, K. Hayashi, et al. 2006. Premorbid Childhood Ocular Alignment Abnormalities and Adult Schizophrenia-Spectrum Disorder. *Schizophr. Res.* **81**:253–260. [01]

Schnack, H. G., M. Nieuwenhuis, E. M. Neeltje, et al. submitted. Can Structural MRI Aid in Clinical Classification? A Machine Learning Study in Two Independent Samples of Patients with Schizophrenia, Bipolar Disorder and Healthy Subjects. [17]

Schneider, K. 1959. Clinical Psychopathology (translated by M. W. Hamilton and E. W. Anderson). New York: Grune & Stratton. [03]

Schork, N. J., T. A. Greenwood, and D. L. Braff. 2007. Statistical Genetics Concepts and Approaches in Schizophrenia and Related Neuropsychiatric Research. *Schizophr. Bull.* **33**:95–104. [03]

Schubert, E. W., K. M. Henriksson, and T. F. McNeil. 2005. A Prospective Study of Offspring of Women with Psychosis: Visual Dysfunction in Early Childhood Predicts Schizophrenia-Spectrum Disorders in Adulthood. *Acta Psychiatr. Scand.* **112**:385–393. [01]

Schultze-Lutter, F. 2009. Subjective Symptoms of Schizophrenia in Research and the Clinic: The Basic Symptom Concept. *Schizophr. Bull.* **35**:5–8. [17]

Schulze, B., and W. Rössler. 2005. Caregiver Burden in Mental Illness: Review of Measurement, Findings and Interventions in 2004–2005. *Curr. Opin. Psychiatry* **18**:684–691. [17]

Schwartz, J. M., P. W. Stoessel, L. R. Baxter, Jr., K. M. Martin, and M. E. Phelps. 1996. Systematic Changes in Cerebral Glucose Metabolic Rate after Successful Behavior Modification Treatment of Obsessive-Compulsive Disorder. *Arch. Gen. Psychiatry* **53**:109–113. [07]

Schwartz, S., and E. Susser. 2006. Relationships among Causes. In: Psychiatric Epidemiology: Searching for the Causes of Mental Disorders, ed. E. Susser et al., pp. 62–74. Oxford: Oxford Univ. Press. [09]

Scott, A., B. A. Wowra, and R. McCarter. 1999. Validation of the Empowerment Scale with an Outpatient Mental Health Population. Psychiatric Services. *Psychiatr. Serv.* **50**:959–961. [16]

Seamans, J. K., D. Durstewitz, B. R. Christie, C. F. Stevens, and T. J. Sejnowski. 2001a. Dopamine D1/D5 Receptor Modulation of Excitatory Synaptic Inputs to Layer V Prefrontal Cortex Neurons. *PNAS* **98**:301–306. [12]

Seamans, J. K., N. Gorelova, D. Durstewitz, and C. R. Yang. 2001b. Bidirectional Dopamine Modulation of Gabaergic Inhibition in Prefrontal Cortical Pyramidal Neurons. *J. Neurosci.* **21**:3628–3638. [12]

Sebat, J., D. L. Levy, and S. E. McCarthy. 2009. Rare Structural Variants in Schizophrenia: One Disorder, Multiple Mutations; One Mutation, Multiple Disorders. *Trends Genet.* **25**:528–535. [01, 07]

Seeman, P. 2011. All Roads to Schizophrenia Lead to Dopamine Supersensitivity and Elevated Dopamine D2(High) Receptors. *CNS Neurosci. Ther.* **17**:118–132. [13]

Seeman, P., and H. C. Guan. 2009. Glutamate Agonist LY404,039 for Treating Schizophrenia Has Affinity for the Dopamine D2(High) Receptor. *Synapse* **63**:935–939. [07]

Seeman, P., J. Schwarz, J. F. Chen, et al. 2006. Psychosis Pathways Converge via D2high Dopamine Receptors. *Synapse* **60**:319–346. [12]

Seidman, L. J., A. J. Giuliano, E. C. Meyer, et al. 2010. Neuropsychology of the Prodrome to Psychosis in the NAPLS Consortium: Relationship to Family History and Conversion to Psychosis. *Arch. Gen. Psychiatry* **67**:578–588. [02, 07]

Seidman, L. J., H. W. Thermenos, R. A. Poldrack, et al. 2006. Altered Brain Activation in Dorsolateral Prefrontal Cortex in Adolescents and Young Adults at Genetic Risk for Schizophrenia: An fMRI Study of Working Memory. *Schizophr. Res.* **85**:58–72. [07]

Seki, T., S. Yuasa, M. Oda, et al. 2010. Generation of Induced Pluripotent Stem Cells from Human Terminally Differentiated Circulating T Cells. *Cell Stem Cell* **7**:11–14. [10]

Selten, J. P., and E. Cantor-Graae. 2005. Social Defeat: Risk Factor for Schizophrenia? *Br. J. Psychiatry* **187**:101–102. [09]

Serretti, A., M. Kato, D. De Ronchi, and T. Kinoshita. 2007. Meta-Analysis of Serotonin Transporter Gene Promoter Polymorphism (5-HTTLPR) Association with Selective Serotonin Reuptake Inhibitor Efficacy in Depressed Patients. *Mol. Psychiatry* **12**:247–257. [08]

Sesack, S. R., and D. B. Carr. 2002. Selective Prefrontal Cortex Inputs to Dopamine Cells: Implications for Schizophrenia. *Physiol. Behav.* **77**:513–517. [07]

Sesack, S. R., D. B. Carr, N. Omelchenko, and A. Pinto. 2003. Anatomical Substrates for Glutamate-Dopamine Interactions: Evidence for Specificity of Connections and Extrasynaptic Actions. *Ann. NY Acad. Sci.* **1003**:36–52. [07]

Seshadri, S., A. Kamiya, Y. Yokota, et al. 2010. Disrupted-in-Schizophrenia-1 Expression Is Regulated by Beta-Site Amyloid Precursor Protein Cleaving Enzyme-1-Neuregulin Cascade. *PNAS* **107**:5622–5627. [08]

Shahbazian, M. D., B. Antalffy, D. L. Armstrong, and H. Y. Zoghbi. 2002. Insight into Rett Syndrome: MeCP2 Levels Display Tissue- and Cell-Specific Differences and Correlate with Neuronal Maturation. *Hum. Mol. Genet.* **11**:115–124. [10]

Sharp, F. R., M. Tomitaka, M. Bernaudin, and S. Tomitaka. 2001. Psychosis: Pathological Activation of Limbic Thalamocortical Circuits by Psychomimetics and Schizophrenia? *Trends Neurosci.* **24**:330–334. [07]

Shaw, P., D. Greenstein, J. Lerch, et al. 2006a. Intellectual Ability and Cortical Development in Children and Adolescents. *Nature* **440**:676–679. [13]

Shaw, P., A. Sporn, N. Gogtay, et al. 2006b. Childhood-Onset Schizophrenia: A Double-Blind, Randomized Clozapine-Olanzapine Comparison. *Arch. Gen. Psychiatry* **63**:721–730. [04]

Sheldon, K. M., T. Kasser, K. Smith, and T. Share. 2002. Personal Goals and Psychological Growth: Testing an Intervention to Enhance Goal Attainment and Personality Integration. *J. Pers.* **70**:5–31. [15]

Shelley, A. M., P. B. Ward, S. V. Catts, et al. 1991. Mismatch Negativity: An Index of a Preattentive Processing Deficit in Schizophrenia. *Biol. Psychiatry* **30**:1059–1062. [07]

Shen, S., B. Lang, C. Nakamoto, et al. 2008. Schizophrenia-Related Neural and Behavioral Phenotypes in Transgenic Mice Expressing Truncated DISC1. *J. Neurosci.* **28**:10,893–10,904. [13]

Sicard, M. N., C. C. Zai, A. K. Tiwari, et al. 2010. Polymorphisms of the HTR2C Gene and Antipsychotic-Induced Weight Gain: An Update and Meta-Analysis. *Pharmacogenomics* **11**:1561–1571. [07]

Siegle, G. J., C. S. Carter, and M. E. Thase. 2006. Use of fMRI to Predict Recovery from Unipolar Depression with Cognitive Behavior Therapy. *Am. J. Psychiatr.* **163**:735–738. [09]

Sigurdsson, T., K. L. Stark, M. Karayiorgou, J. A. Gogos, and J. A. Gordon. 2010. Impaired Hippocampal-Prefrontal Synchrony in a Genetic Mouse Model of Schizophrenia. *Nature* **464**:763–767. [09, 11, 13]

Sikich, L., R. Hamer, R. Bashford, B. Sheitman, and J. Lieberman. 2004. A Pilot Study of Risperidone, Olanzapine and Haloperidol in Psychotic Youth: A Double-Blind, Randomized, 8-Week Trial. *Neuropsychopharmacology* **29**:133–145. [04]

Silverstein, S. M. 1993. Methodological and Empirical Considerations in Assessing the Validity of Psychoanalytic Theories of Hypnosis. *Genet. Soc. Gen. Psychol. Monogr.* **119**:5–54. [01]

———. 2000. Psychiatric Rehabilitation of Schizophrenia: Unresolved Issues, Current Trends and Future Directions. *Appl. Prev. Psychol.* **9**:227–248. [06]

———. 2008. Measuring Specific, Rather Than Generalized, Cognitive Deficits and Maximizing between-Group Effect Size in Studies of Cognition and Cognitive Change. *Schizophr. Bull.* **34**:645–655. [01, 09]

———. 2010. Failures of Dynamic Coordination in Disease States, and Their Implications for Normal Brain Function. In: Dynamic Coordination in the Brain: From Molecules to Mind, ed. C. von der Malsburg et al., Strüngmann Forum Report, J. Lupp, series ed., vol. 5, pp. 245–265. Cambridge, MA: MIT Press. [01]

Silverstein, S. M., S. D. All, J. L. Thompson, et al. 2012a. Absolute Level of Gamma Synchrony Is Increased in First Episode Schizophrenia During Face Processing. *J. Exp. Psychopath.* **3**:702–723. [04]

Silverstein, S. M., and A. S. Bellack. 2008. A Scientific Agenda for the Concept of Recovery as It Applies to Schizophrenia. *Clin. Psychol. Rev.* **28**:1108–1124. [01, 06, 15, 17]

Silverstein, S. M., S. Berten, B. Essex, et al. 2009a. An fMRI Examination of Visual Integration in Schizophrenia. *J. Integr. Neurosci.* **8**:175–202. [01, 02]

Silverstein, S. M., M. Hatashita-Wong, B. A. Solak, et al. 2005. Effectiveness of a Two-Phase Cognitive Rehabilitation Intervention for Severely Impaired Schizophrenia Patients. *Psychol. Med.* **35**:829–837. [01]

Silverstein, S. M., M. Hatashita-Wong, S. M. Wilkniss, et al. 2006a. Behavioral Rehabilitation of the Treatment-Refractory Schizophrenia Patient: Conceptual Foundations, Interventions and Outcome Data. *Psychol. Serv.* **3**:145–169. [17]

Silverstein, S. M., and B. P. Keane. 2009. Perceptual Organization in Schizophrenia: Plasticity and State-Related Change. *Learn. Percept.* **1**:229–261. [01]

———. 2011. Perceptual Organization Impairment in Schizophrenia and Associated Brain Mechanisms; Review of Research from 2005–2010. *Schizophr. Bull.* **37**:690–699. [07]

Silverstein, S. M., B. P. Keane, D. M. Barch, et al. 2012b. Optimization and Validation of a Visual Integration Test for Schizophrenia Research. *Schizophr. Bull.* **38**:125–134. [02]

Silverstein, S. M., B. P. Keane, Y. Wang, et al. 2013a. Visual Context Sensitivity Predicts Clinical State and Short-Term Treatment Responsiveness in Schizophrenia. *Front. Psychol.*, in press. [01]

Silverstein, S. M., I. Kovacs, R. Corry, and C. Valone. 2000. Perceptual Organization, the Disorganization Syndrome and Context Processing in Chronic Schizophrenia. *Schizophr. Res.* **43**:11–20. [02]

Silverstein, S. M., and D. R. Palumbo. 1995. Nonverbal Perceptual Organization Output Disability and Schizophrenia Spectrum Symptomatology. *Psychiatry* **58**:66–81. [01]

Silverstein, S. M., W. D. Spaulding, A. A. Menditto, et al. 2009b. Attention Shaping: A Reward-Based Learning Method to Enhance Skills Training Outcomes in Schizophrenia. *Schizophr. Bull.* **35**:222–232. [15]

Silverstein, S. M., P. J. Uhlhaas, B. Essex, et al. 2006b. Perceptual Organization in First Episode Schizophrenia and Ultra-High-Risk States. *Schizophr. Res.* **83**:41–52. [01]

Silverstein, S. M., Y. Wang, and B. P. Keane. 2012c. Cognitive and Neuroplasticity Mechanisms by Which Congenital or Early Blindness May Confer a Protective Effect against Schizophrenia. *Front Psychol.* **3**:624. [01]

Silverstein, S. M., Y. Wang, and M. W. Roché. 2013b. Base Rates, Blindness, and Schizophrenia. *Front. Psychol.* **4**:157. [01]

Simmers, A. J., and P. J. Bex. 2001. Deficit of Visual Contour Integration in Dyslexia. *Invest. Ophthalmol. Vis. Sci.* **42**:2737–2742. [01]

Simon, H. A. 1973. The Organization of Complex Systems. In: Hierarchy Theory: The Challenge of Complex Systems, ed. H. H. Pattee, pp. 1–27. New York: George Braziller. [01]

Singh, F., J. Pineda, and K. S. Cadenhead. 2011. Association of Impaired EEG Mu Wave Suppression, Negative Symptoms and Social Functioning in Biological Motion Processing in First Episode of Psychosis. *Schizophr. Res.* **130**:182–186. [07]

Singh, S. M., P. McDonald, B. Murphy, and R. O'Reilly. 2004. Incidental Neurodevelopmental Episodes in the Etiology of Schizophrenia: An Expanded Model Involving Epigenetics and Development. *Clin. Genet.* **65**:435–440. [13]

Sklar, P., S. Ripke, L. J. Scott, et al. 2011. Large-Scale Genome-Wide Association Analysis of Bipolar Disorder Identifies a New Susceptibility Locus near Odz4. *Nat. Genet.* **43**:977–983. [09]

Sluzki, C. E. 2007. Interfaces: Toward a New Generation of Systemic Models in Family Research and Practice. *Fam. Process.* **46**:173–184. [09]

Smith, R. G., R. L. Kember, J. Mill, et al. 2009. Advancing Paternal Age Is Associated with Deficits in Social and Exploratory Behaviors in the Offspring: A Mouse Model. *PLoS ONE* **4**:e8456. [08]

Smith, S. M., and P. T. Yanos. 2009. Psychotherapy for Schizophrenia in an Act Team Context. *Clinical Case Stud.* **8**:454–462. [01]

Smith, T. E., J. W. Hull, S. Romanelli, E. Fertuck, and K. A. Weiss. 1999. Symptoms and Neurocognition as Rate Limiters in Skills Training for Psychotic Patients. *Am. J. Psychiatr.* **156**:1817–1818. [15]

Snitz, B. E., A. W. MacDonald, III, and C. S. Carter. 2006. Cognitive Deficits in Unaffected First-Degree Relatives of Schizophrenia Patients: A Meta-Analytic Review of Putative Endophenotypes. *Schizophr. Bull.* **32**:179–194. [02, 04]

Soares, D. C., B. C. Carlyle, N. J. Bradshaw, and D. J. Porteous. 2011. DISC1: Structure, Function, and Therapeutic Potential for Major Mental Illness. *ACS Chem. Neurosci.* **2**:609–632. [13]

Soldner, F., D. Hockemeyer, C. Beard, et al. 2009. Parkinson's Disease Patient-Derived Induced Pluripotent Stem Cells Free of Viral Reprogramming Factors. *Cell* **136**:964–977. [10]

Son, E. Y., J. K. Ichida, B. J. Wainger, et al. 2011. Conversion of Mouse and Human Fibroblasts into Functional Spinal Motor Neurons. *Cell Stem Cell* **9**:205–218. [10]

Sørensen, H. J., D. Saebuy, A. Urfer-Parnas, E. L. Mortensen, and J. Parnas. 2012. Premorbid Intelligence and Educational Level in Bipolar and Unipolar Disorders: A Danish Draft Board Study. *J. Affect. Disord.* **136**:1188–1191. [14, 17]

Spaulding, W. D., M. E. Sullivan, and J. S. Poland. 2003. Treatment and Rehabilitation of Severe Mental Illness. New York: Guilford Press. [15, 17]

Spearman, C. 1904. "General Intelligence," Objectively Determined and Measured. *Am. J. Psychol.* **15**:201–293. [02]

Spencer, E., V. Kafantaris, M. Padron-Gaylo, C. Rosenberg, and M. Campell. 1992. Haloperidol in Schizophrenia Children: Early Findings from a Study in Progress. *Psychopharmacol. Bull.* **28**:183–186. [04]

Spencer, K. M., P. G. Nestor, R. Perlmutter, et al. 2004. Neural Synchrony Indexes Disordered Perception and Cognition in Schizophrenia. *PNAS* **101**:17,288–17,293. [04]

Spieker, E. A., R. S. Astur, J. T. West, J. A. Griego, and L. M. Rowland. 2012. Spatial Memory Deficits in a Virtual Reality Eight-Arm Radial Maze in Schizophrenia. *Schizophr. Res.* **135**:84–89. [17]

Staerk, J., M. M. Dawlaty, Q. Gao, et al. 2010. Reprogramming of Human Peripheral Blood Cells to Induced Pluripotent Stem Cells. *Cell Stem Cell* **7**:20–24. [10]

Stangl, D., and S. Thuret. 2009. Impact of Diet on Adult Hippocampal Neurogenesis. *Genes Nutr.* **4**:271–282. [17]

Steen, R. G., C. Mull, R. McClure, R. M. Hamer, and J. A. Lieberman. 2006. Brain Volume in First-Episode Schizophrenia: Systematic Review and Meta-Analysis of Magnetic Resonance Imaging Studies. *Br. J. Psychiatry* **188**:510–518. [09]

Stein, L. I., and A. B. Santos. 1998. Assertive Community Treatment of Persons with Severe Mental Illness. New York: Norton. [15]

Steindler, D. A., and D. W. Pincus. 2002. Stem Cells and Neuropoiesis in the Adult Human Brain. *Lancet* **359**:1047–1054. [07]

Stephan, K. E., K. J. Friston, and C. D. Frith. 2009. Dysconnection in Schizophrenia: From Abnormal Synaptic Plasticity to Failures of Self-Monitoring. *Schizophr. Bull.* **35**:509–527. [13]

Stirling, J., C. White, S. Lewis, et al. 2003. Neurocognitive Function and Outcome in First-Episode Schizophrenia: A 10-Year Follow-up of an Epidemiological Cohort. *Schizophr. Res.* **65**:75–86. [04]

Stone, J. M., F. Day, H. Tsagaraki, et al. 2009. Glutamate Dysfunction in People with Prodromal Symptoms of Psychosis: Relationship to Gray Matter Volume. *Biol. Psychiatry* **66**:533–539. [07]

Stone, W. S., and L. Iguchi. 2011. Do Apparent Overlaps between Schizophrenia and Autistic Spectrum Disorders Reflect Superficial Similarities or Etiological Commonalities? *N. Am. J. Med. Sci.* **4**:124–133. [01]

Strauss, J. S. 1969. Hallucinations and Delusions as Points on Contunua Function. Rating Scale Evidence. *Arch. Gen. Psychiatry* **21**:581–586. [03]

Strauss, J. S., and W. T. Carpenter, Jr. 1972. The Prediction of Outcome in Schizophrenia I. Characteristics of Outcome. *Arch. Gen. Psychiatry* **27**:739–746. [15, 17]

———. 1975. The Key Clinical Dimensions of the Functional Psychoses. *Publ. Assoc. Res. Nerv. Ment. Dis.* **54**:9–17. [03]

———. 1977. Prediction of Outcome in Schizophrenia III. Five-Year Outcome and Its Predictors. *Arch. Gen. Psychiatry* **34**:159–163. [15]

———. 1981. Schizophrenia. New York: Plenum Medical Books. [03]

Strauss, J. S., W. T. Carpenter, Jr., and J. J. Bartko. 1974. The Diagnosis and Understanding of Schizophrenia. Part III. Speculations on the Processes That Underlie Schizophrenic Symptoms and Signs. *Schizophr. Bull.* **Winter**:61–69. [03]

Strauss, M. E. 2001. Demonstrating Specific Cognitive Deficits: A Psychometric Perspective. *J. Abnorm. Psychol.* **110**:6–14. [02]

Sturgeon, D., G. Turpin, L. Kuipers, R. Berkowitz, and J. Leff. 1984. Physiological Responses of Schizophrenic Patients to High and Low Expressed Emotion Relatives: A Follow-up Study. *Br. J. Psychiatry* **145**:62–69. [01]

Sturman, D. A., D. R. Mandell, and B. Moghaddam. 2010. Adolescents Exhibit Behavioral Differences from Adults During Instrumental Learning and Extinction. *Behav. Neurosci.* **124**:16–25. [13]

Sturman, D. A., and B. Moghaddam. 2011. The Neurobiology of Adolescence: Changes in Brain Architecture, Functional Dynamics, and Behavioral Tendencies. *Neurosci. Biobehav. Rev.* **35**:1704–1712. [13]

———. 2012. Striatum Processes Reward Differently in Adolescents Versus Adults. *PNAS* **109**:1719–1724. [13]

Sullivan, G., N. Duan, S. Mukherjee, et al. 2005. The Role of Services Researchers in Facilitating Intervention Research. *Psychiatr. Serv.* **56**:537–542. [15]

Sullivan, P. F. 2007. Spurious Genetic Associations. *Biol. Psychiatry* **61**:1121–1126. [08]

Sullivan, P. F., M. J. Daly, and M. O'Donovan. 2012a. Genetic Architectures of Psychiatric Disorders: The Emerging Picture and Its Implications. *Nat. Rev. Genet.* **13**:537–551. [09, 13]

Sullivan, P. F., K. S. Kendler, and M. C. Neale. 2003. Schizophrenia as a Complex Trait: Evidence from a Meta-Analysis of Twin Studies. *Arch. Gen. Psychiatry* **60**:1187–1192. [07, 08]

Sullivan, P. F., C. Magnusson, A. Reichenberg, et al. 2012b. Family History of Schizophrenia and Bipolar Disorder as Risk Factors for Autism. *Arch. Gen. Psychiatry* **69**:1099–1103. [05, 06]

Sullivan, P. F., M. C. Neale, and K. S. Kendler. 2000. Genetic Epidemiology of Major Depression: Review and Meta-Analysis. *Am. J. Psychiatr.* **157**:1552–1562. [08]

Swerdlow, N. R. 2011. Are We Studying and Treating Schizophrenia Correctly? *Schizophr. Res.* **130**:1–10. [07]

Swerdlow, N. R., D. L. Braff, and M. A. Geyer. 1999. Cross-Species Studies of Sensorimotor Gating of the Startle Reflex. *Ann. NY Acad. Sci.* **877**:202–216. [07]

Swerdlow, N. R., S. Caine, D. Braff, and M. Geyer. 1992. The Neural Substrates of Sensorimotor Gating of the Startle Reflex: A Review of Recent Findings and Their Implications. *J. Psychopharmacol.* **6**:176–190. [07]

Swerdlow, N. R., N. Halim, F. M. Hanlon, A. Platten, and P. P. Auerbach. 2001. Lesion Size and Amphetamine Hyperlocomotion after Neonatal Ventral Hippocampal Lesions: More Is Less. *Brain Res. Bull.* **55**:71–77. [11]

Swerdlow, N. R., S. A. Lelham, A. N. Sutherland Owens, et al. 2009a. Pramipexole Effects on Startle Gating in Rats and Normal Men. *Psychopharmacology* **205**:689–698. [07]

Swerdlow, N. R., G. A. Light, K. S. Cadenhead, et al. 2006. Startle Gating Deficits in a Large Cohort of Patients with Schizophrenia: Relationship to Medications, Symptoms, Neurocognition, and Level of Function. *Arch. Gen. Psychiatry* **63**:1325–1335. [07]

Swerdlow, N. R., N. Stephany, J. M. Shoemaker, et al. 2002. Effects of Amantadine and Bromocriptine on Startle and Sensorimotor Gating: Parametric Studies and Cross-Species Comparisons. *Psychopharmacology* **164**:82–92. [07]

Swerdlow, N. R., D. P. van Bergeijk, F. Bergsma, E. Weber, and J. Talledo. 2009b. The Effects of Memantine on Prepulse Inhibition. *Neuropsychopharmacology* **34**:1854–1864. [07]

Swerdlow, N. R., M. Weber, Y. Qu, G. A. Light, and D. L. Braff. 2008. Realistic Expectations of Prepulse Inhibition in Translational Models for Schizophrenia Research. *Psychopharmacology* **199**:331–388. [07]

Symond, M. P., A. W. Harris, E. Gordon, and L. M. Williams. 2005. "Gamma Synchrony" in First-Episode Schizophrenia: A Disorder of Temporal Connectivity? *Am. J. Psychiatr.* **162**:459–465. [04, 07]

Szoke, A., A. Trandafir, M.-E. Dupont, et al. 2008. Longitudinal Studies of Cognition in Schizophrenia: Meta-Analysis. *Br. J. Psychiatry* **192**:248–257. [04]

Tackett, J. L., A. L. Silberschmidt, R. F. Krueger, and S. R. Sponheim. 2008. A Dimensional Model of Personality Disorder: Incorporating DSM Cluster A Characteristics. *J. Abnorm. Psychol.* **117**:454–459. [02]

Tajinda, K., K. Ishizuka, C. Colantuoni, et al. 2010. Neuronal Biomarkers from Patients with Mental Illnesses: A Novel Method through Nasal Biopsy Combined with Laser-Captured Microdissection. *Mol. Psychiatry* **15**:231–232. [10]

Takahashi, K., and S. Yamanaka. 2006. Induction of Pluripotent Stem Cells from Mouse Embryonic and Adult Fibroblast Cultures by Defined Factors. *Cell* **126**:663–676. [10]

Talamo, B. R., R. Rudel, K. S. Kosik, et al. 1989. Pathological Changes in Olfactory Neurons in Patients with Alzheimer's Disease. *Nature* **337**:736–739. [10]

Talledo, J. A., A. N. Sutherland Owens, T. Schortinghuis, and N. R. Swerdlow. 2009. Amphetamine Effects on Startle Gating in Normal Women and Female Rats. *Psychopharmacology* **204**:165–175. [07]

Tan, P.-N., M. Steinbach, and V. Kumar. 2005. Introduction to Data Mining: Addison-Wesley. [05]
Tan, Z. S., W. S. Harris, A. S. Beiser, et al. 2012. Red Blood Cell Omega-3 Fatty Acid Levels and Markers of Accelerated Brain Aging. *Neurology* **78**:658–664. [07]
Tandon, R., and W. T. Carpenter, Jr. 2012. DSM-5 Status of Psychotic Disorders. *Schizophr. Bull.* **38**:369–370. [02]
Tandon, R., M. S. Keshavan, and H. Nasrallah. 2008. Schizophrenia, "Just the Facts": What We Know in 2008 Part 1: Overview. *Schizophr. Res.* **100**:4–19. [05, 13]
Tandon, R., H. A. Nasrallah, and M. S. Keshavan. 2011. "Just the Facts": Meandering in Schizophrenia's Many Forests. *Schizophr. Res.* **128**:5–6. [07]
Tarasenko, M., M. Sullivan, A. Ritchie, and W. Spauding. 2012. Effects of Eliminating Psychiatric Rehabilitation from the Secure Levels of a Mental Health Service System. *Psychol. Serv.* Epub ahead of print. [17]
Tarrier, N., C. Barrowclough, G. Haddock, and J. McGovern. 1999. The Dissemination of Innovative Cognitive-Behavioural Psychosocial Treatments for Schizophrenia. *J. Ment. Health* **8**:569–582. [15]
Tarrier, N., R. Beckett, S. Harwood, et al. 1993. A Trial of Two Cognitive Behavioural Methods of Treating Drug-Resistant Residual Psychotic Symptoms in Schizophrenic Patients: I. Outcome. *Br. J. Psychiatry* **162**:524–532. [15]
Taylor, M. J., S. Sen, and Z. Bhagwagar. 2010. Antidepressant Response and the Serotonin Transporter Gene-Linked Polymorphic Region. *Biol. Psychiatry* **68**:536–543. [08]
Tayoshi, S., S. Sumitani, K. Taniguchi, et al. 2009. Metabolite Changes and Gender Differences in Schizophrenia Using 3-Tesla Proton Magnetic Resonance Spectroscopy (1H-MRS). *Schizophr. Res.* **108**:69–77. [07]
Teague, G. B., K. T. Mueser, and C. A. Rapp. 2012. Advances in Fidelity Measurement for Mental Health Services Research: Four Measures. *Psychiatr. Serv.* **63**:765–771. [15]
Thaker, G. 2008. Neurophysiological Endophenotypes across Bipolar and Schizophrenia Psychosis. *Schizophr. Bull.* **34**:760–773. [03]
Theberge, J., Y. Al-Semaan, P. C. Williamson, et al. 2003. Glutamate and Glutamine in the Anterior Cingulate and Thalamus of Medicated Patients with Chronic Schizophrenia and Healthy Comparison Subjects Measured with 4.0-T Proton MRS. *Am. J. Psychiatr.* **160**:2231–2233. [07]
Theodoridou, A., F. Schlatter, V. Ajdacic, W. Rössler, and M. Jäger. 2012. Therapeutic Relationship in the Context of Perceived Coercion in a Psychiatric Population. *Psychiatry Res.* **200**:939–944. [16]
Thermenos, H. W., L. J. Seidman, H. Breiter, et al. 2004. Functional Magnetic Resonance Imaging During Auditory Verbal Working Memory in Nonpsychotic Relatives of Persons with Schizophrenia: A Pilot Study. *Biol. Psychiatry* **55**:490–500. [07]
Thornberry, T. P. 1987. Toward an Interactional Theory of Delinquency. *Criminology* **25**:863–891. [01]
Tienari, P., L. C. Wynne, A. Sorri, et al. 2004. Genotype-Environment Interaction in Schizophrenia-Spectrum Disorder: Long-Term Follow-up Study of Finnish Adoptees. *Br. J. Psychiatry* **184**:216–222. [01, 09]
Tiihonen, J., J. Lönnqvist, K. Wahlbeck, et al. 2009. 11-Year Follow-up of Mortality in Patients with Schizophrenia: A Population-Based Cohort Study (FIN11 Study). *Lancet* **374**:620–627. [05]
Ting, J. T., J. Peca, and G. Feng. 2012. Functional Consequences of Mutations in Postsynaptic Scaffolding Proteins and Relevance to Psychiatric Disorders. *Annu. Rev. Neurosci.* **35**:49–71. [13]

Toch, H., and K. Adams. 1989. The Disturbed Violent Offender. New Haven: Yale Univ. Press. [01]
Tost, H., and A. Meyer-Lindenberg. 2012. Puzzling over Schizophrenia: Schizophrenia, Social Environment and the Brain. *Nat. Med.* **18**:211–213. [09, 12]
Toulopoulou, T., T. E. Goldberg, I. R. Mesa, et al. 2010. Impaired Intellect and Memory: A Missing Link between Genetic Risk and Schizophrenia? *Arch. Gen. Psychiatry* **67**:905–913. [14]
Townsend, L. A., and R. M. Norman. 2004. Course of Cognitive Functioning in First Episode Schizophrenia Spectrum Disorders. *Expert. Rev. Neurother.* **4**:61–68. [04]
Toyokawa, S., M. Uddin, K. C. Koenen, and S. Galea. 2012. How Does the Social Environment "Get into the Mind"? Epigenetics at the Intersection of Social and Psychiatric Epidemiology. *Soc. Sci. Med.* **74**:67–74. [09]
Trantham-Davidson, H., L. C. Neely, A. Lavin, and J. K. Seamans. 2004. Mechanisms Underlying Differential D1 Versus D2 Dopamine Receptor Regulation of Inhibition in Prefrontal Cortex. *J. Neurosci.* **24**:10652–10659. [12]
Trikalinos, T. A., E. E. Ntzani, D. G. Contopoulos-Ioannidis, and J. P. Ioannidis. 2004. Establishment of Genetic Associations for Complex Diseases Is Independent of Early Study Findings. *Eur. J. Hum. Genet.* **12**:762–769. [08]
Trojanowski, J. Q., P. D. Newman, W. D. Hill, and V. M. Lee. 1991. Human Olfactory Epithelium in Normal Aging, Alzheimer's Disease, and Other Neurodegenerative Disorders. *J. Comp. Neurol.* **310**:365–376. [10]
Tryson, G. S., and G. Winograd. 2001. Goal Consensus and Collaboration. *Psychotherapy* **38**:385–389. [15]
Tseng, K. Y., R. A. Chambers, and B. K. Lipska. 2009. The Neonatal Ventral Hippocampal Lesion as a Heuristic Neurodevelopmental Model of Schizophrenia. *Behav. Brain Res.* **204**:295–305. [11]
Tseng, K. Y., B. L. Lewis, T. Hashimoto, et al. 2008. A Neonatal Ventral Hippocampal Lesion Causes Functional Deficits in Adult Prefrontal Cortical Interneurons. *J. Neurosci.* **28**:12691–12699. [11]
Tseng, K. Y., and P. O'Donnell. 2007. Dopamine Modulation of Prefrontal Cortical Interneurons Changes During Adolescence. *Cereb. Cortex* **17**:1235–1240. [13]
Tsukada, H., H. Ohba, S. Nishiyama, and T. Kakiuchi. 2011. Differential Effects of Stress on [(1)(1)C]Raclopride and [(1)(1)C]MNPA Binding to Striatal D(2)/D(3) Dopamine Receptors: A PET Study in Conscious Monkeys. *Synapse* **65**:84–89. [07]
Tulsky, D. S., and L. R. Price. 2003. The Joint WAIS-III and WMS-III Factor Structure: Development and Cross-Validation of a Six-Factor Model of Cognitive Functioning. *Psychol. Assess.* **15**:149–162. [04]
Turetsky, B. I., M. E. Calkins, G. A. Light, et al. 2007. Neurophysiological Endophenotypes of Schizophrenia: The Viability of Selected Candidate Measures. *Schizophr. Bull.* **33**:69–94. [03]
Turetsky, B. I., C. G. Hahn, K. Borgmann-Winter, and P. J. Moberg. 2009. Scents and Nonsense: Olfactory Dysfunction in Schizophrenia. *Schizophr. Bull.* **35**:1117–1131. [10]
Turetsky, B. I., and P. J. Moberg. 2009. An Odor-Specific Threshold Deficit Implicates Abnormal Intracellular Cyclic Amp Signaling in Schizophrenia. *Am. J. Psychiatr.* **166**:226–233. [10]
Turner, E. E., N. Fedtsova, and D. V. Jeste. 1997. Cellular and Molecular Neuropathology of Schizophrenia: New Directions from Developmental Neurobiology. *Schizophr. Res.* **27**:169–180. [12]

Twamley, E. W., C. Z. Burton, and L. Vella. 2011. Compensatory Cognitive Training for Psychosis: Who Benefits? Who Stays in Treatment? *Schizophr. Bull.* **37(Suppl. 2)**:S55–62. [07]

Twamley, E. W., G. N. Savla, C. H. Zurhellen, R. K. Heaton, and D. V. Jeste. 2008. Development and Pilot Testing of a Novel Compensatory Cognitive Training Intervention for People with Psychosis. *Am. J. Psychiatr. Rehabil.* **11**:144–163. [07]

Uhlhaas, P. J., F. Roux, W. Singer, et al. 2009. The Development of Neural Synchrony Reflects Late Maturation and Restructuring of Functional Networks in Humans. *PNAS* **106**:9866–9871. [13]

Uhlhaas, P. J., S. M. Silverstein, W. A. Phillips, and P. G. Lovell. 2004. Evidence for Impaired Visual Context Processing in Schizotypy with Thought Disorder. *Schizophr. Res.* **68**:249–260. [02]

Uhlhaas, P. J., and W. Singer. 2010. Abnormal Neural Oscillations and Synchrony in Schizophrenia. *Nat. Rev. Neurosci.* **11**:100–113. [04, 12]

———. 2012. Neuronal Dynamics and Neuropsychiatric Disorders: Toward a Translational Paradigm for Dysfunctional Large-Scale Networks. *Neuron* **75**:963–980. [13]

Umbricht, D., D. Javitt, G. Novak, et al. 1998. Effects of Clozapine on Auditory Event-Related Potentials in Schizophrenia. *Biol. Psychiatry* **44**:716–725. [07]

———. 1999. Effects of Risperidone on Auditory Event-Related Potentials in Schizophrenia. *Int. J. Neuropsychopharmacol.* **2**:299–304. [07]

Umbricht, D. S., J. A. Bates, J. A. Lieberman, J. M. Kane, and D. C. Javitt. 2006. Electrophysiological Indices of Automatic and Controlled Auditory Information Processing in First-Episode, Recent-Onset and Chronic Schizophrenia. *Biol. Psychiatry* **59**:762–772. [07]

Vacic, C., S. McCarthy, Malhotra D, et al. 2011. Duplications of the Neuropeptide Receptor Gene VIPR2 Confer Significant Risk for Schizophrenia. *Nature* **471**:499–503. [05]

Vaidya, V. A., G. J. Marek, G. K. Aghajanian, and R. S. Duman. 1997. 5-HT2A Receptor-Mediated Regulation of Brain-Derived Neurotrophic Factor mRNA in the Hippocampus and the Neocortex. *J. Neurosci.* **17**:2785–2795. [07]

Valkonen-Korhonen, M., M. Purhonen, I. M. Tarkka, et al. 2003. Altered Auditory Processing in Acutely Psychotic Never-Medicated First-Episode Patients. *Brain Res. Cogn. Brain Res.* **17**:747–758. [07]

van Berckel, B. N., B. Oranje, J. M. van Ree, M. N. Verbaten, and R. S. Kahn. 1998. The Effects of Low Dose Ketamine on Sensory Gating, Neuroendocrine Secretion and Behavior in Healthy Human Subjects. *Psychopharmacology* **137**:271–281. [07]

Vancampfort, D., M. Probst, L. Helvik Skjaerven, et al. 2012. Systematic Review of the Benefits of Physical Therapy within a Multidisciplinary Care Approach for People with Schizophrenia. *Phys. Ther.* **92**:11–23. [07]

Van Craenenbroeck, K., K. De Bosscher, W. Vanden Berghe, P. Vanhoenacker, and G. Haegeman. 2005. Role of Glucocorticoids in Dopamine-Related Neuropsychiatric Disorders. *Mol. Cell Endocrinol.* **245**:10–22. [07]

Van Den Bossche, M. J., M. Johnstone, M. Strazisar, et al. 2012. Rare Copy Number Variants in Neuropsychiatric Disorders: Specific Phenotype or Not? *Am. J. Med. Genet. B Neuropsychiatr. Genet.* **159B**:812–822. [09]

van den Buuse, M. 2010. Modeling the Positive Symptoms of Schizophrenia in Genetically Modified Mice: Pharmacology and Methodology Aspects. *Schizophr. Bull.* **36**:246–270. [13]

van der Gaag, M., D. H. Nieman, J. Rietdijk, et al. 2012. Cognitive Behavioral Therapy for Subjects at Ultrahigh Risk for Developing Psychosis: A Randomized Controlled Clinical Trial. *Schizophr. Bull.* **38**:1180–1188. [01]

Van Oel, J., M. M. Sitskoorn, M. P. M. Cremer, and R. S. Kahn. 2002. School Performance as a Premorbid Marker for Schizophrenia: A Twin Study. *Schizophr. Bull.* **28**:401–414. [14, 17]

van Os, J. 2004. Does the Urban Environment Cause Psychosis? *Br. J. Psychiatry* **184**:287–288. [07]

van Os, J., and S. Kapur. 2009. Schizophrenia. *Lancet* **374**:635–645. [15]

van Os, J., G. Kenis, and B. P. Rutten. 2010. The Environment and Schizophrenia. *Nature* **468**:203–212. [09]

van Os, J., R. J. Linscott, I. Myin-Germeys, P. Delespaul, and L. Krabbendam. 2009. A Systematic Review and Meta-Analysis of the Psychosis Continuum: Evidence for a Psychosis Proneness-Persistence-Impairment Model of Psychotic Disorder. *Psychol. Med.* **39**:179–195. [05, 09, 16]

van Os, J., C. B. Pedersen, and P. B. Mortensen. 2004. Confirmation of Synergy between Urbanicity and Familial Liability in the Causation of Psychosis. *Am. J. Psychiatr.* **161**:2312–2314. [09]

van Os, J., B. P. Rutten, and R. Poulton. 2008. Gene–Environment Interactions in Schizophrenia: Review of Epidemiological Findings and Future Directions. *Schizophr. Bull.* **34**:1066–1082. [08, 09]

van Winkel, R. 2011. Family-Based Analysis of Genetic Variation Underlying Psychosis-Inducing Effects of Cannabis: Sibling Analysis and Proband Follow-Up. *Arch. Gen. Psychiatry* **68**:148–157. [09]

van Winkel, R., B. P. Rutten, O. Peerbooms, et al. 2010. Mthfr and Risk of Metabolic Syndrome in Patients with Schizophrenia. *Schizophr. Res.* **121**:193–198. [17]

Varese, F., F. Smeets, M. Drukker, et al. 2012. Childhood Adversities Increase the Risk of Psychosis: A Meta-Analysis of Patient-Control, Prospective- and Cross-Sectional Cohort Studies. *Schizophr. Bull.* **38**:661–671. [09]

Varghese, D., J. Scott, J. Welham, et al. 2011. Psychotic-Like Experiences in Major Depression and Anxiety Disorders: A Population-Based Survey in Young Adults. *Schizophr. Bull.* **37**:389–393. [09]

Varker, T., and G. J. Devilly. 2012. An Analogue Trial of Inoculation/Resilience Training for Emergency Services Personnel: Proof of Concept. *J. Anxiety Disord.* **26**:696–701. [06]

Vauth, R., P. W. Corrigan, M. Clauss, et al. 2005. Cognitive Strategies Versus Self-Management Skills as Adjunct to Vocational Rehabilitation. *Schizophr. Bull.* **31**:55–66. [15]

Velligan, D. I., P. M. Diamond, N. J. Maples, et al. 2008. Comparing the Efficacy of Interventions That Use Environmental Supports to Improve Outcomes in Patients with Schizophrenia. *Schizophr. Res.* **102**:312–319. [07]

Velligan, D. I., P. M. Diamond, J. Mintz, et al. 2006. The Use of Individually Tailored Environmental Supports to Improve Medication Adherence and Outcomes in Schizophrenia. *Schizophr. Bull.* **32**:483–489. [15]

Velligan, D. I., T. J. Prihoda, J. L. Ritch, et al. 2002. A Randomized Single-Blind Pilot Study of Compensatory Strategies in Schizophrenia Outpatients. *Schizophr. Bull.* **28**:283–292. [15]

Ventura, J., S. A. Wilson, R. C. Wood, and G. S. Hellemann. 2013. Cognitive Training at Home in Schizophrenia Is Feasible. *Schizophr. Res.* **143**:397–398. [01]

Verdoux, H., and J. van Os. 2002. Psychotic Symptoms in Non-Clinical Populations and the Continuum of Psychosis. *Schizophr. Res.* **54**:59–65. [14]

Verhaeghe, N., J. De Maeseneer, L. Maes, C. Van Heeringen, and L. Annemans. 2011. Effectiveness and Cost-Effectiveness of Lifestyle Interventions on Physical Activity and Eating Habits in Persons with Severe Mental Disorders: A Systematic Review. *Int. J. Behav. Nutr. Phys. Act.* **8**:28. [17]

Vértes, P. E., A. F. Alexander-Bloch, N. Gogtay, et al. 2012. Simple Models of Human Brain Functional Networks. *PNAS* **109**:5868–5873. [06]

Vesell, E. S. 1978. Twin Studies in Pharmacogenetics. *Hum. Genet. Suppl.* 19–30. [07]

Vierbuchen, T., A. Ostermeier, Z. P. Pang, et al. 2010. Direct Conversion of Fibroblasts to Functional Neurons by Defined Factors. *Nature* **463**:1035–1041. [10]

Vijayraghavan, S., M. Wang, S. G. Birnbaum, G. V. Williams, and A. F. Arnsten. 2007. Inverted-U Dopamine D1 Receptor Actions on Prefrontal Neurons Engaged in Working Memory. *Nat. Neurosci.* **10**:376–384. [09, 12]

Vinogradov, S., M. Fisher, and E. de Villers-Sidani. 2012. Cognitive Training for Impaired Neural Systems in Neuropsychiatric Illness. *Neuropsychopharmacology* **37**:43–76. [07]

Volavka, J., and L. Citrome. 2011. Pathways to Aggression in Schizophrenia Affect Results of Treatment. *Schizophr. Bull.* **37**:921–929. [01]

Vollenweider, F. X., M. Barro, P. A. Csomor, and J. Feldon. 2006. Clozapine Enhances Prepulse Inhibition in Healthy Humans with Low but Not with High Prepulse Inhibition Levels. *Biol. Psychiatry* **60**:597–603. [07]

Vonk, R., A. C. van der Schot, G. C. M. van Baal, et al. 2012. Premorbid School Performance in Twins Concordant and Discordant for Bipolar Disorder. *J. Affect. Disord.* **136**:294–303. [14, 17]

Waddington, J. L. 1993. Schizophrenia: Developmental Neuroscience and Pathobiology. *Lancet* **341**:531–536. [11]

Walker, E., N. Lewis, R. Loewy, and S. Palyo. 1999. Motor Dysfunction and Risk for Schizophrenia. *Dev. Psychopathol.* **11**:509–523. [01]

Walker, E., V. Mittal, and K. Tessner. 2008. Stress and the Hypothalamic Pituitary Adrenal Axis in the Developmental Course of Schizophrenia. *Annu. Rev. Clin. Psychol.* **4**:189–216. [07]

Walker, E. F., P. A. Brennan, M. Esterberg, et al. 2010. Longitudinal Changes in Cortisol Secretion and Conversion to Psychosis in At-Risk Youth. *J. Abnorm. Psychol.* **119**:401–408. [07]

Walker, E. F., D. J. Walder, and F. Reynolds. 2001. Developmental Changes in Cortisol Secretion in Normal and at-Risk Youth. *Dev. Psychopathol.* **13**:721–732. [07]

Wallace, C. J., R. Tauber, and J. Wilde. 1999. Teaching Fundamental Workplace Skills to Persons with Serious Mental Illness. *Psychiatr. Serv.* **50**:1147–1153. [15]

Walport, M., and P. Brest. 2011. Sharing Research Data to Improve Public Health. *Lancet* **377**:537–539. [09]

Wand, G. S., L. M. Oswald, M. E. McCaul, et al. 2007. Association of Amphetamine-Induced Striatal Dopamine Release and Cortisol Responses to Psychological Stress. *Neuropsychopharmacology* **32**:2310–2320. [07]

Wang, S. J., H. M. Hung, and R. O'Neill. 2011. Regulatory Perspectives on Multiplicity in Adaptive Design Clinical Trials Throughout a Drug Development Program. *J. Biopharm. Stat.* **21**:846–859. [17]

Warren, L., P. D. Manos, T. Ahfeldt, et al. 2010. Highly Efficient Reprogramming to Pluripotency and Directed Differentiation of Human Cells with Synthetic Modified mRNA. *Cell Stem Cell* **7**:618–630. [10]

Weickert, T. W., T. E. Goldberg, A. Mishara, et al. 2004. Catechol-O-Methyltransferase Val(108/158)Met Genotype Predicts Working Memory Response to Antipsychotic Medications. *Biol. Psychiatry* **56**:677–682. [07]

Weinberger, D. R. 1987. Implications of Normal Brain Development for the Pathogenesis of Schizophrenia. *Arch. Gen. Psychiatry* **44**:660–669. [03, 07, 08, 09, 12]

Weiser, M., and S. Noy. 2005. Interpreting the Association between Cannabis Use and Increased Risk for Schizophrenia. *Dialog. Clin. Neurosci.* **7**:81–85. [07]

Welham, J., M. Isohanni, P. Jones, and J. McGrath. 2009. The Antecedents of Schizophrenia: A Review of Birth Cohort Studies. *Schizophr. Bull.* **35**:603–623. [09]

Wells, K., J. Miranda, M. L. Bruce, M. Alegria, and N. Wallerstein. 2004. Bridging Community Intervention and Mental Health Services Research. *Am. J. Psychiatr.* **161**:955–963. [15]

White, J. G., E. Southgate, J. N. Thomson, and S. Brenner. 1986. The Structure of the Nervous System of the Nematode *Caenorhabditis Elegans*. *Phil. Trans. R. Soc. Lond. B* **314**:1–340. [10]

Whitford, T. J., T. F. D. Farrow, C. J. Rennie, et al. 2006. Longitudinal Changes in Regional Grey Matter Volume and Corresponding EEG Power in First-Episode Psychosis. *Aust. NZ J. Psychiatry* **40**:A122–A123. [04]

Whyte, J. 1998. Distinctive Methodological Challenges. In: Assessing Medical Rehabilitation Practices : The Promise of Outcomes Research, ed. M. J. Fuhrer, pp. 43–59. Baltimore: Paul H. Brookes. [01]

Whyte, M. C., C. Brett, L. K. Harrison, et al. 2006. Neuropsychological Performance over Time in People at High Risk of Developing Schizophrenia and Controls. *Biol. Psychiatry* **59**:730–739. [07]

Wicks, S., A. Hjern, and C. Dalman. 2010. Social Risk or Genetic Liability for Psychosis? A Study of Children Born in Sweden and Reared by Adoptive Parents. *Am. J. Psychiatr.* **167**:1240–1246. [09]

Wicks, S., A. Hjern, D. Gunnell, G. Lewis, and C. Dalman. 2005. Social Adversity in Childhood and the Risk of Developing Psychosis: A National Cohort Study. *Am. J. Psychiatr.* **162**:1652–1657. [09]

Wienberg, M., B. Y. Glenthoj, K. S. Jensen, and B. Oranje. 2010. A Single High Dose of Escitalopram Increases Mismatch Negativity without Affecting Processing Negativity or P300 Amplitude in Healthy Volunteers. *J. Psychopharmacol.* **24**:1183–1192. [07]

Wiersma, D., F. J. Nienhuis, C. J. Slooff, and R. Giel. 1998. Natural Course of Schizophrenic Disorders: A 15-Year Follow-up of a Dutch Incidence Cohort. *Schizophr. Bull.* **24**:75–85. [06]

Wilke, R. A., and M. E. Dolan. 2011. Genetics and Variable Drug Response. *JAMA* **306**:306–307. [07]

Williams, G. V., and P. S. Goldman-Rakic. 1995. Modulation of Memory Fields by Dopamine D1 Receptors in Prefrontal Cortex. *Nature* **376**:572–575. [09]

Williams, L. M., P. Das, B. J. Liddell, et al. 2006. Mode of Functional Connectivity in Amygdala Pathways Dissociates Level of Awareness for Signals of Fear. *J. Neurosci.* **26**:9264–9271. [04]

———. 2007a. Fronto-Limbic and Autonomic Disjunctions to Negative Emotion Distinguish Schizophrenia Subtypes. *Psychiatry Res.* **155**:29–44. [04]

Williams, L. M., T. J. Whitford, G. Flynn, et al. 2007b. General and Social Cognition in First Episode Schizophrenia: Identification of Separable Factors and Prediction of Functional Outcome Using the Integneuro Test Battery. *Schizophr. Res.* **99**:182–191. [04]

Williams, L. M., T. J. Whitford, E. Gordon, et al. 2009a. Neural Synchrony in Patients with a First Episode of Schizophrenia: Tracking Relations with Grey Matter and Symptom Profile. *J. Psychiatry Neurosci.* **34**:21–29. [07]

Williams, L. M., T. J. Whitford, M. Nagy, et al. 2009b. Emotion-Elicited Gamma Synchrony in Patients with First-Episode Schizophrenia: A Neural Correlate of Social Cognition Outcomes. *J. Psychiatry Neurosci.* **34**:303–313. [04, 07]

Wilson, B. A. 1991. Long-Term Prognosis of Patients with Severe Memory Disorders. *Neuropsychol. Rehabil.* **1**:117–134. [01]

———. 1997. Cognitive Rehabilitation : How It Is and How It Might Be. *J. Int. Neuropsychol. Soc.* **3**:487–496. [01]

Wittchen, H. U. 2000. Epidemiological Research in Mental Disorders: Lessons for the Next Decade of Research: The Nape Lecture 1999. *Acta Psychiatr. Scand.* **101**:2–10. [16]

Wolf, R. C., N. Vasic, F. Sambataro, et al. 2009. Temporally Anticorrelated Brain Networks During Working Memory Performance Reveal Aberrant Prefrontal and Hippocampal Connectivity in Patients with Schizophrenia. *Prog. Neuropsychopharmacol. Biol. Psychiatry* **33**:1464–1473. [09]

Wolf, R. C., N. Vasic, and H. Walter. 2006. The Concept of Working Memory in Schizophrenia: Current Evidence and Future Perspectives. *Fortschr. Neurol. Psychiatr.* **74**:449–468. [01]

Wolfson, M., S. E. Wallace, N. Masca, et al. 2010. Datashield: Resolving a Conflict in Contemporary Bioscience: Performing a Pooled Analysis of Individual-Level Data without Sharing the Data. *Int. J. Epidemiol.* **39**:1372–1382. [09]

Wood, J., Y. Kim, and B. Moghaddam. 2012. Disruption of Prefrontal Cortex Large Scale Neuronal Activity by Different Classes of Psychotomimetic Drugs. *J. Neurosci.* **32**:3022–3031. [13]

Wood, S., A. Yung, P. McGorry, and C. Pantelis. 2011. Neuroimaging and Treatment Evidence for Clinical Staging in Psychotic Disorders: From the At-Risk Mental State to Chronic Schizophrenia. *Biol. Psychiatry* **70**:619–625. [05]

Woodberry, K. A., A. J. Giuliano, and L. J. Seidman. 2008. Premorbid IQ in Schizophrenia: A Meta-Analytic Review. *Am. J. Psychiatr.* **165**:579–587. [14]

Woodward, N. D., K. Jayathilake, and H. Y. Meltzer. 2007. COMT Val108/158met Genotype, Cognitive Function, and Cognitive Improvement with Clozapine in Schizophrenia. *Schizophr. Res.* **90**:86–96. [07]

Woodward, N. D., S. E. Purdon, H. Y. Meltzer, and D. H. Zald. 2005. A Meta-Analysis of Neuropsychological Change to Clozapine, Olanzapine, Quetiapine, and Risperidone in Schizophrenia. *Int. J. Neuropsychopharmacol.* **8**:457–472. [04]

Woolf, C. M. 1997. Does the Genotype for Schizophrenia Often Remain Unexpressed Because of Canalization and Stochastic Events During Development? *Psychol. Med.* **27**:659–668. [13]

World Health Organization. 2010. International Classification of Diseases (ICD-10). Geneva: World Health Organization. [05]

Wright, A. G. C., K. M. Thomas, C. J. Hopwood, et al. 2012. The Hierarchical Structure of DSM-5 Pathological Personality Traits. *J. Abnorm. Psychol.* **121**:951–917. [02]

Wright, E. R. 1997. The Impact of Organizational Factors on Mental Health Professionals' Involvement with Families. *Psychiatr. Serv.* **48**:921–927. [15]

Wu, E. Q., H. G. Birnbaum, L. Shi, et al. 2005. The Economic Burden of Schizophrenia in the United States in 2002. *J. Clin. Psychiatry* **66**:1122–1129. [01]

Wyatt, R. J. 1991. Neuroleptics and the Natural Course of Schizophrenia. *Schizophr. Bull.* **17**:325–351. [04]

Wykes, T. 1998. What Are We Changing with Neurocognitive Rehabilitation? Illustrations from Two Single Cases of Changes in Neuropsychological Performance and Brain Systems as Measured by SPECT. *Schizophr. Res.* **34**:77–86. [01]

———. 2010. Cognitive Remediation Therapy Needs Funding. *Nature* **468**:165–166. [04]

Wykes, T., M. Brammerm, J. Mellers, et al. 2002. The Effects on the Brain of a Psychological Treatment, Cognitive Remediation Therapy (CRT): An fMRI Study. *Br. J. Psychiatry* **181**:144–152. [01]

Wykes, T., V. Huddy, C. Cellard, S. R. McGurk, and P. Czobor. 2011. A Meta-Analysis of Cognitive Remediation for Schizophrenia: Methodology and Effect Sizes. *Am. J. Psychiatr.* **168**:472–485. [01, 04, 05, 06, 07, 15, 17]

Wykes, T., E. Newton, S. Landau, et al. 2007. Cognitive Remediation Therapy (CRT) for Young Early Onset Patients with Schizophrenia: An Exploratory Randomized Controlled Trial. *Schizophr. Res.* **94**:221–230. [07]

Wykes, T., and C. Reeder. 2005. Cognitive Remediation Therapy for Schizophrenia: Theory and Practice. London: Routledge. [15]

Wykes, T., C. Reeder, C. Huddy, et al. 2012. Developing Models of How Cognitive Improvements Change Functioning: Mediation, Moderation and Moderated Mediation. *Schizophr. Res.* **138**:88–93. [01]

Wykes, T., C. Steel, B. Everitt, and N. Tarrier. 2008. Cognitive Behavior Therapy for Schizophrenia: Effect Sizes, Clinical Models, and Methodological Rigor. *Schizophr. Bull.* **34**:523–537. [01, 05, 15]

Wynn, J. K., M. F. Green, J. Sprock, et al. 2007. Effects of Olanzapine, Risperidone and Haloperidol on Prepulse Inhibition in Schizophrenia Patients: A Double-Blind, Randomized Controlled Trial. *Schizophr. Res.* **95**:134–142. [07]

Xu, B., I. Ionita-Laza, J. L. Roos, et al. 2012. De Novo Gene Mutations Highlight Patterns of Genetic and Neural Complexity in Schizophrenia. *Nat. Genet.* [09]

Yan, J. 2013. NIMH Tries to Jumpstart Drug Innovations. *Psychiatric News* **48**:8–10. [17]

Yang, C. R., and J. K. Seamans. 1996. Dopamine D1 Receptor Actions in Layer V–VI Rat Prefrontal Cortex Neurons in Vitro: Modulation of Dendritic-Somatic Signal Integration. *J. Neurosci* **16**:1922–1935. [12]

Yang, N., Y. H. Ng, Z. P. Pang, T. C. Sudhof, and M. Wernig. 2011. Induced Neuronal Cells: How to Make and Define a Neuron. *Cell Stem Cell* **9**:517–525. [10]

Yanos, P. T., D. Roe, M. L. West, S. M. Smith, and P. H. Lysaker. 2012. Group-Based Treatment for Internalized Stigma among Persons with Severe Mental Illness: Findings from a Randomized Controlled Trial. *Psychol. Serv.* **9**:248–258. [15]

Yazawa, M., B. Hsueh, X. Jia, et al. 2011. Using Induced Pluripotent Stem Cells to Investigate Cardiac Phenotypes in Timothy Syndrome. *Nature* **471**:230–234. [10]

Yoo, A. S., A. X. Sun, L. Li, et al. 2011. Microrna-Mediated Conversion of Human Fibroblasts to Neurons. *Nature* **476**:228–231. [10]

Young, C. E., and C. R. Yang. 2004. Dopamine D1/D5 Receptor Modulates State-Dependent Switching of Soma-Dendritic Ca^{2+} Potentials via Differential Protein kinase A and C Activation in Rat Prefrontal Cortical Neurons. *J. Neurosci.* **24**:8–23. [12]

Young, J. W., X. Zhou, and M. A. Geyer. 2010. Animal Models of Schizophrenia. *Curr. Top. Behav. Neurosci.* **4**:391–433. [13]

Yu, J., K. Hu, K. Smuga-Otto, et al. 2009. Human Induced Pluripotent Stem Cells Free of Vector and Transgene Sequences. *Science* **324**:797–801. [10]

Yu, J., M. A. Vodyanik, K. Smuga-Otto, et al. 2007. Induced Pluripotent Stem Cell Lines Derived from Human Somatic Cells. *Science* **318**:1917–1920. [10]

Yung, A., L. Phillips, P. McGorry, et al. 2002. Comprehensive Assessment of At-Risk Mental States (CAARMS). Melbourne: The PACE Clinic, University of Melbourne. [07]

Yung, A. R., and P. D. McGorry. 1996. The Initial Prodrome in Psychosis: Descriptive and Qualitative Aspects. *Aust. NZ J. Psychiatry* **30**:587–599. [09]

Yung, A. R., and B. Nelson. 2011. Young People at Ultra High Risk for Psychosis: A Research Update. *Early Interv. Psychiatry* **5(Suppl. 1)**:52–57. [01, 09]

Yung, A. R., B. Nelson, C. Stanford, et al. 2008. Validation of "Prodromal" Criteria to Detect Individuals at Ultra High Risk of Psychosis: 2 Year Follow-Up. *Schizophr. Res.* **105**:10–17. [02]

Yung, A. R., H. P. Yuen, G. Berger, et al. 2007. Declining Transition Rate in Ultra High Risk (Prodromal) Services: Dilution or Reduction of Risk? *Schizophr. Bull.* **33**:673–681. [02]

Zald, D. H., and J. V. Pardo. 1997. Emotion, Olfaction, and the Human Amygdala: Amygdala Activation During Aversive Olfactory Stimulation. *PNAS* **94**:4119–4124. [10]

Zammit, S., P. Allebeck, A. S. David, et al. 2004. A Longitudinal Study of Premorbid IQ Score and Risk of Developing Schizophrenia, Bipolar Disorder, Severe Depression, and Other Nonaffective Psychoses. *Arch. Gen. Psychiatry* **61**:354–360. [14, 17]

Zammit, S., G. Lewis, C. Dalman, and P. Allebeck. 2010a. Examining Interactions between Risk Factors for Psychosis. *Br. J. Psychiatry* **197**:207–211. [09]

Zammit, S., G. Lewis, J. Rasbash, et al. 2010b. Individuals, Schools, and Neighborhood: A Multilevel Longitudinal Study of Variation in Incidence of Psychotic Disorders. *Arch. Gen. Psychiatry* **67**:914–922. [08]

Zanelli, J., A. Reichenberg, K. Morgan, et al. 2010. Specific and Generalized Neuropsychological Deficits: A Comparison of Patients with Various First-Episode Psychosis Presentations. *Am. J. Psychiatr.* **167**:78–85. [14]

Zhang, J. P., T. Lencz, and A. K. Malhotra. 2010. D2 Receptor Genetic Variation and Clinical Response to Antipsychotic Drug Treatment: A Meta-Analysis. *Am. J. Psychiatr.* **167**:763–772. [07]

Zhang, X., K. M. Klueber, Z. Guo, et al. 2006. Induction of Neuronal Differentiation of Adult Human Olfactory Neuroepithelial-Derived Progenitors. *Brain. Res.* **1073-1074**:109–119. [10]

Zhang, X., K. M. Klueber, Z. Guo, C. Lu, and F. J. Roisen. 2004. Adult Human Olfactory Neural Progenitors Cultured in Defined Medium. *Exp. Neurol.* **186**:112–123. [10]

Zheng, P., X. X. Zhang, B. S. Bunney, and W. X. Shi. 1999. Opposite Modulation by Cortical N-Methyl-D-Aspartate Receptor-Mediated Responses by Low and High Concentrations of Dopamine. *Neuroscience* **91**:527–535. [12]

Ziermans, T., P. Schothorst, M. Magnee, H. van Engeland, and C. Kemner. 2011. Reduced Prepulse Inhibition in Adolescents at Risk for Psychosis: A 2-Year Follow-up Study. *J. Psychiatry Neurosci.* **36**:127–134. [07]

Zuardi, A. W., J. E. Hallak, S. M. Dursun, et al. 2006. Cannabidiol Monotherapy for Treatment-Resistant Schizophrenia. *J. Psychopharmacol.* **20**:683–686. [07]

Zuckerman, L., M. Rehavi, R. Nachman, and I. Weiner. 2003. Immune Activation During Pregnancy in Rats Leads to a Postpubertal Emergence of Disrupted Latent Inhibition, Dopaminergic Hyperfunction, and Altered Limbic Morphology in the Offspring: A Novel Neurodevelopmental Model of Schizophrenia. *Neuropsychopharmacology* **28**:1778–1789. [11]

Subject Index

22q11 78, 83, 190, 191, 216

adolescence 188, 219, 227
  cognitive decline in 229–233, 277–279
  early 218, 233, 277–279
  late 2, 14, 102, 104, 108, 186, 218, 233
affective dysregulation 52, 143, 204
agranulocytosis, clozapine-induced 104, 293
altered sense of self 6, 10, 13, 87
Alzheimer's disease 29, 36, 37, 55, 148
  OE technology 177, 178
amygdala 72, 108, 132, 161
anhedonia 27, 53, 60, 86
animal models 10–12, 83, 106, 147, 209, 217, 274, 275, 288
  hypothesis-based 183–194, 219
  limitations of 150, 168, 192
  link to human cell models 179, 180
  rodent 115, 117, 129, 133, 169, 180, 184, 185
  role in novel therapeutics 192–194
  validity in 184, 185, 189, 194, 210
anterior cingulate cortex (ACC) 111, 120, 121, 130, 131, 152
anti-inflammatory interventions 8, 106, 107, 124. *See also* omega-3 fatty acids
antipsychotic medication 29, 41, 79, 82, 96, 110, 118, 265, 283, 293
  effect on BOLD signal 121, 122
  effect on prepulse inhibition 116
  impact on cognition 73, 193, 232
anxiety 15, 34, 61, 81, 83, 89, 97, 98, 138, 141, 142, 158, 159, 214, 242, 263
aspirin 106, 123, 161
assertive community treatment 13, 246, 247, 271
assessment 15, 40, 71, 96, 147, 151, 153, 156, 157, 163
  CARE program 116, 153
  multidimensional 86–88

  personalized treatment 294, 295
  virtual reality techniques 15, 290, 291
associative thought disorder 27, 37
at-risk mental state 8, 25, 34, 35, 108, 160, 161
  secondary prevention 276, 277, 281, 282
attachment 30, 61, 95, 149
attention 17, 61, 63–66, 70–72, 120, 217, 244, 253
  impaired 16, 37, 38, 157
  selective 38, 39
attractor states 21, 195, 198–200, 204, 220, 221
autism 5, 19, 52, 83, 84, 94, 120, 129, 139, 193, 212, 213, 228
  iPS cell characterization 172
avolition 50–54, 58, 60, 77, 85, 87
Avon Longitudinal Study of Parents and Children 154, 155
AX continuous performance task 38, 39

behavioral construct 6, 15, 49, 51, 56, 60, 61, 81, 86, 88
biomarkers 55, 56, 59, 62, 79, 80, 102, 115, 121, 125, 142, 160, 287
  abnormal gamma synchrony 88, 118, 221
  mu rhythm suppression 119, 120
  role of animal models 185, 191–194
bipolar disorder 5, 6, 25, 29, 33, 35–37, 62, 83, 129, 151, 193, 228
  cognitive function in 231–233, 278
  OE technology 177, 178
  predicting 146
  risk factors 141, 142, 148
Bleuler, Eugen 25, 27, 38, 52–54, 78, 228
Bleuler, Manfred 94
BOLD signal 121, 122
bonding 149
brain-derived neurotrophic factor (BDNF) 105, 123, 144

brain function 3, 5, 6, 11, 20, 21, 63, 64, 70, 72, 131, 150
   computational models 196
brain imaging 2, 71, 74, 79, 180, 294. *See also* neuroimaging

CACNA1C 132, 205
cancer 6, 84, 142
candidate genes 61, 62, 103, 113, 131, 132, 150
cannabis 4, 96, 104, 107, 140, 143, 145, 164
CARE. *See* Cognitive Assessment and Risk Evaluation (CARE) program
case formulation approach 254, 294
catechol-O-methyl transferase (COMT) 96, 107, 112, 113, 115, 201, 205
category fluency 66, 67, 231
celecoxib 106
cellular models. *See* human cellular models
childhood abuse 4, 8, 19, 79, 141, 154–156, 279, 291
chlorpromazine 55, 227, 232
classification 55–58, 77–79, 227, 263, 264, 279, 282
   Kleist–Leonhardt 290
clinical symptoms 77, 81, 86–88, 119, 214
   heterogeneity in 211
clinical trials 13, 16, 63–67, 125, 288, 289
   barriers to 161, 194
closed concept/construct 28–30, 36, 37, 41, 78, 85, 213, 277
clozapine 55, 118, 123, 297
   agranulocytosis 104, 293
CNTNAP2 213, 215
CNTRICS. *See* Cognitive Neuroscience Treatment Research to Improve Cognition in Schizophrenia
cognition 2, 16–18, 39, 40, 63, 96, 97, 109, 113, 142, 216, 228, 237, 294. *See also* social cognition
   impairments in 4, 17, 37–40, 58, 81, 85, 280, 282
   modeling 198–200

Cognitive Assessment and Risk Evaluation (CARE) program 116, 153
cognitive behavioral therapy (CBT) 8, 15, 82, 108, 109, 114, 122–125, 235, 237, 240–244, 253, 254, 257, 258, 271, 294–296, 303
cognitive decline 79, 188, 189, 228–233, 278
   identifying 277–280
cognitive–emotional impairment 63–73
Cognitive Neuroscience Treatment Research to Improve Cognition in Schizophrenia (CNTRICS) 71, 115, 121
cognitive remediation 16, 73, 74, 82, 96, 97, 109, 114, 122–125, 235, 237, 245, 251–253, 258, 282, 283, 286, 294
   computer-based 29, 69, 287
   defined 244
   target age 280
comorbidity 35, 44–46, 79, 81, 143, 154, 264
computational models 9, 12, 195–210, 220, 221
computational neuroscience 195–197, 206, 275
congenital blindness 19, 20
connectivity. *See* neural connectivity
construct validity 71, 115, 119, 184
context processing 5, 39, 40, 202
contingent reinforcement 235, 258
continuous performance tests (CPT) 38, 39, 64, 66, 67, 71, 72
copy number variants (CNVs) 83, 103, 129, 139, 140, 148, 190, 213, 218
cortical disinhibition 110, 185–187, 191–194
cortical morphogenesis 212
cortisol 105, 147, 154
COX-2 inhibitors 106, 123, 124
cross-disciplinary translational research 127–136

data mining 45, 86–89
data sharing 22, 88, 155, 156
deep brain stimulation 287, 295

deficit schizophrenia 56, 58, 59, 61, 82, 87, 270
deinstitutionalization 239, 261, 304
delusions 2, 20, 29–31, 51–54, 58, 60, 77, 78, 82, 85, 97, 146, 228, 242
dementia 6, 55, 113, 280
    Alzheimer's 29, 37, 148, 177, 178
    praecox 36, 52, 54, 77, 79, 93, 98, 227
de novo mutations 129, 130
depression 34, 37, 60, 83, 85, 97, 98, 138, 141, 142, 157, 158, 161, 265, 288, 304
    MDD 58, 131, 132, 263
    treatment 105, 239, 242
developmental models 56, 57, 163, 186–189
diabetes 142, 145, 255, 283, 284
diagnosis 2, 28, 31, 35, 49–62, 77–79, 81–84, 295
Diagnostic Statistical Manual (DSM) 28, 51, 79, 82, 263, 282
    A criteria 51, 53, 54
    DSM-IV 30, 53, 54, 83
    DSM-5 46, 56, 58–61, 77
diet 255, 282, 286
dimensional psychopathology 290, 291
direct cell conversion 174–176, 222
DISC1 128, 180, 190, 213, 215, 218
discrimination 4, 13, 130, 268, 302, 304
disease entity 52–58, 78, 80, 81, 84
    defined 50
disorganization 40, 54, 77, 81, 85
    social 20, 21
    thought 51–53, 58, 60, 85
dopamine 6, 103, 107–111, 115–117, 121, 144, 147, 187, 206, 218, 219
    animal models 192
    antagonists 41, 54, 81
    dual-state theory 204
    dysfunction 189, 201
    hypothesis 13, 109, 124, 184, 192
dorsolateral prefrontal cortex (DLPFC) 120, 150, 151, 160
drug abuse. *See* substance abuse
drug development 3, 10, 13, 65, 168, 275, 287, 288, 289
    me-too 54, 55
drug screening 11, 169, 171, 189, 210

DSM. *See* Diagnostic Statistical Manual
dual-state theory of prefrontal dopamine function 204
Dunedin birth cohort 154, 155, 229
dysbindin 190

early adolescence 218, 233
    cognitive decline in 277–279
early adulthood 2, 14, 102, 188, 218, 233
early intervention 63, 73, 105, 123, 158–161, 281, 291
early psychosis 101–126
    neurocognition in 112–114
    treatment of 104–107
eccentricity 30, 32
EEG gamma synchrony 71, 72
eicosapentaenoic acid (EPA) 111, 113
electrophysiology 115–120
emotion 53, 68, 87, 241, 255, 278. *See also* cognitive–emotional impairment, social cognition
    impaired 63, 70–73
    recognition 117, 120
    relevance to olfaction 178
empathy, impaired 53
employment 2, 14, 70, 281
    supported 235–237, 241, 242, 244, 247, 258, 271, 303, 304
empowerment 238, 245, 268, 270, 296
endophenotype 16, 17, 27, 42, 55, 132, 155, 185, 191, 192
environmental models 189, 190
environmental risk factors 101, 104, 150, 163, 190, 216, 217, 270, 274, 280, 291. *See also* urban environment
poverty 8, 9, 98, 281
Environmental Risk Longitudinal Twin Study 154, 155
Epidemiological Catchment Area Study 263
epidemiology 4, 8, 9, 97, 127, 129, 133, 139, 140, 151, 152, 155, 213
    genetic 25, 26, 36, 46, 83
epigenetics 8, 101, 104–107, 137, 149, 291

epilepsy 6, 11, 18, 83, 84, 139, 212, 213, 217
equipotentiality 27, 44, 45
ethnicity 105, 130, 140, 145, 156, 158, 163, 281
etiology 2, 14, 18, 28, 50, 81, 84–86, 89, 209
  models for understanding 42–45
  role of animal models 194
  systemic view 284
  traditional view of 283, 284
etiology–pathophysiology–symptoms (E–P–S) framework 211–214, 219, 220, 223
event-related potentials (ERPs) 118, 147, 151
evidence-based medicine 240, 268
evidence-based treatment 13, 14, 271, 295, 297–299
  National Implementing Evidence-Based Practices Project 247, 248
excitation-inhibition balance 184, 186, 188, 193, 212, 218, 219
executive function 17, 63, 64, 66, 88, 120, 132, 244
exercise. *See* physical activity
externalizing 30–34, 37

face validity 11, 184, 210
family history 34, 79, 89, 94, 95, 103, 140, 146, 154
family intervention 235, 236, 240, 241, 244, 247–249, 251, 253, 271, 294, 304
  training for 257, 258
fault tree analysis 25, 43–46
functional capacity 63, 70, 72, 124
functional connectivity 19, 71, 132, 150, 151, 170. *See also* neural connectivity
functional imaging 115–122
  fMRI 3, 4, 15, 70, 71, 130, 131, 150
functional outcome 64, 72, 77, 95, 114, 238. *See also* outcome
  determinants of 69, 70, 96, 112, 227, 228, 232
  improving 124, 245, 283

future research 6, 21, 33, 59, 71, 73, 81, 88, 114, 124, 134, 139, 152, 157, 163, 173, 179, 215, 251, 259, 275, 279

GABA 5, 88, 110, 111, 119, 180, 184, 192, 202
gamma synchrony 71, 72, 119, 220, 221
  abnormal 88, 118
gene–environment interactions 10, 94, 98, 134, 143, 149, 151, 152, 170, 279
generalized deficit 17, 25, 38–40
genetic epidemiology 25, 26, 36, 46, 83
genetic models 147, 173, 183, 190, 191, 288
genetic risk 7, 34–37, 79, 83, 94, 97, 103, 140, 147, 150, 151, 155, 213, 217, 228, 229, 232, 274
genetics 3, 6, 13, 77, 127, 137, 139, 140, 148–150, 155, 164
  imaging 131, 132, 150, 274
genome-wide association (GWA) studies 103, 131, 132, 146, 151, 213
genotype 96, 113, 216, 222, 293
glutamate 109–111, 115, 116, 121, 124, 148, 180, 192
goal setting 235, 254, 255, 258, 266, 283
gray matter loss 108–112, 117, 119, 124, 125, 141, 146

hallucinations 2, 20, 30, 50–54, 58, 60, 77, 78, 85, 204, 228, 242
haloperidol 73, 122
Head Start 280, 282
health economics 2, 268, 269, 299, 300
health policy 14, 79, 247, 248, 261, 268, 269, 292, 299, 303
help-seeking behavior 261, 264–269, 276
heritability 102, 131, 132, 217
heterogeneity 3, 5–7, 11, 21, 33, 49, 51, 54, 69, 77–90, 102, 128, 130, 137, 139, 141, 142, 193, 211, 216, 273, 289, 294
  conceptual paradigms 84–88
  in iPS cells 170, 173
  reducing 56, 84, 88, 213

hippocampus 108, 109, 111, 121, 123, 150, 151, 187, 189, 200, 218
  neurogenesis 286
  prefrontal connectivity 214, 216
homeorhesis 284, 285
human cellular models 12, 167–182, 209, 210, 221–223
  defined 167
  translation of 180, 181
human olfactory neurons 176–179
Huntington's disease 184
hypoferremia 106
hypokrisia 27
hypothalamic-pituitary-adrenal (HPA) axis 9, 105, 130, 141, 147
  dysregulation 108, 130
hypoxia 104, 140, 190

imaging genetics 131, 132, 150, 274
immune activation 107, 186, 188, 189, 192
  maternal 104, 106, 132, 133, 216
impaired insight 53, 54, 81, 303
implementation science 250, 251, 259, 297, 298
independent living 70, 236, 243, 255, 258, 268, 283
individualized treatment 102, 103, 125
induced neuronal (iN) cells 167, 170, 174–176, 180, 181
induced pluripotent stem (iPS) cells 167, 170–173, 175, 180, 181, 222, 274, 288
information processing 198–200
Integrated Psychological Therapy (IPT) 253
intellectual ability 96–98, 129, 231
  deterioration 69, 110, 227
  low IQ 8, 228–231, 233, 278
intellectual disability 83–85, 212, 213, 279
interactive developmental systems model 56, 57
interleukin-6 186, 214
internalizing 30–34, 37
International Classification of Diseases (ICD) 28, 51, 79

ICD-10 77, 83, 282
International Schizophrenia Consortium (ISC) 146
interventions 14, 15, 21, 61, 80
  psychosocial 13, 79
  school-based 9
IQ. *See* intellectual ability

Jaspers, Karl 53, 79

ketamine 105, 115, 116, 119, 185
ketoconazole 105
Kraepelin, Emil 25, 27, 38, 52, 53, 77, 78, 81, 93, 98, 227, 228, 233

labeling theory 302
Langfeldt, Gabriel 52, 53
late adolescence 2, 14, 102, 104, 108, 186, 218, 219, 233
learning disabilities 19, 80, 85, 139
life-crisis model 270, 303
life expectancy 3, 79, 282, 285, 286
lithium 109, 123, 124
low intelligence (IQ) 8, 228–231, 233, 278
LY404039 111

major depression disorder (MDD) 58, 131, 132, 263
mania 30, 32, 58, 60, 85
maternal age 140
maternal immune activation 104, 106, 132, 133, 189, 216
MATRICS 64, 69, 71, 72, 113, 115
medical model 51, 54, 298
medication management 304
Meehl, Paul 26–28, 35, 37, 42, 45
melanocortin 293
memantine 116, 118
memory 17, 55, 63, 70, 157, 202, 244. *See also* verbal memory, working memory
  epigenetic 170, 173, 175

Subject Index

mental disorder, defined 50
mental health care
  access to service 262, 270
  cost-benefit ratio 268, 269, 300
  patient-therapist relationship 261, 267, 270, 296
mental health policy. *See* health policy
mental health professionals 270, 298
  training 235, 248, 249, 256–259, 299
mental health promotion 281, 282, 286
mental illness, defined 50
mesolimbic dopaminergic system 130, 141
metabolic syndrome 285, 291
methionine 144
methylazoxymethanol acetate (MAM) 187–189, 215
migrant status 47, 104, 105, 130, 140, 164
minocycline 106, 113, 124, 161
mismatch negativity 117, 118
modeling 10–12, 81, 186–189, 209–224, 217. *See also* computational models
  neonatal ventral hippocampal lesion (NVHL) model 187–189
  time course 218–220
molindone 73
mood disorders 81, 87, 131, 154, 159
mortality 3, 282, 285, 286
motivation 4, 6, 16, 61, 87, 217, 255, 283, 286
  relation to olfaction 178
motivational interviewing 283, 296
motor development 7, 98
mu wave suppression 119, 120

N-acetylcysteine 116, 118
National Comorbidity Survey (NCS) 263
National Implementing Evidence-Based Practices Project 247, 248
negative symptoms 15, 51, 53, 54, 58, 77, 78, 81, 85, 287
  treating 82, 106, 111, 113, 117, 124, 160
neonatal ventral hippocampal lesion 187–189, 215, 216

neural connectivity 93, 98, 99, 108, 112, 121, 122, 125, 132, 150, 151, 170
  measure of 70, 71
neural synchrony 71, 72, 88, 118, 119, 151, 191, 216, 218, 220, 221
neural system dynamics 198–200
Neurexin 81, 84
neurobiology 9, 133, 150–152
neurocognition 111–113, 118, 124
neurofeedback training 120
neurogenesis 109, 124, 286
neuroimaging 86, 115–122, 137, 150, 151, 154, 287, 289. *See also* brain imaging
neuroinflammation 105, 106, 111
neuroticism 132
neurotoxicity 101, 111, 121, 124
New York High-Risk Project 38
nicotinic receptors 113, 118
NMDA 186, 202, 214
  antagonists 111, 116, 117
  blockade 185
  receptor hypofunction 109, 111
nondeficit schizophrenia 56, 58, 87
nonspecificity 47, 137, 139, 142
nosology 26, 36, 41, 46, 84
  defined 50
NR3C1 130
NRXN1 215
nutrition 98, 104, 149, 157, 159, 160, 276, 280–282, 285

obsessive compulsive disorder (OCD) 108, 124
obstetric complications 104, 140, 141, 154
olanzapine 73, 192
olfactory epithelium (OE) 167, 170, 177, 180, 181
  biopsy 167, 176, 178, 222
  technology 179
omega-3 fatty acids 8, 106, 107, 111, 113, 123, 124, 282
open concept/construct 6, 18, 29, 30, 33, 36, 41, 47, 78, 79, 85, 210, 277
optogenetics 198, 204, 288

outcome 10, 42, 44, 56, 73, 74, 209–224, 270. *See also* functional outcome
  evaluation 235, 247, 254, 279, 290, 299
  heterogeneity 63, 77, 138, 139, 268, 273, 289
  predictors 39, 71, 228, 232, 235, 261, 267, 296
oxidative stress 106, 178, 179, 184, 186, 188, 192, 214
oxytocin 117, 120, 295

pandysmaturation 7
paranoia 30, 82, 86
Parkinson's disease 29, 177, 178, 184
paternal age 79, 104, 129, 130, 140, 141
paternal antisocial personality traits 143, 144
pathogenesis 4, 137, 147, 209, 211, 212, 219, 223
pathophysiology 3, 6, 8, 12, 51, 64, 77, 81, 85, 86, 89, 209
patient. *See also* empowerment
  help-seeking behavior 261, 264–269, 276
  relationship to therapist 261, 267, 296
Patient Outcomes Research Team (PORT) 239, 271
peer support 13, 80
perception 17, 61, 63, 72, 157, 216
  treatment effects on 111–122
perigenual cingulate cortex (pACC) 151, 161
personalized medicine 6, 21, 169, 254–256, 288, 292–295
  defined 293
  patient's perspective 296, 297
pharmacogenomics 103, 104
pharmacological models 185, 186
phase 2 clinical trials 288, 289
phencyclidine 115, 118, 119, 185
phenotypic convergence 12, 211, 212, 215, 221, 222
physical abuse 4, 8, 19, 140. *See also* childhood abuse
physical activity 99, 105, 125, 159, 255, 280, 282, 285, 286, 290

pleiotropy 83, 217, 218
pluripotent risk state 8, 10, 86, 89, 142, 158, 159, 163
polygenic risk scores 146
poor rapport 53, 54
population-based sampling 97, 129, 153–155, 161, 163
positive symptoms 58, 63, 79, 81, 82, 120, 161, 204, 282, 297
  link with dopamine 109, 192
positron emission tomography (PET) 71, 121, 154, 181, 280
postmortem brain analysis 58, 168–170, 187, 215
posttraumatic stress disorder 15, 29, 41, 95, 141, 270
poverty 8, 9, 98, 281
practice-based medicine 256, 259
predictive validity 146, 184, 210
prefrontal cortex 108, 109, 120–122, 144, 151, 188, 200–204, 206, 215
  cortical thickness 219
  dorsolateral 120, 150, 151, 160
  medial 110
prenatal factors 47, 79, 98, 106, 133, 138, 150, 159
prepulse inhibition (PPI) 11, 115–117, 151, 187, 189, 191, 214
prevention 8, 14, 62, 97–99, 143, 158–161, 273, 280–282
  primary 60, 89, 99, 276–278
  secondary 60, 61, 89, 276–278, 281, 282
  tertiary 14, 277, 281
Problem-Oriented Medical Information Systems (PROMIS) 15, 294
problem solving 64, 68, 70, 112, 244, 304
protection 94, 97, 102, 107, 109, 114, 123, 125, 140
proton magnetic resonance spectroscopy (^{1}H-MRS) 109–111, 121
psychiatric rehabilitation. *See* psychosocial interventions
  defined 236
psychomotor impairment 51, 53, 54, 58, 77, 85, 244

psychopathology 26, 56, 58, 61, 62, 97, 99, 282, 290
  development of 34, 35
  dissociative 52, 78
  domains 49, 50, 53, 59
  meta-structure of 30–34
psychopharmacology 55, 265, 295
psychosis 33, 47, 52, 54, 82, 97, 139–141, 213, 233, 276
  cognitive behavioral therapy for 242, 243
  impact of cognitive decline 228–230, 277
  treatments for early 101–126
  watershed analogy 42–44
psychosocial functioning 238, 243–245, 250, 253, 254, 276, 282, 283
  improvements in 241
psychosocial interventions 3, 13, 14, 93, 96, 124, 125, 161, 235–260
  access to 246–251
  implementing 249–251
  personalizing 254–256
  training for 256–258
psychotherapy 108, 134, 144, 256, 262, 264, 265
public awareness 13, 265, 268–270, 301
public health interventions 98, 145, 158, 159, 276, 280–282, 286

quality of life 122, 266–268, 273, 276, 297, 302, 303

reality distortion 52–54, 58, 78
real-time biofeedback 15
reasoning/problem solving 64, 68, 70
recovery 237–246, 296
  new conceptualization 238, 254, 258, 275, 276
recovery movement 238, 245, 258, 275, 276, 296
rehabilitation 232, 236–239, 245, 246, 251, 252, 258, 283, 297
  vocational 241, 303
reinforcement learning 87
relapse prevention 238, 244, 245, 253
reliability engineering 41–45

remission 231, 235, 238, 258, 275, 276
Research Domain Criteria (RdOC) 27, 46, 56, 60, 61, 86, 123, 291
resilience 7–10, 62, 89, 93–100, 131, 134, 137, 149, 152, 154, 158
  defined 93–95, 140
restricted affect 53, 54, 58, 60, 85
Rett syndrome 168, 172
reward processing 217, 219
  modeling 199
reward system 286
rheumatoid arthritis 19, 20
risk 4, 7–10, 19, 36, 89, 102–107, 115, 116, 130, 134, 140, 286
  estimating 153–155
  factors 47, 55, 89, 94, 98, 101, 137–164, 228–231
  role of animal models 185, 194
  intervention 280, 282
  phenotype 150, 233
  prediction 103, 144–147, 158
  specificity 141–143
risperidone 73
rodent models 115, 117, 129, 133, 169, 180, 184, 185
  limitations of 168, 192

sampling 152–157, 289
schizoaffective disorder 35, 54
schizophrenia
  classification 227, 263, 264, 279, 282
  clinical syndrome 49–51, 54, 55, 60
  cognitive disorder 227–234
  deficit 56–61, 82, 87, 270
  defined 81, 234, 285
  interactive developmental systems model 56, 57
  nondeficit 56, 58, 87
schizophrenia construct 78–89
  defined 1, 5, 6, 18, 51, 63, 77
  models of 78
schizotaxia 26, 27, 37, 42
schizotypy 19, 26, 27, 35, 37, 42, 155
Schneider, Kurt 52–54, 78
school-based interventions 9, 14, 159, 163, 237, 276, 277

school performance 70, 142, 159, 228–232, 278, 280, 291
secondary prevention 60, 61, 89, 276–278, 281, 282
selective attention 38, 39
self-care 70, 243, 246, 255, 283, 294
self-esteem 268, 296, 303
self-management 237, 243, 244, 253, 255, 258, 286
   training 235, 245
self-report measures 28, 32–34, 73, 153
self-stigma 252, 255, 270, 302, 303
serotonin 103, 111, 115, 132
   reuptake inhibitors (SSRI) 109, 118, 123, 124
service delivery 161–163, 235, 247, 273, 285, 292, 297, 298
   business model for 299
sexual abuse 4, 8, 19, 143, 154
SHANK 213, 215
shared decision making 238, 245, 258, 275, 296–299
shared research platforms 127, 134. *See also* data sharing
single nucleotide polymorphisms (SNPs) 103, 104, 112, 115, 140
social adversity 9, 94, 140–143, 158, 280
social cognition 2, 64, 68, 71, 72, 85, 117–120, 237, 253, 296
social defeat 4, 8, 9, 105, 157
social-emotional processing 130, 152, 159
social isolation 187, 204, 214, 255, 286, 302
social learning theory 243, 254
social skills training 124, 235, 237, 243–245, 252, 253, 257, 258, 295, 304, 305
socioeconomic status 47, 130, 156–158, 161
sociology 9, 127, 128, 157
SRGAP2 gene 168
staging 56, 59–62, 89
stakeholders 266–268, 299, 300
steeling effect 93–95

stigma 13, 54, 80, 159, 242, 252, 255, 262, 268–270, 280, 281, 285, 301–303
stress 4, 15, 104–108, 130, 131, 147, 151, 164, 169, 216, 217, 304
   management 237, 241, 257
   oxidative 178, 179, 184, 186, 188, 192, 214
   periadolescent psychosocial 89
   posttraumatic 15, 29, 41, 95, 141, 270
striatum 200, 215, 218, 219
   hyperdopaminergia 121
substance abuse 8, 9, 34, 36, 47, 81, 104, 108, 141, 142, 146, 282, 283. *See also* drug abuse
suicide 270, 282, 286
supported employment 235–237, 241–244, 247, 258, 271, 303, 304
   effectiveness of 252–254
   individual placement and support program 242
   training for 257
symptomatology 204, 209, 223
synaptic connectivity 179, 200, 212, 213, 221
synaptogenesis 109, 123
syndrome 29, 30, 52
   defined 50
systemic disorders 172, 284, 285

tardive dyskinesia 104, 291
TEOSS trial 73
tertiary prevention 14, 277, 281
tetrahydrocannabinol (THC) 105, 107, 140
thought disorder 20, 25, 26, 30–33, 40–42, 78, 87
   associative 27, 37
   heterogeneity of 45–47
Timothy syndrome 172
token economy 235, 240, 258, 271
transcranial magnetic stimulation 15, 160, 287
transduction methods 171, 173, 217
translation
   impact of heterogeneity on 5–7, 77
   impediments to 80–84

translation (continued)
  research 127–136, 275, 291, 292
trauma exposure 142–145, 156, 164
treatment 3, 12, 84, 261–272, 291, 293, 297. *See also* psychosocial interventions
  barriers 269–271
  costs 2, 14, 268, 269, 299, 300
  eclectic approach to 253, 282, 294
  effects on neurocognition 111–122
  evidence-based 13, 14, 271, 295–299
  goal setting 235, 254, 255, 258, 266
  improving outcomes 301–303
  outcome predictors 104, 123, 124, 235, 252, 261, 267, 279
  personalized 6, 102, 103, 125, 254–256, 292–296
  systemic view 284
  traditional view 283, 284
  tricyclic antidepressant 105
  validation 239, 240, 290
treatment development 79, 93–99, 103, 122–124, 160, 205–207, 223, 239–241, 273, 287–294
  framework for 291–293
  investment in 299–301
treatment outcome. *See* outcome
tryptophan depletion 118
twins 40, 94, 216, 229, 231, 278
  phenotypic heterogeneity 216

unemployment 2, 80, 142, 255, 302

urban environment 9, 15, 47, 98, 104, 105, 130, 131, 140, 143, 151, 152, 156, 158, 161, 163, 247, 280, 291

validity 54, 88, 183, 185, 188, 189, 194
  construct 71, 115, 119, 184
  face 11, 184, 210
  predictive 146, 184, 210
valine (Val) 144
verbal learning 64–71
  disabilities 19, 85
verbal memory 39, 112, 120–122, 231
VIPR2 84
viral infection 4, 9, 89, 141, 171
virtual reality tools 290, 291
visual learning 64, 65, 68, 69
visual processing 17–20, 39, 40, 82, 85
  contour recognition 217
vitamin D 98, 190
vitamin E 123
vocational rehabilitation 124, 241, 242, 252, 303. *See also* supported employment
volition 228

watershed model of psychosis 42–44
weight gain 73, 83, 104, 282–286, 293
working memory 11, 17, 37, 39, 61, 64–66, 88, 97, 119–122, 150, 195, 200, 202, 204, 216–218

zinc deficiency 106
ZNF804A 132, 151, 152